Ontology Matching

Jérôme Euzenat · Pavel Shvaiko

Ontology Matching

With 67 Figures and 18 Tables

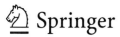

Authors

Jérôme Euzenat
INRIA Rhône-Alpes
655, avenue de l'Europe
Montbonnot St Martin
38334 Saint-Ismier cedex
France
Jerome.Euzenat@inrialpes.fr

Pavel Shvaiko
Department of Information and Communication Technology
University of Trento
via Sommarive 14
38050, Povo, Trento (TN)
Italy
pavel@dit.unitn.it

Library of Congress Control Number: 2007926257

ACM Computing Classification (1998): H.3, H.4, I.2, F.4

ISBN 978-3-540-49611-3 Springer Berlin Heidelberg New York

This work is subject to copyright. All rights are reserved, whether the whole or part of the material is concerned, specifically the rights of translation, reprinting, reuse of illustrations, recitation, broadcasting, reproduction on microfilm or in any other way, and storage in data banks. Duplication of this publication or parts thereof is permitted only under the provisions of the German Copyright Law of September 9, 1965, in its current version, and permission for use must always be obtained from Springer. Violations are liable for prosecution under the German Copyright Law.

Springer is a part of Springer Science+Business Media

springer.com

© Springer-Verlag Berlin Heidelberg 2007

The use of general descriptive names, registered names, trademarks, etc. in this publication does not imply, even in the absence of a specific statement, that such names are exempt from the relevant protective laws and regulations and therefore free for general use.

Typeset by the authors
Production: LE-TeX Jelonek, Schmidt & Vöckler GbR, Leipzig
Cover design: KünkelLopka Werbeagentur, Heidelberg

Printed on acid-free paper   45/3180/YL - 5 4 3 2 1 0

À mes parents et grand-parents qui ont poussé derrière,
à Anton et Johanna qui tirent par devant,
et à Jutta qui marche à mes côtés.

Jérôme

Моим родителям, Ларисе и Леониду,
за понимание и поддержку.
A Marlene, il mio amore, che mi è sempre stata vicino.

Pavel

Contents

Introduction 1

Part I The matching problem

1 Applications 9
  1.1 Ontology engineering 9
  1.2 Information integration 11
  1.3 Peer-to-peer information sharing 16
  1.4 Web service composition 19
  1.5 Autonomous communication systems 20
  1.6 Navigation and query answering on the web 22
  1.7 Summary 24

2 The matching problem 29
  2.1 Vocabularies, schemas and ontologies 29
  2.2 Ontology language 36
  2.3 Types of heterogeneity 40
  2.4 Terminology 42
  2.5 The ontology matching problem 44
  2.6 Summary 56

Part II Ontology matching techniques

3 Classifications of ontology matching techniques 61
  3.1 Matching dimensions 61
  3.2 Classification of matching approaches 63
  3.3 Other classifications 70
  3.4 Summary 72

4 Basic techniques ... 73
- 4.1 Similarity, distances and other measures ... 73
- 4.2 Name-based techniques ... 74
- 4.3 Structure-based techniques ... 92
- 4.4 Extensional techniques ... 105
- 4.5 Semantic-based techniques ... 110
- 4.6 Summary ... 115

5 Matching strategies ... 117
- 5.1 Matcher composition ... 117
- 5.2 Similarity aggregation ... 121
- 5.3 Global similarity computation ... 126
- 5.4 Learning methods ... 133
- 5.5 Probabilistic methods ... 141
- 5.6 User involvement and dynamic composition ... 142
- 5.7 Alignment extraction ... 144
- 5.8 Summary ... 149

Part III Systems and evaluation

6 Overview of matching systems ... 153
- 6.1 Schema-based systems ... 154
- 6.2 Instance-based systems ... 169
- 6.3 Mixed, schema-based and instance-based systems ... 176
- 6.4 Meta-matching systems ... 184
- 6.5 Summary ... 186

7 Evaluation of matching systems ... 193
- 7.1 Evaluation principles ... 193
- 7.2 Data sets for evaluation ... 198
- 7.3 Evaluation measures ... 203
- 7.4 Application-specific evaluation ... 213
- 7.5 Summary ... 216

Part IV Representing, explaining, and processing alignments

8 Frameworks and formats: representing alignments ... 219
- 8.1 Alignment formats ... 219
- 8.2 Alignment frameworks ... 235
- 8.3 Ontology editors with alignment manipulation capabilities ... 241
- 8.4 Summary ... 243

9 Explaining alignments ... 245
9.1 Justifications ... 245
9.2 Explanation approaches ... 247
9.3 A default explanation ... 249
9.4 Explaining basic matchers ... 251
9.5 Explaining the matching process ... 252
9.6 Arguing about correspondences ... 255
9.7 Summary ... 257

10 Processing alignments ... 259
10.1 Ontology merging ... 260
10.2 Ontology transformation ... 261
10.3 Data translation ... 261
10.4 Mediation ... 262
10.5 Reasoning ... 264
10.6 Towards an alignment service ... 264
10.7 Summary ... 265

Part V Conclusions

11 Conclusions ... 269
11.1 A brief outlook of the trends in the field ... 269
11.2 Future challenges ... 270
11.3 Final words ... 274

Appendix A: Legends of figures ... 275

Appendix B: Running example ... 277

Appendix C: Exercises ... 289

References ... 297

Index ... 323

Introduction

An ontology typically provides a vocabulary describing a domain of interest and a specification of the meaning of terms in that vocabulary. Depending on the precision of this specification, the notion of ontology encompasses several data or conceptual models, e.g., classifications, database schemas, fully axiomatised theories. Ontologies tend to be everywhere. They are viewed as the silver bullet for many applications, such as database integration, peer-to-peer systems, e-commerce, semantic web services, social networks [Fensel, 2004]. They are, indeed, a practical means to conceptualise what is expressed in a computer format [Brodie *et al.*, 1984]. However, in open or evolving systems, such as the semantic web, different parties would, in general, adopt different ontologies. Thus, merely using ontologies, like using XML, does not reduce heterogeneity: it raises heterogeneity problems to a higher level.

For instance, imagine two organisations dealing with books: one is a cultural product electronic commerce site (which sells books, music, movies, etc.) and the other is a university library. The activities of both organisations deal with some related products, the books, but are concerned with different aspects of these: the seller is concerned by the margin, the publisher or the type of binding. The library, in turn, pays more attention to the topic, the size and the year of publication. Both are concerned by the price and the author. Yet they may consider these differently, because the price can include tax and shipping fees or not and being expressed in different currencies or because the authors can be denoted by individual objects or by the character string of their names. Moreover, the seller may organise the books according to their commercial types and the library according to their literary types. In summary, these two organisations will obviously have different and heterogeneous ontologies.

These two institutions may have to interact, for example, because the second one wants to order books to the first one or because the first one wants to digitise the collections of the second one. In order to do so seamlessly, they need to find the correspondences between the entities in their respective ontologies. The correspondences may express that what is called a book in the ontology of the seller stands for what is called a volume in that of the library. Furthermore, the price in the seller ontology should be multiplied by a tax rate for obtaining the corresponding price in the library ontology. The process of finding these correspondences is called 'ontology matching'.

This book is devoted to ontology matching as a solution to the semantic heterogeneity problem faced by computer systems. Ontology matching aims at finding correspondences between semantically related entities of different ontologies. These correspondences may stand for equivalence as well as other relations, such as consequence, subsumption, or disjointness, between ontology entities. Ontology entities, in turn, usually denote the named entities of ontologies, such as classes, properties or individuals. However, these entities can also be more complex expressions, such as formulas, concept definitions, queries or term building expressions. Ontology matching results, called alignments, can thus express with various degrees of precision the relations between the ontologies under consideration.

Alignments can be used for various tasks, such as ontology merging, query answering, data translation or for browsing the semantic web. In the above mentioned example, the library can take advantage of alignments for automatically ordering a book and the seller can use them for checking the availability of a reference by the library. Matching ontologies enables the knowledge and data expressed in the matched ontologies to interoperate. It is thus of utmost importance for the above mentioned applications whose interoperability is jeopardised by heterogeneous ontologies.

Many different matching solutions have been proposed so far from various viewpoints, e.g., databases, information systems, artificial intelligence. They take advantage of various properties of ontologies, e.g., structures, data instances, semantics, or labels, and use techniques from different fields, e.g., statistics and data analysis, machine learning, automated reasoning, and linguistics. These solutions share some techniques and tackle similar problems, but differ in the way they combine and exploit their results. As a consequence, they are quite difficult to compare and describe, lacking a uniform framework.

About *Ontology Matching*

Ontology Matching aims at being a reference book that presents currently available work in the topic in a uniform framework. In particular, though we use the word ontology, the work and the techniques considered in this book can equally be applied to database schema matching, catalogue integration, XML schema matching and other related problems. The objectives of the book include presenting (i) the state of the art and (ii) the latest research results in ontology matching by providing a detailed account of matching techniques and matching systems in a systematic way from theoretical, practical and application perspectives. The main emphasis of this book is thus on technical solutions for matching.

We have aimed at a sufficiently comprehensive and documented book so that readers can find and learn about almost any subject related to ontology matching and be referred to further reading. Several topics are not covered in full depth but presented only in some salient details for completeness purpose.

It is not the goal of this book to advocate one approach to ontology matching against the others, but rather to show the variety of approaches and their adequacy in different contexts. We are convinced that there is not one unique approach to ontology matching. We concentrate, however, on automatic solutions for matching. Many applications require submitting matching results to user scrutiny and control before using them, but the better the automated part of the task, the easier the control.

This book provides a comprehensive coverage of ontology matching for the researcher and the practitioner. In particular, it reconsiders the former frameworks and classifications, broadening their scope and accounting for more solutions. It goes as far as describing in detail basic techniques used in matching systems, reviewing available systems, providing a framework for their evaluation and discussing their applications. This unified view of ontology matching techniques and solutions aims at being the starting point to implementing matching solutions dedicated to a particular application context or developing new techniques. So readers should find in this book a starting point for implementing and understanding matching, they should not expect the ultimate matching solution to be unveiled.

Ontology Matching is not meant to be a textbook, though it features exercises for a selected number of chapters. These exercises can help readers in evaluating their understanding of some technical concepts. This book is also complemented by a web site[1] which features additional information and resources.

Outline of the book

This book is organised in five parts.

Part I is dedicated to the motivation and the definition of the ontology matching problem. The motivation is given in Chap. 1 through various applications that can take advantage of matching ontologies and the presentation of how matching contributes to these applications. In Chap. 2, the ontology matching problem is technically defined in various instances of ontology matching occurring in different contexts, such as folkosomies, classifications, databases, XML and entity–relationship schemas and finally formal ontologies. It justifies the emphasis of this book on ontology matching and provides definitions for the vocabulary used. Finally, it technically defines the ontology expression languages, the ontology matching process and its result: the alignment.

Part II provides a comprehensive coverage of the techniques currently used for ontology matching. It is the main part of the book. Chap. 3 defines a classification of matching approaches which will be used in the subsequent chapters. Chap. 4 presents the basic methods that can be used for assessing the similarity or dissimilarity of ontology entities. These techniques are the basis of most, if not all, current ontology matchers. The composition of an ontology matching system from these basic techniques is considered in Chap. 5, which presents the high-level tools for designing ontology matching systems.

Part III is devoted to packaged matching systems that can be used in applications. Chap. 6 presents a large panel of state of the art matching systems. The reader will find that the basic techniques presented before can lead to a large diversity of

---

[1] http://book.ontologymatching.org

systems. Chap. 7 is dedicated to the evaluation of matching solutions. It presents techniques for discriminating empirically among these systems and evaluating their suitability to a particular application.

Part IV is devoted to the use of the ontology matching results in applications once they have been obtained. Chap. 8 considers how alignments can be expressed either for being stored or for being communicated between systems. This chapter also presents frameworks in which alignments can be both obtained and used in various ways. Chap. 9 deals with the explanation of matching results to users. This manipulation is important when matching is not expected to be automatic. Finally, Chap. 10 addresses the ultimate use of ontology matching results through their implementation as an effective procedure, e.g., rules, articulation axioms, mediators that can be used within applications.

Part V concludes the book, summarising the current state of ontology matching and emphasising remaining problems that will have to be addressed by further research.

A graphical representation of this organisation is presented below. The arrows offer different independent reading paths through the book.

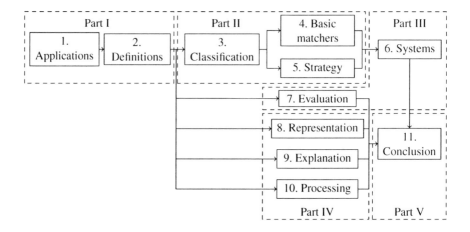

Readership and lecture guide

This book is intended for researchers and practitioners of information and ontology engineering.

The book outline provides a progressive presentation of the ontology matching field and can be read in its entirety. However, each chapter considers ontology matching under a different perspective and can be read in isolation (though it is advised to read the first part before any other). Those who are only interested in getting acquainted with ontology matching can start by reading Chaps. 1, 2 and 11.

For researchers and students dealing with the problem of semantic heterogeneity, we provide not only a comprehensive overview of the state of the art in ontology matching, but also present in detail recent research developments. They show how ontology matching technologies are going to evolve, indicating which research topics are in the academic agenda and which of them represent the scientific challenges. A course on ontology matching should take some motivations from Chap. 1, explain the concepts introduced in Chap. 2, use the classification of Chap. 3 for exposing Chaps. 4 and 5 and certainly provide some insights from Chap. 7.

For information technology practitioners, both from industry and academia, who want to implement an ontology matching component, this book will help take advantage of state of the art solutions. These readers will take more profit in Chaps. 4, 5, 6, 8, 9 and 10.

For professionals in the areas of e-commerce and knowledge management, the book provides decision support on the use of ontology matching technologies, information about potential problems, and guidelines for the successful application of existing approaches. These readers will take more profit in Chaps. 1, 2, 3, 6, 7, 8 and 10.

We only expect from readers a basic knowledge about data and conceptual modelling and graph theory. Knowledge about logics can also be helpful, thought not strictly necessary.

Acknowledgements

The work presented in this book has been partly supported by the Knowledge Web network of excellence (IST-2004-507482) of the European Commission 6th Framework Programme for Research and Technological Development. We emphasise the crucial role played by the European networks of excellence in providing support for cooperative research on important and emerging topics such as this one. This book testifies to the rich working atmosphere these networks contributed to create.

We thank all the participants of the Heterogeneity workpackage of Knowledge Web and, in particular, Than-Le Bach, Jesus Barrasa, Paolo Bouquet, Jan De Bo, Jos De Bruijn, Rose Dieng-Kuntz, Enrico Franconi, Raúl García Castro, Manfred Hauswirth, Pascal Hitzler, Mustafa Jarrar, Markus Krötzsch, Ruben Lara, Malgorzata Mochol, Amedeo Napoli, Luciano Serafini, François Sharffe, Giorgos Stamou, Heiner Stuckenschmidt, York Sure, Vojtěch Svátek, Valentina Tamma, Sergio Tessaris, Paolo Traverso, Raphaël Troncy, Sven van Acker, Frank van Harmelen, and Ilya Zaihrayeu.

Some people had a particular impact on the book through many fruitful discussions, detailed technical feedback on various ontology matching themes, joint work and continuous support during the time we have been elaborating on it. We are very grateful for this to Marc Ehrig, Fausto Giunchiglia, Loredana Laera, Diana Maynard, Deborah McGuinness, Petko Valchev, Mikalai Yatskevich, and Antoine Zimmermann.

We also thank Amedeo Napoli for his careful reading. We are indebted to Fiona McNeill for the time she kindly spent on a first complete draft of this book and her insightful suggestions.

Finally, we are grateful to our Springer Verlag editor, Ralph Gerstner, for his belief that we had material for such a book and for his kind patience during its production.

Part I

The matching problem

1
Applications

Matching models is an important operation in traditional applications, such as ontology integration, schema integration, or data warehouses. Typically, these applications are characterised by heterogeneous structural models that are analysed and matched either manually or semi-automatically at design time. In such applications matching is a prerequisite of running the actual system.

A line of applications that can be characterised by their dynamics, e.g., agents, peer-to-peer systems, web services, is emerging. Such applications, contrary to traditional ones, require (ultimately) a run time matching operation and take advantage of more explicit conceptual models.

In this chapter we first present some well-known applications where matching has been recognised as a plausible solution for a long time. These are ontology engineering (§1.1) and information integration, including schema integration, catalogue integration, data warehouses and data integration (§1.2). Then, we discuss some recently emerged new applications, such as peer-to-peer information sharing (§1.3), web service composition (§1.4), autonomous communication systems, including agents and mobile devices communication (§1.5), and navigation and query answering on the web (§1.6). Finally, the legends to the figures illustrating scenarios under consideration can be found in Appendix A.

1.1 Ontology engineering

A context where users are confronted with heterogeneous ontologies is ontology engineering, and, more generally, the task of designing, implementing and maintaining ontology-based applications. This activity requires support of ontology matching because ontology engineering has to deal with multiple, distributed and evolving ontologies.

1.1.1 Ontology editing and import

Ontology heterogeneity may be first faced while designing an ontology for a domain of interest. Ontology-based system designers often have to integrate different ontologies, either for the sake of enforcing reuse, and thus not multiplying ontologies on the same topic, or because it is necessary for interconnecting various relevant resources.

It is often the case that application engineering requires an external set of ontologies to be put together. For instance, building a library cataloguing ontology may require assembling ontologies for people, books and topics as well as ontologies for measurement units, geographic coordinates, book identification numbers, metadata ontology, etc. These ontologies share related concepts: for instance, the friend-of-a-friend (FOAF[1]) ontology (which can be used as a starting point for modelling people) offers a document concept that has to be related to the classes of the book identification numbers ontology.

Ontology engineers need support for (i) identifying the relevant ontologies and (ii) matching and recording the relations between the entities in these ontologies. Additionally, they may want to import the identified ontologies and merge them (in which case, they will use some axioms generated from the result of the matching phase) or to use data expressed under another ontology in the application (in which case, they will generate a mediator from the matching result).

The scenario under consideration is simple because it is static. In fact, ontologies are encountered at design time and mediators can be built at that moment. Thus, the application developer can find the correspondences and design the necessary transformations manually. Some tools provide support for finding the correspondences, for example, Protégé through the Prompt suite of tools [Noy and Musen, 2000]. Newer ontology development environments will have to take into account, from the beginning, the existence of multiple ontologies and the need for the mediators between them.

1.1.2 Ontology evolution and versioning

It is natural that domains of interest, application requirements and the way in which knowledge engineers conceptualise those by means of ontologies undergo changes and evolve over time. Moreover, ontology development, similar to software development, is often performed in a distributed and collaborative manner. Therefore, multiple versions of the same ontology, e.g., the Gene ontology[2], often exist. Some applications keep their ontologies up to date, while others may continue to use old ontology versions and update them on their own. These situations arise because knowledge engineers and developers usually do not have a global view of how and where the ontologies have changed. In fact, change logs may not always be available (which is often the case in distributed ontology development). Therefore, developers need to manage and maintain the different versions of their ontologies.

---

[1] http://www.foaf-project.org
[2] http://www.geneontology.org

The matching operation is of help here, see Fig. 1.1. Its main focus is on discovering the differences, e.g., what ontology entities have been added, deleted or renamed, between two ontology versions [Roddick, 1995, Noy and Klein, 2004, Noy and Musen, 2002b, Noy and Musen, 2004].

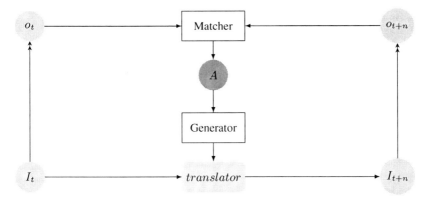

Fig. 1.1. Ontology evolution scenario. In this scenario it is useful to: (1) *match* the old version o_t and the new version o_{t+n} of the ontology, thus resulting in a set of correspondences (A) between these versions, (2) *generate* a transformation by using these correspondences and (3) *translate* the underlying data instances I_t to I_{t+n}.

1.2 Information integration

Information integration is one of the oldest classes of applications where matching is viewed as a plausible solution. Under the information integration heading, we gather here such problems as schema integration [Batini et al., 1986, Sheth and Larson, 1990, Spaccapietra and Parent, 1991, Parent and Spaccapietra, 1998], data warehousing [Bernstein and Rahm, 2000], data integration (also known as enterprise information integration) [Chawathe et al., 1994, Wache et al., 2001, Draper et al., 2001, Halevy et al., 2005], and catalogue integration [Agrawal and Srikant, 2001, Ichise et al., 2003, Bouquet et al., 2003c, Giunchiglia et al., 2005a].

A general information integration scenario is presented in Fig. 1.2: given a set of local information sources (local ontologies $LO_1, \ldots LO_n$) potentially storing their data in different formats, e.g., SQL DDL, XML, or RDF, provide users with a uniform query interface via the mediated (or global) ontology CO, to all the local information sources. This allows users to avoid querying the local information sources one by one, and to obtain a result from them just by querying a common ontology.

For example, if users pose queries like *find a book about Logics* to a common ontology, then, an information integration system communicates with information

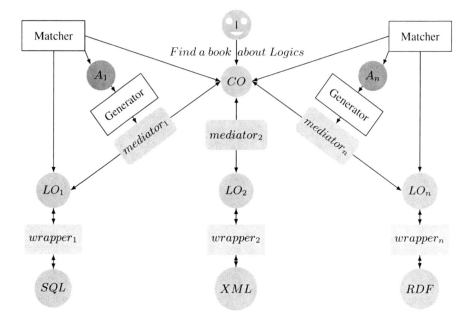

Fig. 1.2. A general (centralised) information integration scenario. The data sources (SQL, RDF, etc.) are wrapped ($wrapper_i$) to ontologies (LO_i) which are *matched* against a common ontology (CO). The alignments (A_i) between these help *generate* mediators ($mediator_i$) which in turn transform queries against the common ontology into a query to the information source and translate the answers in the other way.

sources, e.g., Amazon, Barnes & Noble, and returns a reconciled result based on the input provided by these sources. In general, the information integration system performs several macro steps. These include:

- interpret (rewrite) the query in terms of the common ontology;
- identify the correspondences between semantically related entities of the local information sources and the common ontology;
- translate the relevant data instances of the local information sources (involved in handling the query) into a knowledge representation formalism of the information integration system;
- reconcile the results obtained from multiple information sources, namely detecting and eliminating, e.g., redundancies, duplications, before returning the final answer.

Identifying the correspondences between semantically related entities of the local information sources and the common ontology is a matching step. Let us limit our vision of matching to the description above for the moment. We will expand it to some extent in the next sections.

In some concrete information integration scenarios, the common ontology can be either physically existing or virtual. Below, we discuss these scenarios in some detail.

1.2.1 Schema integration

Schema integration is the oldest scenario [Batini *et al.*, 1986, Sheth and Larson, 1990, Parent and Spaccapietra, 1998]. Suppose that two (or more) enterprises want to perform either a merger or an acquisition among them. Ultimately, these enterprises have to integrate their databases into a single one. Usually, a first technical step is to identify correspondences between semantically related entities of the schemas before merging the databases. This step, known as matching, is required even if the databases to be integrated are coming from the same domain of interest, e.g., book selling, car rentals. This is because the schemas have been designed and developed independently. In fact, people follow diverse modelling principles and patterns, even if they have to encode the same real-world object. Finally, the schemas to be integrated might have been developed according to different business goals. This makes the matching problem even harder.

Under the schema integration heading we can classify some other scenarios. For example, (tightly-coupled) federated databases [Sheth and Larson, 1990]. These typically have one global schema providing a unified access to the federation of component databases. Component databases, in turn, are autonomous. Thus, in this application when, for example, one component schema of the federated database is changed, the federated (global) schema has consequently to be also reconsidered. Matching can help in identifying those changes.

Finally, it is worth noting the applications which we are not discussing here, e.g., distributed databases systems [Özsu and Valduriez, 1999]. These are usually designed in a centralised way, e.g., by a database administrator, and therefore, semantic heterogeneity does not exist there by construction [Elmagarmid *et al.*, 1999].

1.2.2 Catalogue integration

In Business-to-Business (B2B) applications, trade partners store information about their products in electronic catalogues. Typical examples of catalogues are product directories of electronic sales portals, such as Amazon or eBay. In order for a merchant to participate in the marketplace, e.g., eBay, it has to determine correspondences between entries of its catalogues and those of the marketplace catalogue (see Fig. 1.3). This process of finding correspondences among entries of the catalogues is referred to as the catalogue matching problem [Bouquet *et al.*, 2003c]. Notice that if we look at this problem from a merchant viewpoint, matching has to be performed for each marketplace it would like to participate. Having identified the correspondences between the entries of the catalogues, they are further analysed in order to generate query expressions that automatically translate data instances between the catalogues. Finally, having matched the catalogues, users of a marketplace have a unified access to the products which are on sale. The above described scenario involving interactions between marketplaces and merchants can be viewed as a typical example of integrating local data sources into a data warehouse, see also [Bernstein and Rahm, 2000].

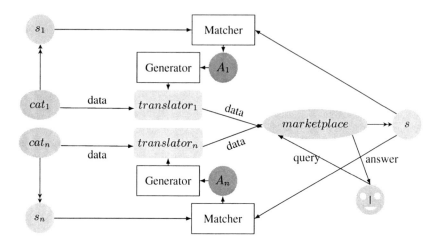

Fig. 1.3. Catalogue integration scenario with matching. Each merchant *matches* its catalogue (s_i) with that of the marketplace (s). From the matching result (A_i) it is *generated* a data translation program ($translator_i$) which is used for loading the catalogue (cat_i) to the *marketplace*. Users can ask queries to the *marketplace* and receive answers based on the integrated catalogue.

Another catalogue integration scenario deals with (typically large-scale) product classifications, such as UNSPSC[3] (The United Nations Standard Products and Services Code) and eCl@ss[4] (Standardised Material and Service Classification). In a sense, we can view this scenario as one which enables interoperability among multiple B2B marketplaces, thus, facilitating product exchange between the enterprises subscribing to different product classifications [Schulten et al., 2001]. This is to be achieved by establishing the correspondences between semantically related entities of the standardised product classifications, which is a matching operation as well.

1.2.3 Data integration

Data integration is an approach where integration of information coming from multiple local sources is performed *without* first loading their data into a central warehouse [Halevy et al., 2005]. This allows interoperation across multiple local sources having access to the up-to-date data. Notice that in the above considered catalogue

---

[3] http://www.unspsc.org
[4] http://www.eclass.de

integration scenario, merchants are those who have to perform updates of the central warehouse of the marketplace. In this scenario the data integration system provides this functionality.

The scenario, depicted in Fig. 1.4, is as follows. First, local information sources participating in the application, e.g., bookstore, library, museum, are identified. Then, a virtual common ontology is built. Queries are posed over the virtual common ontology, and are then reformulated into queries over the local information sources, e.g., in cultural heritage applications, these might be museums, such as Iconclass[5] and Rijksmuseum[6]. In order to enable semantics-preserving query answering, correspondences between semantically related entities of the local information sources and the virtual ontology are to be established. Establishing these correspondences is known as a matching.

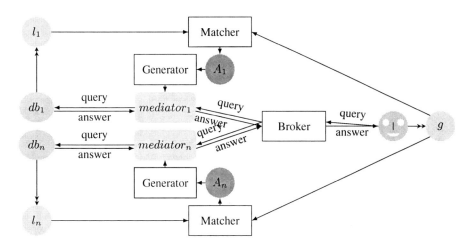

Fig. 1.4. Data integration scenario with matching. Depending on if the global schema (g) is considered as *matched* against existing local schemas (l_i) or the other way around, this describes GAV or LAV approach, respectively. Usually, the matching phase, resulting in alignments (A_i), is done off-line, *generating* mediators ($mediator_i$) for each local database. The query is sent to a *broker* calling the necessary mediators. They translate the query, evaluate it against the database and translate the answer before returning it.

Query answering is then performed by using these correspondences (mappings) within the Local-as-View (LAV), Global-as-View (GAV), or Global-Local-as-View (GLAV) settings [Lenzerini, 2002]. In the LAV approach, local schemas are defined in terms of the global schema, i.e., the mapping is specified by defining each local schema construct as a view over global schema constructs. Queries are processed by means of an inference mechanism that re-expresses the atoms of the global schema in terms of atoms of the local schemas. In GAV, a global schema is defined in terms of

---

[5] http://icontest.iconclass.nl/libertas/ic?style=index.xsl
[6] http://www.rijksmuseum.nl/aria/aria_catalogs/index?lang=en

the local schemas, i.e., the mapping is specified by writing a definition of each global schema construct as a view over local schema constructs. Queries are processed by means of unfolding, i.e., by expanding the atoms according to their definitions (so as to come up with local schema relations). GLAV, in turn, is a mixed approach. We can think of it as a variation of the LAV approach that allows the head of the view definition to contain any query on the local schemas.

Finally, as noticed in [Lenzerini, 2002], the main task in these applications is to establish the mappings, i.e., perform the matching operation. Besides using matching results for creating the global (respectively local) views, it can also be used for maintaining them when schemas evolve.

1.3 Peer-to-peer information sharing

Peer-to-Peer (P2P) is a distributed communication model in which parties (also called peers) have equivalent functional capabilities in providing each other with data and services [Zaihrayeu, 2006]. P2P networks became popular through a file, e.g., pictures, music, videos, books, sharing paradigm. There exists several widely used P2P file sharing systems, e.g., Kazaa, Edonkey, and BitTorrent. These applications describe file contents by a simple schema (set of attributes, such as title of a song, its author, etc.) to which all the peers in the network have to subscribe. These schemas cannot be modified locally by a single peer. Therefore, in the above mentioned systems the semantic heterogeneity problem (at the schema level) does not exist by construction. The use of a single system schema violates the *total autonomy* of peers. Although robust P2P systems allow peers to connect to and disconnect from the network at any time, thereby respecting some forms of peers autonomy, such as *participation autonomy*, they still restrict the *design autonomy* of peers, in matters such as how to describe the data and what constraints apply on the data [Zaihrayeu, 2006].

If peers are meant to be totally autonomous, they may use different terminologies and metadata models in order to represent their data, even if they refer to the same domain of interest. Thus, in order to establish (meaningful) information exchange between peers, one of the steps is to identify and characterise relationships between their ontologies. This is a matching operation. Having identified the relationships between ontologies, these can be used for the purpose of query answering, e.g., using techniques applied in data integration systems, see Sect. 1.2.

1.3.1 Semantic P2P systems

Semantic P2P systems [Staab and Stuckenschmidt, 2006] use more complex specifications of their contents, such as database schemas [Bernstein *et al.*, 2002], or formal ontologies [Rousset *et al.*, 2006], than the classical P2P systems mentioned above. The main idea behind this is to improve the search accuracy by providing a finer-grained description of items. For example, users who want to share their book library with their friends may index them by authors, topics, and years of publication.

1.3 Peer-to-peer information sharing

This tagging approach will benefit from using some ontological descriptions, e.g., for retrieving books on mathematics written by Cambridge authors before 1920 as opposed to books by Bertrand Russell in 1908 on logic. For instance, the BibSter system [Haase et al., 2004] uses a bibliographic ontology expressed in RDF. Such systems as BibSter still follow a single ontology approach, thereby limiting the design autonomy of peers, and thus, the semantic heterogeneity problem, at the schema level, does not exist by construction.

More advanced semantic P2P systems relax the homogeneity requirement of classical P2P systems: they allow peers to use independent schemas and ontologies, see Fig. 1.5.

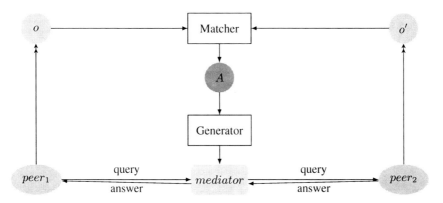

Fig. 1.5. P2P query answering. In this scenario, it is useful to: (1) *match* relevant parts of ontologies o and o', thus resulting in alignment A, (2) *generate* a *mediator* between $peer_1$ and $peer_2$ for translating queries and sometimes for translating answers.

Such applications pose additional requirements on matching solutions. In P2P settings which respect total autonomy of peers, an assumption that all the peers rely on one global schema, as in data integration, cannot be made because the global schema may need to be updated any time the system evolves [Giunchiglia and Zaihrayeu, 2002]. While in the case of data integration schema matching can be performed at design time, in P2P applications peers need to coordinate their databases on-the-fly, therefore ultimately requiring run time schema matching. Finally, incomplete and approximate answers, as long as they are good enough for the application, are also acceptable in such settings. This is the case because some mappings involved in query answering may become temporarily unavailable or invalid [Shvaiko et al., 2006b].

Some examples of various P2P scenarios which rely on different peer metadata models, including relational database schemas, XML schemas, RDF schemas, or OWL ontologies are described in [Bernstein et al., 2002, Zaihrayeu, 2006, Ives et al., 2004, Nejdl et al., 2002, Rousset et al., 2006]. For example, applications like SomeWhere [Rousset et al., 2006] integrate peer databases and connect them through mappings expressed in Horn clauses from one database to another. When a peer needs to answer a query, the system computes possible expansions of the query with regard to these mappings, i.e., it follows the LAV approach [Lenzerini, 2002]. Then it sends to each relevant peer the queries that can help answer the initial query and answers are returned and integrated as soon as they arrive. This approach assumes that peer database schemas have been matched off-line beforehand. Thus, only the query answering part of the system takes into account the dynamics of the P2P environment.

1.3.2 Emergent semantics between peers

Emergent semantics [Aberer et al., 2004b, Aberer et al., 2004a] is the process by which a set of peers gradually converges towards a consensus ontology through constantly interacting and negotiating the meaning of the terms they use. This process mimics to some extent the one exhibited by a society of humans and may never reach an end but at least it improves discourse understanding. Since consensus is built incrementally, emerging from different local point-to-point peer agreements, an alignment between ontologies of peers is viewed as a practical means for establishing those local agreements. Thus peers will have to constantly update the relations between their ontologies. These updates can be achieved by a matching operation. The process of emerging semantics between two peers is illustrated in Fig. 1.6.

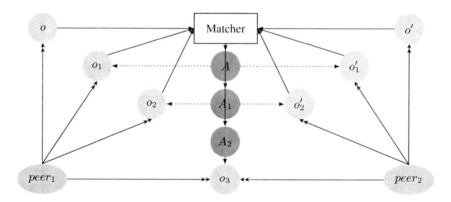

Fig. 1.6. Peer-to-peer and emergent semantics: after a first *matching* between ontologies o and o', the resulting alignment A causes (dotted line) the peers ($peer_1$ and $peer_2$) to evolve their ontologies into o_1 and o'_1, respectively. In turn, these ontologies (o_1 and o'_1) can again be *matched*, thus resulting in alignment A_1, and so on and so forth. Ultimately, the peers may converge to a common ontology (o_3).

Constantly matching ontologies can trigger the confrontation and revision of these ontologies themselves. In fact, users may want to establish more consensual ontologies from this confrontation [Zhdanova et al., 2005]. There are several ways in which alignments can help here:

- Alignments provide a basis from which the negotiation between peers can start (like agent protocols for arguing about correspondences, see Sect. 1.5.1).
- Matching algorithms are very often able to compute a distance between ontologies. This is useful when, for instance, a peer wants to find the 'closest' ontology.
- By building a network of ontologies together with alignments between them and by exploiting, with the help of social network analysis techniques, the distance between the ontologies, it is also possible to determine the proximity between users or agents. This, in turn, facilitates customising the query answering process, and even the consensus building.

These results will help users and communities in consolidating their ontologies by achieving agreements gradually with similar domain representations as well as for determining the most central ontology (in social network analysis terms) for the domain of interest.

1.4 Web service composition

Web services are processes that expose their interfaces to the web so that users can invoke them. Semantic web services provide a richer and more precise way to describe the services through the use of knowledge representation languages and ontologies [Fensel *et al.*, 2007]. Web service discovery and integration is the process of finding a web service able to deliver a particular service and composing several services in order to achieve a particular goal [Paolucci *et al.*, 2002, Medjahed and Bouguettaya, 2005, Oundhakar *et al.*, 2005].

Web services have been designed for being independent and replaceable. So web service processors are able to incorporate new services in their workflows, and therefore customers can dynamically choose new and more promising services. For that purpose, they must be able to compare the descriptions of these services (in order to know if they are indeed relevant) and to route the knowledge they process in order to compose different services by routing the output of some service to the input of another service.

However, in the case of semantic web services, which can be described with regard to ontologies, imposing a central common ontology (like in single ontology P2P systems), as real-world experiences demonstrate, is not realistic and would freeze the evolution of such services. Henceforth, both for finding the adequate service and for interfacing services, a data mediator comes into play as a bridge between different vocabularies [Bussler *et al.*, 2002, Roman *et al.*, 2004]. From the correspondences between the terms of the descriptions, mediators must be able to translate the output of one service into a suitable input for another service, see Fig. 1.7.

Thus, the core part of a mediator definition is an alignment between two ontologies. This, in turn, can be provided through matching the corresponding ontologies either off-line when someone is designing a preliminary service composition, or dynamically (on-line) [Giunchiglia *et al.*, 2006b, Robertson *et al.*, 2006], when new services are sought for completing a request.

20   1 Applications

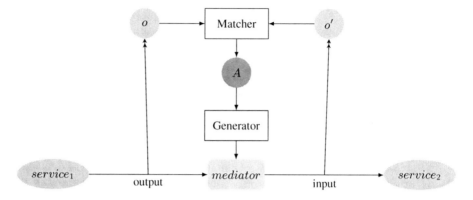

Fig. 1.7. Web service composition. In this scenario it is useful to: (1) *match* relevant parts of ontologies o and o', thus resulting in alignment A, (2) *generate* a *mediator* between $service_1$ and $service_2$ in order to enable transformation of the actual data.

For instance, suppose some on-line library service provides its output description in some ontology and a parcel shipping service uses a second ontology for describing its input. Matching these ontologies is useful for: (i) checking that what is delivered by the first service, e.g., a Book, matches what is expected by the second one, e.g., an Object, (ii) verifying preconditions of the second service, e.g., size in centimetres against dimensions in inches, and (iii) generating a mediator able to transform the output of the first service in order to be input to the second one (see Fig. 1.7).

1.5 Autonomous communication systems

Other kinds of applications also involve autonomous entities that can meet on a network and which have been designed independently. When these entities are software programs, they have been considered as agents for a long time (§1.5.1). However, if they are a combination of hardware and software they are a matter of ambient computing (§1.5.2). Obviously, as we have already discussed in previous sections, such entities cannot share a common ontology. Thus, if they want to communicate, it is useful to match their ontologies.

1.5.1 Multi-agent communication

Agents are software entities characterised by their autonomy and capacity of interaction. They are often divided into cognitive agents and reactive agents. Reactive agents implement a simple behaviour and the strength of these agents is their capacity to let a global behaviour emerge from the individual behaviour of many such agents. Cognitive agents have a rather more elaborate behaviour often characterised as the ability to pursue goals, to plan their actions and to negotiate with other agents in order to achieve their goals.

1.5 Autonomous communication systems

Agents communicate by exchanging messages expressed in an agent communication languages, such as the FIPA Agent Communication Language [FIPA0061, 2002, FIPA0037, 2002]. These languages determine the 'envelope' of the messages and enable agents to position themselves within a particular interaction context. However, they do not specify the actual content of the message, which is often expressed with respect to some ontology accessible to the agent. Current standards for expressing these messages provide slots for declaring the content language and the ontology used.

As a consequence, when two autonomous and independently designed agents meet, they have the opportunity to exchange messages but little chance to understand one another if they do not share the same content language and ontology. It is thus useful to help these agents to match their ontologies in order to either translate their messages or integrate bridge axioms in their own models. Several proposals have been made to assess the correspondences between the terms of the ontologies [van Eijk *et al.*, 2001, Wiesman *et al.*, 2002, Bailin and Truszkowski, 2002, Wang and Gasser, 2002, Euzenat *et al.*, 2005a].

Agents confronted with heterogeneous ontologies have to find the correspondences between these ontologies in order to start understanding each other's messages. They can perform ontology matching by themselves or by taking advantage of alignment libraries or matching services. Once an alignment is obtained, agents can start a negotiation phase [Laera *et al.*, 2006] in which they exchange arguments for or against correspondences. When they find a mutual agreement they can transform the resulting alignment in a program that translates the exchanged messages or in axioms which, once integrated in the agent knowledge, enable interpretation of messages, see Fig. 1.8.

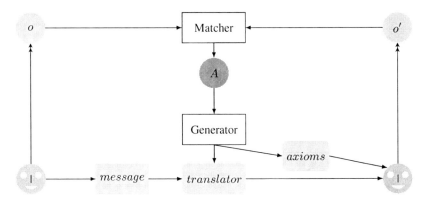

Fig. 1.8. Agent communication. In this scenario it is useful to: (1) *match* relevant parts of ontologies o and o' used by each of the agents, thus resulting in alignment A, (2) *generate* bridge *axioms* between two ontologies, and (3) incorporate the *axioms* into o'. Alternatively, the process can (2′) *generate* a *message translator* from ontology o to ontology o' and (3′) apply this *translator* to the *message*.

1.5.2 Matching contexts in ambient computing

In ambient computing, applications running on mobile devices take advantage of the surrounding environment for providing services to users. Naturally, this environment undergo changes, e.g., with regard to user locations, and applications must always keep track of these changes, including newly appearing devices and sensors. Characterising the context in ambient computing goes through finding the information about the current situation in the environment by using various devices available in that environment, e.g., sensors. By doing so, applications provide context-aware solutions. If one wants to design flexible and smart ambient computing applications, it is useful to take advantage of the ontologies of these various devices, those of sensors available in the environments and their capabilities [Coutaz *et al.*, 2005]. Similar to web service descriptions, these ontologies will provide descriptions of the devices, even of abstract devices, such as a temperature service, and the way to interact with them.

Once again, it is expected that device providers will develop different ontologies adapted to their products or will extend some standard ontologies. Moreover, since applications evolve in ever changing environments in which devices can fail and new ones can appear, there is no way to freeze once and for all the ontologies that are relevant and available at a particular moment.

Therefore, in order to properly operate in ambient computing environments, applications have to be expressed in terms of generic features that are matched against the actual environment. This matching process can take advantage of ontology matching, since similar devices are likely to be used by similar applications. Thus, providing a service for reconciling various ontologies and storing the results obtained from previous interactions should help these applications in sharing and reusing the established alignments.

1.6 Navigation and query answering on the web

This section presents several applications some of which extend the web experience to the semantic web by using resources such as formal ontologies. Operating in an open environment, these applications most often require matching. In particular, the applications under consideration include: navigation on the semantic web (§1.6.1), query answering on the web (§1.6.2), and query answering on the deep web (§1.6.3).

1.6.1 Navigation on the semantics web

Browsers such as Magpie [Dzbor *et al.*, 2003, Dzbor *et al.*, 2004] are designed to take advantage of semantic annotations associated with web pages. For instance, Magpie can recognise manifestation of instances of an ontology in a web page, display these instances specifically (different colours for different classes) and add services such as linking to the instance web page.

In open web browsing, the key point is to be able to select, at run time, the appropriate ontologies for the given browsing context. Indeed, the web pages are linked to other web pages whose content may notably differ from that of the source page. In order to improve the user experience, it is necessary to take new ontologies into account dynamically and to be able to connect them to the current ontologies. Thus, ontology matching is needed to match between a set of terms that describes the topic of the current page and the relevant on-line ontologies.

Let us consider an example [Sabou et al., 2006b]. The following short news story is about both trips to exotic locations and talks.

> For April and May 2005, adventurer Lorenzo Gariano was part of a ten-man collaborative expedition between 7summits.com and the 7summits club from Russia, led by Alex Abramov and Harry Kikstra, to the North Face of Everest. This evening he will present a talk on his experiences, together with some of the fantastic photos he took.

An ontology that covers such concepts as adventurer, expedition, talk and photos should be selected or discovered from the web. This requires that the above mentioned concepts are matched to the corresponding concepts from the available on-line ontologies. In addition, if some of the search terms cannot be found in an ontology, correspondences with more or less general concepts in the ontology are acceptable. Finally, not all the entities of the ontology need to be involved in matching. It is sufficient to consider only those entities that are similar to the terms found on the web page.

1.6.2 Query answering on the web

Contrary to the information integration scenario (§1.2), information on the web is not described by a global schema over which queries can be expressed. Moreover, users are used to query the web using their own terminology. Then a semantic query answering system on the web has to rewrite the query with respect to available ontologies in order to use reasoning for providing answers.

For instance, a query answering system such as AquaLog [Lopez et al., 2005] is aware of an ontology about academic life which has been populated to describe knowledge related to some university [Sabou et al., 2006b]. For answering a query such as: *Which projects are related to researchers working with ontologies?*, AquaLog interprets it in terms of entities available in the system ontology. For this, it first translates this query into the following triples: ⟨projects, related to, researchers⟩ and ⟨researchers, working, ontologies⟩. Then it attempts to match these triples to the concepts of the underlying ontology. For example, the term projects should be identified to be equivalent to the ontology concept Project and ontologies is assumed equivalent to the ontologies instance of the Research-Area concept. If Action is a subclass of Project, the system will be able to take actions into account in its answers.

Currently, the scope of AquaLog is limited by the amount of knowledge encoded in the ontology of the system. A new version of AquaLog, called Power-Aqua [Lopez et al., 2006], extends its predecessor, as well as some other systems with similar goals, such as Observer [Mena *et al.*, 1996], towards open query answering. PowerAqua aims at selecting and aggregating information derived from multiple heterogeneous ontologies on the web. Matching constitutes the core of this selection task. Unlike AquaLog, matching is now performed between the triples and many on-line ontologies (not just the single ontology of the system). It is not necessary to match all query triples within one ontology. When no ontology concept is found for an element of a triple, the use of more general concepts is also acceptable. Moreover, it is not necessary to try to match the whole ontology against the query, but only the relevant fragments.

1.6.3 Query answering on the deep web

The so-called *deep web* is made of the web sites searchable via query interfaces (HTML forms) giving access to one or more back-end web databases. It is believed that it contains much more information [Chang *et al.*, 2004] than the billions of static HTML pages. At the moment, search engines are not very effective at crawling and indexing the deep web, since they cannot handle meaningfully the query interfaces. For example, according to [Chang *et al.*, 2004], Google and Yahoo both manage to index 32% of the existing deep web objects. Hence, the deep web remains largely unexplored, in spite of containing a huge number of on-line databases, which may be of use.

Thus, users have difficulties, first in discovering the relevant deep web resources and then in querying them. A standard use case includes buying a book with the lowest price among multiple on-line book stores. Query interfaces can be viewed as simple schemas (sets of terms). For example, in the book selling domain, the query interface of an on-line booksore can be considered as a schema represented as a set of concept attributes, namely Author, Title, Subject, ISBN, Publisher. Thus, in order to enable query answering from multiple sources on the deep web, it is necessary to identify semantic correspondences between the attributes of the query interfaces of the web sites. This correspondence identification is a matching operation. Ultimately, these correspondences are used for the on-the-fly translation of a query between interfaces of the web databases.

1.7 Summary

The above panorama shows a widespread need for ontology (in a wide sense) matching. Moreover, the need for matching is not limited to one particular application. In fact, it exists in any application that communicates through ontologies. Thus, it is natural that in the future more examples of applications requiring matching will appear, e.g., ontology repair [McNeill, 2006].

Since semantic heterogeneity is an intrinsic problem of any application involving more than one party, it is reasonable to consider ontology matching as a unified object of study. However, there are notable differences in the way these applications use matching. The application related differences must be clearly identified in order to provide the best suited solution in each case.

These applications can be ordered according to their *dynamics*, namely autonomy of parties participating in an application and rate of changes in an application.

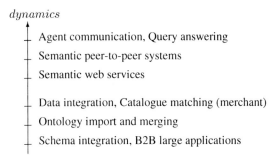

Fig. 1.9. Distribution of some applications with regard to their dynamics.

Fig. 1.9 orders the applications based on dynamics. It shows that agent communication and query answering have a more dynamic profile compared to other applications. In fact, agents, besides having the ability to enter or leave the network or to change their ontologies at any moment (as in the peer-to-peer case), are also able to negotiate the alignments and potential mismatches. Data integration and merchant catalogue matching, due to multiple new merchants being willing to participate in marketplaces, have a higher dynamics than ontology import and schema integration, where typically only a small and limited number of parties participate. Finally, the three bottom applications represent traditional applications, while the three top applications can be considered as dynamic applications. The uneven step in the middle of the dynamics axis in Fig. 1.9 is used to stress the above mentioned distinction.

Another dimension along which these applications differ is the purpose for which they perform matching:

- ontology engineering requires the ability to *transform* relevant ontologies or some parts of these ontologies into an ontology focusing on a domain of interest being modelled or to generate a set of bridge axioms that will help in identifying corresponding concepts (the transformations apply at the ontological level);
- schema integration requires the ability to *merge* the schemas under consideration into a single schema (the transformations apply at the ontological level and instance translation applies at the data level);
- data integration requires the ability to *translate data* instances residing in multiple local schemas according to a global schema definition in order to enable query answering over the global schema (this involves query translation at the ontological level and data translation at the data level);
- peer-to-peer systems and more generally query answering systems require bidirectional *mediators* able to transform queries (ontological level) and translate back answers (data level);

Table 1.1. Summary of application requirements.

Application	instances	run time	automatic	correct	complete	operation
Ontology evolution (§1.1)	✓			✓	✓	transformation
Schema integration (§1.2)	✓			✓	✓	merging
Catalogue integration (§1.2)	✓			✓	✓	data translation
Data integration (§1.2)	✓			✓	✓	query mediation
P2P information sharing (§1.3)		✓				query mediation
Web service composition (§1.4)		✓	✓	✓		data mediation
Multi agent communication (§1.5)		✓	✓	✓	✓	data translation
Context matching in ambient computing (§1.5)		✓	✓	✓		data translation
Semantic web browsing (§1.6)	✓	✓				navigation
Query answering (§1.6)	✓	✓		✓		query reformulation

- agent communication requires *translators* for messages sent from one agent to another, which apply at the data level; similarly, semantic web services require one-way data translations for composing services.

This leads to different requirements for different applications. These requirements concern:

- the type of available input a matching system can rely on, such as schema or instance information. There are cases when data instances are not available, for instance due to security reasons [Clifton et al., 1997] or when there are no instances given beforehand. Therefore, these applications require only a matching solution able to work without instances (here schema-based method).
- some specific behaviour of matching, such as requirements of (i) being *automatic*, i.e., not relying on user feedback; (ii) being *correct*, i.e., not delivering incorrect matches; (iii) being *complete*, i.e., delivering all the matches; and (iv) being performed at *run time*.
- the use of the matching result as described above. In particular, how the identified alignment is going to be processed, e.g., by merging the data or conceptual models under consideration or by translating data instances among them.

Table 1.1 summarises what we found to be the most important requirements for matching solutions in the applications considered in this chapter. This is obviously a general approximation that must be adapted to each particular application.

Some of these hard requirements can be derived into comparative (or non functional) requirements such as *speed*, resource consumption (in particular memory requirements), degree of correctness or completeness. They are useful for comparing solutions on a scale instead of an absolute (yes/no) comparison. Moreover, they allow trading a requirement, e.g., completeness, for another more important one, e.g., speed.

These general requirements for applications will be used in Chap. 6 for assessing the capacity of matching systems to be applied to particular applications, in Chap. 7 for designing evaluation procedures related to applications, and in Chap. 10 to classify the operations performed after matching.

As this brief overview indicates, there are many different applications which require or can take advantage of matching ontologies. However, in spite of a common need for matching, the application matching requirements are quite different. In particular, one can distinguish between traditional and dynamic applications both from the dynamics standpoint and the requirement standpoint. These two observations justify both the unified treatment of matching that we take in this book and the position of considering matching being a separate operation, as opposed to considering merging or mediating being the primitive ones.

The next chapter will go deeper in providing a more precise definition to this unified view of matching.

2

The matching problem

In a distributed and open system, such as the semantic web and many other applications presented in the previous chapter, heterogeneity cannot be avoided. Different actors have different interests and habits, use different tools and knowledge, and most often, at different levels of detail. These various reasons for heterogeneity lead to diverse forms of heterogeneity, and, therefore, should be carefully taken into consideration.

In this chapter we first present various existing ways of expressing knowledge that are found in diverse applications (§2.1). We then discuss in more detail ontologies and ontology languages as knowledge representation formalisms (§2.2). We introduce several justifications for heterogeneity (§2.3). These should help in designing a matching strategy with respect to the kind of heterogeneity that has to be faced. Then, we briefly review some terminology related to matching and alignment as well as provide the meaning that will be used for these terms in this book (§2.4). Finally, we give a formal account of the matching problem by defining a semantics for the matching result, i.e., the alignment (§2.5).

Our goal here is not to close the debate by providing some ultimate semantics for alignments or by settling the definitive meaning of terms, but rather to give definitions that help the reader in understanding better the matching solutions that are presented in this book, as well as the results they produce.

2.1 Vocabularies, schemas and ontologies

So far we have considered ontologies without being precise about their meaning. An ontology can be viewed as a set of assertions that are meant to model some particular domain. Usually, the ontology defines a vocabulary used by a particular application. In various areas of computer science there are different data and conceptual models that can be thought of as ontologies. These are, for instance, folksonomies, database schemas, UML models, directories, thesauri, XML schemas and formal ontologies (axiomatised theories). These and other examples are given in decreasing order of formality in Fig. 2.1. Thus, a top level ontology is supposed to have an explicit well defined semantics, whereas the interpretation of directories in a file system is mostly implicit. In fact, it depends only on what its creator had in mind, i.e., the meaning of labels, the background knowledge, and the context in which those labels occur are all implicit, and therefore, these are not a part of the directory specification.

	'Ordinary' glossaries	Data dictionaries	Structured glossaries	Principled, informal hierarchies	XML schemas	Formal taxonomies	Description logics	
Terms	Ad hoc hierarchies	Thesauri	XML DTDs	Database schemas	Entity-relationship models	Frames	Logics	$expressivity$
	Glossaries and data dictionaries		**Thesauri and taxonomies**		**Metadata and data models**		**Formal ontologies**	

Fig. 2.1. Various forms of ontologies ordered by their expressivity (adapted from [Uschold and Gruninger, 2004]).

We provide below examples of various forms of ontologies of Fig. 2.1 and illustrate some heterogeneity problems encountered in these forms.

2.1.1 Tags and folksonomies

Tags and *folksonomies* are used as very simple ways to describe a corpus of knowledge by just giving names, called tags, to them. This is used in popular web sites, such as del.icio.us[1] for web site annotation, or Flickr[2] for annotating pictures. An example of tags for books and book collections is given in Fig. 2.2.

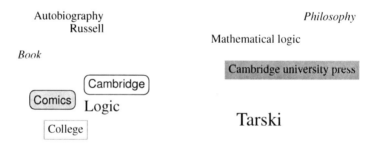

Fig. 2.2. Fragments of two folksonomies.

---

[1] http://del.icio.us
[2] http://www.flickr.com

Obviously, different users use different tags. Even if these tags remain internally coherent for the user who created them, this internal structure is not explicit for the machine. It is difficult to find relations between the tags of two folksonomies. Moreover, the fact that these tags do not have direct relations with each other (in one folksonomy) makes that problem even harder. However, there has been work aiming at inducing a structure between the tags, e.g., the Flickr clusters. These are based mostly on the set of objects, e.g., pictures, web sites, that are indexed by the corresponding tags.

2.1.2 Directories

A taxonomy is a partially ordered set of taxons (classes) in which one taxon is greater than another one only if what it denotes includes what is denoted by the other. Directories or classifications are taxonomies that are used by companies for presenting goods on sale, by libraries for storing books, or by individuals to classify files on a personal computer. Some well-known examples of directories include those of Google[3], Yahoo[4] and the Open Directory Project[5]. These directories are hierarchies of folders identified by labels and containing items, such as bookmarks, or goods. The semantics of these folders is given by the items they ultimately contain [Giunchiglia et al., 2006a]. Of course, each independent entity tends to develop its own directory based on its own needs and tastes, see Fig. 2.3.

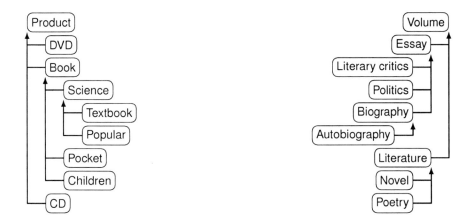

Fig. 2.3. Fragments of two directories.

In Fig. 2.3, the directory on the left represents the set of items of a bookstore or a cultural good seller, while the one on the right is the directory of a person that

---

[3] http://www.google.com/dirhp
[4] http://www.yahoo.com
[5] http://dmoz.org

illustrates the content of his or her personal library. As we can see, these directories encode the domain under consideration at different levels of details, since these directories have been designed independently and for different purposes, i.e., selling versus classifying.

Finally, there exist some consensus classifications. In library science, the Dewey classification has been used for more than a century for classifying books by topics [Chan et al., 1996]. In natural sciences, the principled classification of species represents another example [Schuh, 1999].

2.1.3 Relational database schemas

Relational databases require the data to be organised in a predefined way as tables or relations. A relational schema specifies the names of the tables as well as their types: the names and types of the columns of each table. The relational model also includes the notion of a key for each table: a subset of the columns that uniquely identifies each row, see Fig. 2.4. Finally, a column in a table may be specified as a foreign key pointing to a column in another table. This is used to keep referential constraints among various entities.

```
                                    book (key: isbn):
                                        isbn    -> int(11)     auto_incr
                                        type    -> varchar(10)  [Volume]
                                        year    -> int(11)
                                        title   -> varchar(100)

item (key: id):                     author (key: firstname, lastname):
    id    -> varchar(30)                firstname  -> varchar(30)
    type  -> varchar(10)                middlename -> varchar(30)
    price -> int(11)   [NULL]           lastname   -> varchar(30)
    name  -> varchar(100)

creator (key: id, author):          writer (key: isbn, firstname, lastname):
    id     -> varchar(30)               isbn      -> int(11)
    author -> varchar(100)              firstname -> varchar(30)
                                        lastname  -> varchar(30)
```

id	type	price	name
89	Pocket	9.95	La chute
134	Popular	60	My life
77	Textbook		Introduction to logic
58	Science		Principia mathematica

id	name
89	Albert Camus
58	Alfred N. Whitehead
77	Alfred Tarski
134	Bertrand Russell
58	Bertrand Russell

isbn	type	year	title
2070360105	Novel	1956	La chute
0415189853	Autobiogr	1969	My life
048628462X	Essay	1941	Introduction to logic

firstname	middlename	lastname
Albert		Camus
Alfred	North	Whitehead
Alfred		Tarski
Bertrand		Russell

isbn	firstname	lastname
2070360105	Albert	Camus
2070394387	Albert	Camus
0521626064	Bertrand	Russell
0521626064	Alfred	Whitehead
0415189853	Bertrand	Russell
048628462X	Alfred	Tarski

Fig. 2.4. Fragments of two populated database schemas.

The schemas of Fig. 2.4 are presented with some data instances in tables. They display similar collections of information about books and authors, however, these are presented in different ways.

Relational databases, in a sense, are relatively restricted: table cells can only contain primitive datatypes, such as string or integer and cannot refer to some individual. For instance, the right-hand side schema of Fig. 2.4, in order to express the relationship between a book and its authors, requires an additional table expressing the authorship relation by joining the keys of both book and writer. Moreover, the relational model lacks the facility to organise data in a taxonomy. In both schemas of Fig. 2.4, tables corresponding to books have a type column assigning their class names to the objects. Several approaches have been proposed for overcoming this expressivity problem. For example, (i) by using a more expressive model, like the entity–relationship model (see Sect. 2.1.5), at design time and by generating a database out of it, or (ii) by using a more elaborate model, such as the object-oriented database model.

Finally, it is worth mentioning widely used languages for specifying relational schemas, such as Structured Query Language (SQL) as well as some of its recent versions, e.g., SQL:1999 and SQL:2003. These support many modelling capabilities, such as user-defined types, aggregation, generalisation, etc.

2.1.4 XML schemas

Document Type Definitions (DTDs) and XML schemas have been introduced for specifying the structure of XML documents. The main ingredients of XML schemas include elements, attributes, and types. In turn, elements can be either complex for specifying nested subelements, or simple for specifying built-in datatypes, such as string, for an element or attribute. XML schemas are rather complementary to directories: instead of describing how things are classified, they describe how things are made from the inside. For instance, the schema at the top of Fig. 2.5 describes the Product element that comprises a name element which is a string, an id which is a URI, a price which is a non negative integer, and topics which are a strings. It also describes a Book element which is a Product that, in addition, has a sequence of authors which, in turn, are Person elements, and exactly one publisher. Even if element definitions can be extended or restricted as subcategories of a classification, the emphasis is on their structure: the extension of an element is made by providing the elements which are modified in this structure. The sequential aspect of XML documents is part of the element specification, though it can be overruled.

In fact, these schemas are a shape according to which future documents are to be created, as opposed to an ontology, which is a description of existing, external objects. The specialisation hierarchy in XML schema is a type hierarchy that defines which kind of elements can occupy the place of another kind, e.g., if a shelf contains books, then putting a biography on this shelf is authorised. In principle, this classification structure does not have to correspond to any natural classification of the objects expressed themselves.

```xml
<schema xmlns:xsd="http://www.w3.org/2001/XMLSchema"
  xmlns="http://www.w3.org/2001/XMLSchema">
 <complexType name="Person">
  <sequence><element name="name" type="xsd:string"/></sequence>
 </complexType>

 <simpleType name="creator"><restriction base="Person"/></simpleType>
 <simpleType name="author"><restriction base="creator"/></simpleType>

 <complexType name="Product">
  <sequence>
   <element ref="creator" minOccurs="1"/>
   <element name="name" type="xsd:string" minOccurs="1"/>
   <element name="id" type="xsd:anyURI" minOccurs="1" maxOccurs="1"/>
   <element name="price" type="xsd:nonNegativeInteger" minOccurs="1"/>
   <element name="topic" type="xsd:string"/>
  </sequence>
 </complexType>

 <complexType name="Book">
  <complexContent>
   <extension base="Product">
     <sequence>
       <element ref="author" type="xsd:any"/>
       <element name="publisher" type="Publisher" minOccurs="1" maxOccurs="1"/>
     </sequence>
   </extension>
  </complexContent>
 </complexType>
</schema>

<schema xmlns:xsd="http://www.w3.org/2001/XMLSchema"
  xmlns="http://www.w3.org/2001/XMLSchema">
 <complexType name="Volume">
  <sequence>
   <element name="author" type="Writer" minOccurs="1"/>
   <element name="title" type="xsd:string" minOccurs="1"/>
   <element name="year" type="xsd:decimal"/>
  </sequence>
   <attribute name="isbn" type="xsd:anyURI"/>
 </complexType>

 <complexType name="Essay">
  <complexContent>
   <extension base="Volume">
     <sequence><element name="subject" type="xsd:any"/></sequence>
   </extension>
  </complexContent>
 </complexType>

 <complexType name="Human">
  <sequence>
   <element name="firstname"  type="xsd:string"/>
   <element name="middlename" type="xsd:string"/>
   <element name="lastname" type="xsd:string"/>
  </sequence>
 </complexType>

 <complexType name="Writer">
  <complexContent><extension base="Human"/></complexContent>
 </complexType>
</schema>
```

Fig. 2.5. Fragments of two XML schemas.

2.1.5 Conceptual models

Often database researchers do not consider directly the relational schema but are rather concerned with the underlying entity–relationship model [Madhavan *et al.*, 2002]. Conceptual models cover what was properly described as such in [Brodie *et al.*, 1984], as well as entity–relationship models [Chen, 1976] that aim at abstracting databases, and UML [Booch *et al.*, 1998] models that aim at abstracting object-oriented programs.

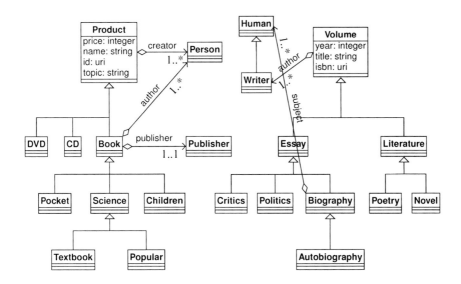

Fig. 2.6. Fragments of two conceptual models as UML class diagrams. Boxes describe entities and their internal structure; Specialisation is expressed by vertical triangular arrows; other relationships are displayed as regular arrows bearing a multiplicity indication.

These models offer a rich way of expressing entities which in this case can be meant as entities of some modelled domain, like people in a database or specification of entities to be created like programs. They offer constructors for organising classes in a hierarchy as well as constructors for describing the internal structure of objects. They thus offer the best of both worlds: directories and databases. For instance, Fig. 2.6 describes two UML class diagrams corresponding to the same sort of models as presented before: a taxonomy of classes from an e-commerce site selling cultural goods on the left and a book library on the right. They both offer a complete description of the items through the specification of their properties and a taxonomy of classes. Moreover, they can express relationships between classes, e.g., that the author of a Book is a Person in the model on the left. The two models of Fig. 2.6 express comparable domains, e.g., a Volume will correspond to a Book, and yet largely different, e.g., there is no Product superclass in the right-hand side model.

2.1.6 Ontologies

It is nowadays common to see directories or conceptual models promoted as ontologies. Ontologies contain most of the features of entity–relationship models, and thus, most parts of the kinds of schemas considered above. The ontologies of Fig. 2.7 syntactically correspond to the models of Fig. 2.6.

The distinctive feature of ontologies is the existence of a model theoretic semantics: ontologies are logic theories. Ontology interpretation is not left to the users that read the diagrams or to the knowledge management systems implementing them, it is specified explicitly. The semantics provides the rules for interpreting the syntax which do not provide the meaning directly but constrains the possible interpretations of what is declared.

It is commonplace in theoretical database research to consider relational databases with a first order semantics. However, this is not part of the official SQL standard [Melton (ed.), 2003]. Moreover, the relational algebra used in database schemas is not very expressive: expressiveness resides in the query language.

For these reasons of rich expressiveness and presence of a model theoretic semantics, we will specifically focus on ontologies. Traditionally, ontologies were considered different from knowledge bases, like a database schema is different from a database that uses it. We will not enforce this distinction here and only use the term 'ontology' as it is common place in logic. We thus discuss ontologies in more detail with the idea that these discussions are for part relevant to other kinds of conceptual models.

The semantics of ontologies can be constrained by additional axioms. This could be, in some languages, the opportunity to add axioms, such as an autobiography is a biography whose topic is the author:

$$\forall x, \mathsf{Autobiography}(x) \Rightarrow \exists y; \mathsf{Person}(y) \land \mathsf{author}(x,y) \land \mathsf{topic}(x,y)$$

For the sake of completeness we give in the next section a syntax and semantics for a minimal ontology language.

2.2 Ontology language

Ontologies are expressed in an ontology language. There are a large variety of languages for expressing ontologies [Staab and Studer, 2004]. Fortunately, most of these languages share the same kinds of entities, often with different names but comparable interpretations. We briefly describe what entities are found in ontology languages. It is not our purpose to commit to one particular language, however this section aims to facilitate the understanding of some of the forthcoming examples which are given in OWL [Smith *et al.*, 2004, Dean and Schreiber (eds.), 2004], an ontology language recommended by the W3C.

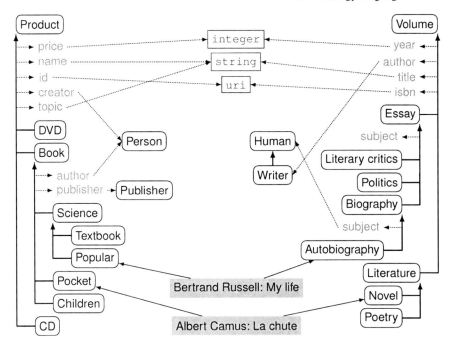

Fig. 2.7. Fragments of two ontologies.

2.2.1 Ontology entities

Ontology languages usually deal with the following kinds of entities:

Classes or concepts are the main entities of an ontology. These are interpreted as a set of individuals in the domain. They are introduced in OWL by the `owl:Class` construct. For example, in Fig. 2.7, Book and Person are classes.

Individuals or objects or instances are interpreted as particular individual of a domain. These are introduced in OWL by the `owl:Thing` construct. For example, in Fig. 2.7, the objects Albert Camus: La chute and Bertrand Russell: My life are individuals.

Relations are the ideal notion of a relation independently to what it applies. Relations are interpreted as a subset of the product of the domain. These are introduced in OWL by the `owl:ObjectProperty` or `owl:DatatypeProperty` construct. For example, in Fig. 2.7, creator and topic are relations.

Datatypes are particular parts of the domain which specify values as opposed to individuals, however, values do not have identities. For example, in Fig. 2.7, String and Integer are datatypes.

Data values are simple values. For example, in Fig. 2.7, the string 'My life' is a data value that can be the title of an Autobiography.

These entities do not have to be named. They can be constructed out of other entities. In OWL, a concept can be created out of the restriction of a relation. For example, this occurs if one defines the class Writer as the set of individuals that have written something:

```
<owl:Restriction>
  <owl:onProperty rdf:resource="#hasWritten" />
  <owl:minCardinality rdf:datatype="&xsd;nonNegativeInteger">1</owl:cardinality>
</owl:Restriction>
```

Alternatively, a new class can also be constructed by combining two other classes. For example, when considering that a low price pocket book for children is a Pocket book that is also a Children and LowPrice book:

```
<owl:intersectionOf>
  <owl:Class rdf:resource="#Pocket" />
  <owl:Class rdf:resource="#Children" />
  <owl:Class rdf:resource="#LowPrice" />
</owl:intersectionOf>
```

Entities can be connected by various kinds of relations, including:

Specialisation between two classes or two properties is interpreted as inclusion of the interpretations. For instance, in Fig. 2.7, the class Book is a specialisation of the class Product. Specialisation is introduced in OWL by the `rdfs:subClassOf` or `rdfs:subPropertyOf` constructs.

Exclusion between two classes or two properties is interpreted as the exclusion of their interpretations, i.e., when their intersection is empty. For instance, in Fig. 2.7, the class Product could be declared to be exclusive to the class Person. Exclusion is introduced in OWL by the `owl:disjointWith` construct.

Instantiation or typing between individuals and classes, property instances and properties, values and datatypes is interpreted as membership. For instance, in Fig. 2.7, the item presented as Bertrand Russell: My life is an instance of the class Popular. Instantiation is expressed in OWL with the `rdf:type` construct.

Example 2.1. The class Book of Fig. 2.7 can be expressed in OWL as follows:

```
<owl:Class rdf:ID="Book">
  <rdfs:subClassOf rdf:resource="#Product" />
  <rdfs:label xml:lang="en">book</rdfs:label>
  <rdfs:comment xml:lang="en">A book.</rdfs:comment>
  <rdfs:subClassOf>
    <owl:Restriction>
      <owl:onProperty rdf:resource="#author" />
      <owl:cardinality rdf:datatype="&xsd;nonNegativeInteger">1</owl:cardinality>
    </owl:Restriction>
  </rdfs:subClassOf>
  <rdfs:subClassOf>
    <owl:Restriction>
      <owl:onProperty rdf:resource="#publisher" />
      <owl:allValuesFrom rdf:resource="#Publisher" />
    </owl:Restriction>
  </rdfs:subClassOf>
</owl:Class>
```

In particular, it defines two classes by restricting the cardinality of the author relation and restricting the range of the publisher relation. It also relates the defined Book class to these classes and the Product class by specialisation (rdfs:subClassOf).

In summary, we can consider an ontology to be characterised as follows.

Definition 2.2 (Ontology). *An ontology is a tuple $o = \langle C, I, R, T, V, \leq, \perp, \in, = \rangle$, such that:*

C is the set of classes;
I is the set of individuals;
R is the set of relations;
T is the set of datatypes;
V is the set of values (C, I, R, T, V being pairwise disjoint);
$\leq$ is a relation on $(C \times C) \cup (R \times R) \cup (T \times T)$ called specialisation;
$\perp$ is a relation on $(C \times C) \cup (R \times R) \cup (T \times T)$ called exclusion;
$\in$ is a relation over $(I \times C) \cup (V \times T)$ called instantiation;
$=$ is a relation over $I \times P \times (I \cup V)$ called assignment.

Many algorithms transform these ontologies into labelled graphs, where nodes are typed. We use such a notation in the diagrams of this book, see for details Fig. A.1 of Appendix A.

2.2.2 Ontology language semantics

The semantics of ontology languages is usually given through model theory. In particular, it defines an interpretation function that maps each ontology entity to a set D called the domain of interpretation .

Definition 2.3 (Interpretation). *Given an ontology $o = \langle C, I, R, T, V, \leq, \perp, \in, = \rangle$, an interpretation of o is a pair $\langle I, D \rangle$, such that D is called the domain of interpretation and I is a function called the interpretation function, such that:*

- $\forall c \in C, I(C) \subseteq D$;
- $\forall r \in R, I(r) \subseteq D \times (D \cup V)$;
- $\forall i \in I, I(i) \in D$;
- $\forall t \in T, I(t) \subseteq V$;
- $\forall v \in V, I(v) \in V$.

An assertion expressed in an ontology language is said to be satisfied by an interpretation if the interpretation is coherent with this assertion.

Definition 2.4 (Satisfiability). *Given an ontology $o = \langle C, I, R, T, V, \leq, \perp, \in, = \rangle$, a formula δ, which is satisfied by an interpretation $\langle I, D \rangle$ of o (denoted as $I \models \delta$), is defined as follows:*

$I \models c \leq c'$	if and only if	$I(c) \subseteq I(c')$
$I \models r \leq r'$	if and only if	$I(r) \subseteq I(r')$
$I \models t \leq t'$	if and only if	$I(t) \subseteq I(t')$
$I \models c \perp c'$	if and only if	$I(c) \cap I(c') = \emptyset$
$I \models r \perp r'$	if and only if	$I(r) \cap I(r') = \emptyset$
$I \models t \perp t'$	if and only if	$I(t) \cap I(t') = \emptyset$
$I \models i \in c$	if and only if	$I(i) \in I(c)$
$I \models v \in t$	if and only if	$I(v) \in I(t)$
$I \models i.r = i'$	if and only if	$\langle I(i), I(i') \rangle \in I(r)$
$I \models i.r = v$	if and only if	$\langle I(i), I(v) \rangle \in I(r)$

Ontology formulas may contain more than these assertions, e.g., quantified assertions or assertions related by logical connectors. We will restrict ourselves to the relations between the entities. An ontology is a set of assertions that selects the set of interpretations which satisfy them. These interpretations are called models. They constitute the possible interpretations of an ontology.

Definition 2.5 (Model). *Given an ontology o, a model of o is an interpretation* $m = \langle I, D \rangle$ *of o, which satisfies all the assertions in o:*

$$\forall \delta \in o, m \models \delta$$

The set of models of an ontology o is denoted as $\mathcal{M}(o)$.

Finally, an important notion is the set of assertions that are consequences of an ontology. These are the assertions implicitly entailed by an ontology and they determine the answers to queries.

Definition 2.6 (Consequence). *Given an ontology formula* δ, δ *is a consequence of an ontology o, if and only if, it is satisfied by all models of o. This is denoted as* $o \models \delta$.

Given a model m, we will denote as $m(e)$ the application of the interpretation function of the model to some ontology entity e.

This digression introduced more precisely, albeit generally, a simplified syntax and semantics of ontologies. This will be useful when considering the meaning of matching ontologies.

2.3 Types of heterogeneity

The goal of matching ontologies is to reduce heterogeneity between them. Heterogeneity does not lie solely in the differences between goals of the applications according to which they have been designed or in the expression formalisms in which ontologies have been encoded. There have been many different classifications to types of heterogeneity [Batini et al., 1986, Sheth and Larson, 1990, Breitbart, 1990, Kim and Seo, 1991,

Goh, 1997, Hull, 1997, Kashyap and Sheth, 1998, Benerecetti *et al.*, 2000, Wache *et al.*, 2001, Klein, 2001, Euzenat, 2001, Corcho, 2004, Hameed *et al.*, 2004, Ghidini and Giunchiglia, 2004, Bouquet *et al.*, 2004a]. Some of them focus on mismatches [Klein, 2001], others rather mention interoperability levels [Euzenat, 2001]. We consider here the most obvious types of heterogeneity:

Syntactic heterogeneity occurs when two ontologies are not expressed in the same ontology language. This obviously happens when comparing, for instance, a directory with a conceptual model. This also happens when two ontologies are modelled by using different knowledge representation formalisms, for instance, OWL and F-logic. This kind of mismatch is generally tackled at the theoretical level when one establishes equivalences between constructs of different languages. Thus, it is sometimes possible to translate ontologies between different ontology languages whilst still preserving the meaning [Euzenat and Stuckenschmidt, 2003].

Terminological heterogeneity occurs due to variations in names when referring to the same entities in different ontologies. This can be caused by the use of different natural languages, e.g., Paper vs. Articulo, different technical sublanguages, e.g., Paper vs. Memo, or the use of synonyms, e.g., Paper vs. Article.

Conceptual heterogeneity, also called semantic heterogeneity in [Euzenat, 2001] and logical mismatch in [Klein, 2001], stands for the differences in modelling the same domain of interest. This can happen due to the use of different (and, sometimes, equivalent) axioms for defining concepts or due to the use of totally different concepts, e.g., geometry axiomatised with points as primitive objects or geometry axiomatised with spheres as primitive objects. As noted in [Klein, 2001] and [Visser *et al.*, 1998], there is a difference between the conceptualisation mismatch, which relies on the differences between modelled concepts, and the explicitation mismatch, which relies on the way these concepts are expressed. [Visser *et al.*, 1998] provides a precise classification of these mismatches.

Finally, in the context of conceptual differences, [Benerecetti *et al.*, 2001] identifies three important reasons for these to hold. We discuss these below and give examples with the help of notion of a geographic map:

- *Difference in coverage* occurs when two ontologies describe different, possibly overlapping, regions of the world at the same level of detail and from a unique perspective. This is obviously the case for two partially overlapping geographic maps.
- *Difference in granularity* occurs when two ontologies describe the same region of the world from the same perspective but at different levels of detail. This applies to geographic maps with different scales, e.g., one displays buildings, while another depicts whole cities as points.
- *Difference in perspective,* also called difference in scope [Chalupsky, 2000], occurs when two ontologies describe the same region of the world, at the same level of detail, but from a different perspective. This occurs for maps

with different purposes: a political map and a geological map do not display the same objects.

Semiotic heterogeneity, also called pragmatic heterogeneity in [Bouquet et al., 2004a], is concerned with how entities are interpreted by people. Indeed, entities which have exactly the same semantic interpretation are often interpreted by humans with regard to the context, for instance, of how they are ultimately used. This kind of heterogeneity is difficult for the computer to detect and even more difficult to solve, because it is out of its reach. The intended use of entities has a great impact on their interpretation, therefore, matching entities which are not meant to be used in the same context is often error-prone. Given the limited grasp that a computer can have on these issues, we do not deal with semiotic heterogeneity here.

Usually, several types of heterogeneity occur together. This book is only concerned with reducing the terminological and conceptual types of heterogeneity. Techniques for dealing with these types individually are presented in Chap. 4, while techniques for considering them together are provided in Chap. 5.

2.4 Terminology

As can be observed from what we have presented so far, in the area of ontology matching, different authors including ourselves use different words to refer to similar concepts, and, vice versa, sometimes different concepts are referred to by the same name [Chalupsky, 2000, Klein, 2001, Euzenat, 2001, Noy and Klein, 2004, Kalfoglou and Schorlemmer, 2003b, Bouquet et al., 2004a]. This is especially confusing since these terms, e.g., mapping, can be used for describing both an action and its result. In this section, we provide a working glossary with the definitions of terms as they are going to be used in this book.

Matching is the process of finding relationships or correspondences between entities of different ontologies.

Alignment is a set of correspondences between two or more (in case of multiple matching) ontologies (by analogy with molecular sequence alignment). The alignment is the output of the matching process.

Correspondence is the relation holding, or supposed to hold according to a particular matching algorithm or individual, between entities of different ontologies. These entities can be as different as classes, individuals, properties or formulas. Some authors use the term mapping instead, however, it will not be used in this sense in this book.

Mapping is the oriented, or directed, version of an alignment: it maps the entities of one ontology to at most one entity of another ontology. This complies with the mathematical definition of a mapping instead of that of a general relation. The mathematical definition would in principle require that the mapped object is equal to its image, i.e., that the relation is an equivalence relation. A mapping can be seen as a collection of mapping rules all oriented in the same direction, i.e., from one ontology to the other, and such that the elements of the source ontology appear at most once.

Mapping rule is a correspondence which maps an entity of one ontology into another one from another ontology.

Ontology merging is the creation of a new ontology from two, possibly overlapping, source ontologies. The initial ontologies remain unaltered. The merged ontology is supposed to contain the knowledge of the initial ontologies, e.g., consequences of each ontology are consequences of the merge. This concept is closely related to that of schema integration in databases.

Ontology integration is the inclusion in one ontology of another ontology and assertions expressing the *glue* between these ontologies, usually as bridge axioms. The integrated ontology is supposed to contain the knowledge of both initial ontologies. Contrary to merging, the first ontology is unaltered while the second one is modified.

Bridge axioms or articulation axioms are formulas, in an ontology language, that express the alignments such that it is possible to integrate the entities of an ontology within one another. Bridge axioms are the basis for ontology merging when the ontologies are expressed in the same language.

Ontology translation is the process of transforming an ontology from one ontology language to another. By extension, it is a program for translating ontologies.

Ontology transformation is the process of expressing the entities of an ontology with respect to the entities of another ontology, i.e., relations between entities of the first ontology and those of the second one are added to the first ontology. So the initial consequences of the first ontology are still consequences of the transformation result. The two initial ontologies are unaltered and a third ontology, the transformation result, is created. By extension, it is a program that transforms ontologies.

Data translation is the process of transforming data or instances from one ontology into corresponding data or instances expressed in another ontology. By extension, it is a program that translates data.

Mediation consists of interfacing two software components by dynamically altering the information stream between these. By extension, a mediator is a program performing mediation. In web service composition a mediator translates the output of a service in the input of another one: it thus performs data translation. In query answering applications it is a dual pair of translations that transforms the query from one ontology to another and that translates the answer back.

Ontology version of an ontology is the ontology resulting from the application of modifications to this ontology.

Ontology reconciliation is a process that harmonises the content of two or more ontologies, typically requiring changes on one of the two sides or even on both sides [Hameed *et al.*, 2004]. In this case, there is no merging of the ontologies but co-evolution. Ontology reconciliation can be performed for the purpose of merging two ontologies or for the purpose of making them independent.

2.5 The ontology matching problem

There have been different formalisations of matching and its result [Bernstein *et al.*, 2000, Lenzerini, 2002, Kalfoglou and Schorlemmer, 2003b, Bouquet *et al.*, 2004a, Zimmermann *et al.*, 2006]. We provide here a general definition. It does not pretend to solve each particular problem nor to strictly cover the complete field. It aims at serving as a guide for this book.

2.5.1 The matching process

The *matching* operation determines the alignment A' for a pair of ontologies o and o'. There are some other parameters that can extend the definition of the matching process, namely: (i) the use of an input alignment A, which is to be completed by the process; (ii) the matching parameters, p, e.g., weights, thresholds; and (iii) external resources used by the matching process, r, e.g., common knowledge and domain specific thesauri.

Technically, this process can be defined as follows:

Definition 2.7 (Matching process). *The matching process can be seen as a function f which, from a pair of ontologies to match o and o', an input alignment A, a set of parameters p and a set of oracles and resources r, returns an alignment A' between these ontologies:*

$$A' = f(o, o', A, p, r)$$

This can be schematically represented as illustrated in Fig. 2.8.

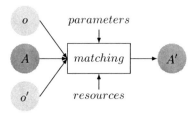

Fig. 2.8. The matching process.

It can be useful to specifically consider the matching of more than two ontologies within the same process. We call this multiple matching.

Definition 2.8 (Multiple matching process). *The multiple matching process can be seen as a function f which, from a set of ontologies to match $\{o_1, \ldots o_n\}$, an input alignment A, a set of parameters p and a set of oracles and resources r, returns an alignment A' between these ontologies:*

$$A' = f(o_1, \ldots o_n, A, p, r)$$

The matching process is the main subject of this book. However, before discussing its internals, let us first consider what it provides: the alignment.

2.5.2 Structure of an alignment

Alignments express the correspondences between entities belonging to different ontologies. All definitions here are given for matching between two ontologies. In case of multiple matching, the definitions can be straightforwardly extended by using n-ary correspondences. A correspondence must consider the two corresponding entities and the relation that is supposed to hold between them. We provide the definition of the alignment following the work in [Euzenat, 2004, Bouquet et al., 2004a].

Since the related entities are an important part of alignments, they have to be defined. We separate the matched entities from the ontology language because it can be desirable to have a different language for identifying the matched entities. Given an ontology language, we use an *entity language* for expressing those entities that will be put in correspondence by matching. The expressions of this language will depend on the ontology on which expressions are defined.

The entity language can be simply made of all the formulas of the ontology language based on the ontology vocabulary. It can restrict its scope to particular kinds of formulas from the language, for instance, atomic formulas, or even to terms of the language, like class expressions. It can also restrict the entities to be only named entities. This is convenient in the context of the semantic web to restrict entities to those identifiable by their URIs. The entity language can also be an extension of the ontology language: this can be a query language, such as SPARQL [Prud'hommeaux and Seaborne (ed.), 2007], adding operations for manipulating ontology entities that are not available in the ontology language itself, like concatenating strings or joining relations. Finally, this entity language can combine both extension and restriction, e.g., by authorising any boolean operations over named ontology entities.

Definition 2.9 (Entity language). *Given an ontology language L, an entity language Q_L is a function from any ontology $o \subseteq L$ which defines the matchable entities of ontology o.*

In the following we will assume that each ontology interpretation can be extended to an interpretation of the entity language associated with the ontology.

The next important component of the alignment is the relation that holds between the entities. We identify a set of relations Θ that is used for expressing the relations between the entities. Matching algorithms primarily use the equivalence relation ($=$) meaning that the matched objects are the same or are equivalent if these are formulas. It is possible to use relations from the ontology language within Θ. For instance, using OWL, it is possible to take advantage of the owl:equivalentClass, owl:disjointWith or rdfs:subClassOf relations in order to relate classes of two ontologies. These relations correspond to set-theoretic relations between classes: *equivalence* ($=$); *disjointness* ($\perp$); *more general* ($\sqsupseteq$). They can be used without reference to any ontology language. Finally, relations can be of any type and are not restricted to relations present within the ontology language, such as fuzzy relations or probability distributions over a complete set of relations, or similarity measures.

For pragmatic reasons, the relationship between two entities is assigned a degree of confidence which can be viewed as a measure of trust in the fact that the correspondence holds – 'I trust 70% the fact that the correspondence is correct or reliable' – and can be compared with the certainty measures provided with meteorological forecasts.

Definition 2.10 (Confidence structure). *A confidence structure is an ordered set of degrees $\langle \Xi, \leq \rangle$ for which there exists a greatest element $\top$ and a smallest element $\bot$.*

The usage of confidence degrees is that the higher the degree with regard to $\leq$, the most likely the relation holds. It is convenient to interpret the greatest element as the boolean true and the smallest element as the boolean false.

The most widely used structure is based on the real number unit interval [0 1], but some systems simply use the boolean lattice. Some other possible structures are fuzzy degrees, probabilities or other lattices. [Gal et al., 2005a] has investigated the structure of fuzzy confidence relations. This structure can be extended, for instance, if one wants to compose alignments. Thus, in this case, it may be necessary to define operations for combining these degrees.

With these ingredients, it is possible to define the correspondences that have to be found by matching algorithms.

Definition 2.11 (Correspondence). *Given two ontologies o and o' with associated entity languages Q_L and $Q_{L'}$, a set of alignment relations Θ and a confidence structure over Ξ, a correspondence is a 5-uple:*

$$\langle id, e, e', r, n \rangle,$$

such that

- *id is a unique identifier of the given correspondence;*
- $e \in Q_L(o)$ *and* $e' \in Q'_{L'}(o')$;
- $r \in \Theta$;
- $n \in \Xi$.

The correspondence $\langle id, e, e', r, n \rangle$ asserts that the relation r holds between the ontology entities e and e' with confidence n.

Example 2.12 (Correspondence). For example, a simple kind of correspondence is as follows:

http://book.ontologymatching.org/example/culture-shop.owl#Book $=$
http://book.ontologymatching.org/example/library.owl#Volume

It asserts the equivalence relation with full confidence between what is denoted by two URIs, namely the Book class in one ontology and the Volume class in another one. Some examples of more complex correspondences are as follows:

$$\text{author}(x, concat(w.\text{firstname}, w.\text{lastname})) \Leftarrow_{.85} \begin{array}{l} \text{Book}(x) \\ \land \text{ writtenBy}(x, w) \\ \land \text{ Writer}(w) \end{array}$$

is a Horn clause expressing that if there exists a Book x written by Writer w, the author of x in the first ontology is identified by the concatenation of the first and last name of w. The confidence in this clause is quantified with a .85 degree.

There can be several possible correspondences for the same entities depending on the language in which correspondences are expressed. For instance, one could have the simple correspondence that speed in one ontology is equivalent to velocity in another one:

$$\text{speed} \equiv \text{velocity}$$

or record that they are expressed in miles per hour and metre per second respectively:

$$\text{speed} = \text{velocity} \times 2.237$$
$$0.447 \times \text{speed} = \text{velocity}$$

Finally, an alignment is defined as a set of correspondences.

Definition 2.13 (Alignment). *Given two ontologies o and o', an* alignment *is made up of a set of correspondences between pairs of entities belonging to $Q_L(o)$ and $Q_{L'}(o')$ respectively.*

Example 2.14 (Alignment). Fig. 2.9 displays a possible alignment for the pair of ontologies of Fig. 2.7. It can be expressed by the following correspondences:

$$\text{Book} =_{1.0} \text{Volume} \qquad \text{name} \geq_{1.0} \text{title}$$
$$\text{id} \geq_{.9} \text{isbn} \qquad \text{author} =_{1.0} \text{author}$$
$$\text{Person} =_{.9} \text{Human} \qquad \text{Science} \leq_{.8} \text{Essay}$$

So far, our alignments are very simple: they are sets of pairs of entities from two ontologies. However, there are, in the literature, at least three types of n:m, multiple or complex alignments:

1. alignments involving more than two ontologies produced by multiple matching, that we may call multialignments,
2. alignments involving correspondences between more than two entities (still belonging to two ontologies),
3. alignments with entities involved in more than one correspondence that are denoted by the use of * (zero-or-more) or + (more-than-zero) in their cardinalities.

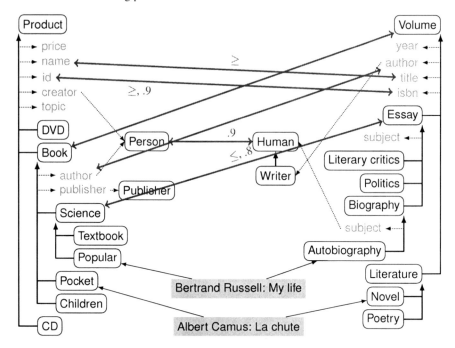

Fig. 2.9. Alignment between the ontologies of Fig. 2.7. Correspondences are expressed by arrows. By default their relation is = and their confidence value is 1.0; otherwise, these are mentioned near the arrows.

In case of multiple matching (1), the alignments must contain correspondences relating more than two entities. The definitions above must then be extended accordingly. This is not covered further here.

The second kind of correspondences (2) can be thought of as using non binary relations. However, given the nature of the problem: matching two ontologies, we will consider that these objects can be grouped by operators in the entity language Q_L which can include operators such as concatenation, arithmetic operations or logical connectors for that purpose. In its simplest expression the only construction can be a set.

Option (3) is related to the multiplicity of the alignment if it is considered as a relation. By analogy with mathematical functions, it is useful to define some properties of the alignments. These apply when the only considered relation is equality (=) and the confidence measures are not taken into account. One can ask for a total alignment with regard to one ontology, i.e., all the entities of one ontology must be successfully mapped to the other one. This property is purposeful whenever thoroughly transcribing knowledge from one ontology to another is the goal: there is no entity that cannot be translated.

One can also require the alignment to be injective with regard to one ontology, i.e., all the entities of the other ontology is part of at most one correspondence. Injectivity is useful in ensuring that entities that are distinct in one ontology remain

2.5 The ontology matching problem 49

distinct in the other one. In particular, this contributes to the reversibility of alignments.

Definition 2.15 (Total alignment, injective alignment). *Given two ontologies o and o', an alignment A over o and o' is called a* total alignment *from o to o' if and only if:*

$$\forall e \in Q_L(o), \exists e' \in Q_{L'}(o'); \langle e, e', = \rangle \in A$$

and, it is called an injective alignment *from o to o' if and only if:*

$$\forall e' \in Q_{L'}(o'), \exists e_1, e_2 \in Q_L(o); \langle e_1, e', = \rangle \in A \land \langle e_2, e', = \rangle \in A \Rightarrow e_1 = e_2$$

These properties heavily depend on the ontology entity languages which are chosen for these alignments.

Usual mathematical properties apply to these alignments. In particular, a total alignment from o to o' is a surjective alignment from o' to o. A total alignment from both o and o', which is injective from one of them, is a bijection. In mathematical English, an injective function is said to be *one-to-one* and a surjective function to be *onto*. Due to the wide use among matching practitioners of the term *one-to-one* for a bijective, i.e., both injective and surjective, alignment, we will only use one-to-one for bijective.

Finally, we can extend these definitions when correspondence relations are *not* equivalence. In such a case, they do not ensure the same properties. For instance, injectivity does not guarantee reversibility of the alignment used as a transformation.

In conceptual models and databases, the term multiplicity denotes the constraints on a relation. Usual notations are 1:1 (one-to-one), 1:m (one-to-many), n:1 (many-to-one) or n:m (many-to-many). If we consider only total and injective properties, denoted as 1 for injective and total, ? for injective, + for total and * for none, and the two possible orientations of the alignments, from o to o' and from o' to o, the multiplicities become: ?:?, ?:1, 1:?, 1:1, ?:+, +:?, 1:+, +:1, +:+, ?:*, *:?, 1:*, *:1, +:*, *:+, *:* [Euzenat, 2003].

Example 2.16 (Alignment multiplicity). The alignment of Example 2.14 is ?:?. If we add the correspondence Product $\geq$ Volume, then it is ?:*. If we now consider relating any entity of the second ontology to another entity of the first one, then it becomes ?:+.

The four pictures below display some of the possible configurations for two ontologies composed of three classes each.

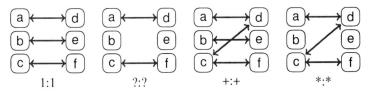

2.5.3 A rough understanding of matching

As an introduction to the ideas behind matching, we present here a very simple and yet powerful description. It relies on the principle that ontologies can approximate other ontologies and that ontologies to be matched are approximation of a common ideal ontology. We give the classical interpretation of it in both model theoretic terms and categorical terms. This is informally presented in Fig. 2.10.

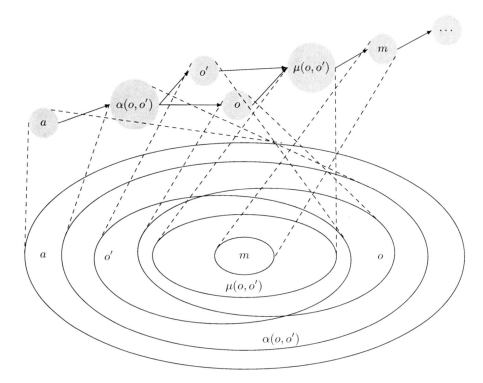

Fig. 2.10. Relations between ontologies, alignment $\alpha(o, o')$ and the corresponding model-theoretic interpretation. Each ontology in the top of the figure is represented by its set of models in the bottom. These ontologies corresponds to an initial (possibly empty) ontology, two ontologies (o and o') approximating some ideal ontology m (itself approximating some even more complete ontology). $\alpha(o, o')$ is an alignment of o and o' with regard to the way it approximates them and $\mu(o, o')$ is their merge with regard to the way o and o' approximate it.

Let approximation be a relation between ontologies which expresses that one ontology a is a representation of at least the same modelled domains as another $\alpha(o, o')$. In logic, this relation corresponds to entailment. In category theory, the ontology is called an *object* and the approximation is called a *morphism*. One can define other relations between ontologies, such as having at least one common approximated ontology. Syntactically, it is possible to provide a set of generators that will complete an ontology, e.g., adding a constraint on a class, classifying an individual, providing an approximated ontology.

Model-theoretic semantics assigns to any ontology the set of its models. If the ontology is correctly designed, the modelled domain is part of these. Model-theoretic semantics provides a formal meaning to the intuitions behind such a notion as approximation: an ontology is approximated by another if its models are also models of the other: this is the standard interpretation of entailment. Thus, the more approximated an ontology, the fewer models it has.

In these very general terms, matching two ontologies o and o' consists of finding a most specific ontology $\alpha(o, o')$ that approximates both ontologies. If one ontology is approximated by another, the result of matching should be the latter $\alpha(o, o') = o$. In model-theoretic terms, it amounts to finding an ontology whose set of models is maximal for inclusion and is included in the intersection of the set of models of two aligned ontologies. In categorical terms, this means that there exists an object $\alpha(o, o')$ and a pair of morphisms from it to the ontologies o and o', and for every other object satisfying these conditions there exists a morphism from it to $\alpha(o, o')$.

This general description of alignment can be compared to three cases explaining reasons for the existence of conceptual heterogeneity that were introduced in Sect. 2.3. In fact, if we take those cases literally:

Coverage mismatch, corresponding to two ontologies modelling totally different domains, resorts to $\alpha(o, o') = \emptyset$, $\emptyset$ being an empty ontology;

Granularity mismatch, corresponding to two ontologies modelling the same domain with different precisions, corresponds to the case where $\alpha(o, o') = o$, i.e., one ontology is an approximation of the other;

Perspective mismatch, corresponding to the representation of different aspects of the same domain at the same granularity, is the general case presented here.

However, in real world matching tasks the most frequent cases do not distinctly belong to one of these cases. In any case, finding the alignment between two ontologies is a useful exercise. For example, if one wants to merge two ontologies, i.e., to find $\mu(o, o')$, it is sufficient to stick the unmatched subparts of one ontology to the matched subpart. This general intuition about matching, similar to that of [Kalfoglou and Schorlemmer, 2003b], strongly resembles a category-theoretic framework. Indeed, this work has been further extended from these lines towards more precise categorical definitions in [Hitzler et al., 2005, Zimmermann et al., 2006]. In particular, in categorical terms the merge corresponds to a push-out construction [Bench-Capon and Malcolm, 1999, Alagic and Bernstein, 2001] or its generalisation as a colimit [Hitzler et al., 2005].

2.5.4 Semantics of alignment

We provide a simple foundation for alignments because it is useful to know what is expected from a matching algorithm. However, as will be demonstrated, the semantics of alignment provides a definition of how alignments must be interpreted and not of how alignments must be found by a matching algorithm. In this respect, we provide only a semantics for interpreting alignment and not for the matching operation.

The usual way of providing a semantics for related conceptual systems is through modal logic of knowledge and belief [Fagin *et al.*, 1995, Wooldridge, 2000]. In the line of the work on data integration, we only give a first-order model theoretic semantics. It depends on the semantics of ontologies but does not interfere with it. In fact, given a set of ontologies and a set of alignments between them, we can evaluate the semantics of the whole system in terms of the semantics of each individual ontology.

The main problem arising is the non compatibility of the domains of interpretation. Given several ontologies, it is possible to consider different positions with regard to the domain of interpretation:

- For all these ontologies, the domain of interpretation D is unique. This approach is useful when ontologies describe a set of well defined entities, like the set of files shared in a peer-to-peer system. This approach has been taken in [Calvanese *et al.*, 2002b, Calvanese *et al.*, 2004].
- For each ontology o, the domain D_o may be different. Domains are related with the help of domain relations $r_{o,o'}$ which map elements of D_o to corresponding elements of domain $D_{o'}$. This approach is used in [Ghidini and Serafini, 1998, Borgida and Serafini, 2003].
- There is no constraint on the domain of interpretation of ontologies. This is the assumption that will be considered here. For dealing with this assumption, we use a universal domain U, that may be defined as the union of all the domains under consideration, and an equalising function γ or rather a set of equalising functions: $\gamma_o : D_o \longrightarrow U$.

[Zimmermann and Euzenat, 2006] considers the implications of these three models. Here, because the models of various ontologies can have different interpretation domains, we use the notion of an equalising function, which helps make these domains commensurate.

Definition 2.17 (Equilising function). *Given a family of interpretations $\langle I_o, D_o \rangle_{o \in \Omega}$ of a set of ontologies Ω, an equalising function for $\langle I_o, D_o \rangle_{o \in \Omega}$ is a family of functions $\gamma = (\gamma_o : D_o \longrightarrow U)_{o \in \Omega}$ from the ontology domains of interpretation to a global domain of interpretation U. The set of all equalising functions is called Γ.*

When it is unambiguous, we will use γ as a function. The goal of this γ function is only to be able to (theoretically) compare elements of the domain of interpretation. It is simpler than the use of domain relations in distributed first-order logics [Ghidini and Serafini, 1998] in the sense that there is one function per domain instead of relations for each pair of domains.

The equalising functions can be different for each ontology. This means, in particular, that even if two ontologies are interpreted over the same domain of interpretation, it is not compulsory that the equalising function maps their elements to the same element of U, though it remains possible. This allows for a loose coupling of the interpretations.

The relations used in correspondences do not necessarily belong to the ontology languages. As such, they do not have to be interpreted by the ontology semantics. Therefore, we have to provide semantics for them.

Definition 2.18 (Interpretation of alignment relations). *Given $r \in \Theta$ an alignment relation and U a global domain of interpretation, r is interpreted as a binary relation over U, i.e., $r^U \subseteq U \times U$.*

For the sake of simplicity, we consider correspondences that are only triples of the following form: $\langle e, e', r \rangle$. The definition of correspondence satisfiability relies on γ and the interpretation of relations. It requires that in the equalised models, the correspondences are satisfied.

Definition 2.19 (Satisfied correspondence). *A correspondence $c = \langle e, e', r \rangle$ is satisfied for an equalising function γ by two models m, m' of o, o' if and only if $\gamma_o \cdot m \in \mathcal{M}(o)$, $\gamma_{o'} \cdot m' \in \mathcal{M}(o')$ and*

$$\langle \gamma_o(m(e)), \gamma_{o'}(m'(e')) \rangle \in r^U$$

This is denoted as $m, m' \models_\gamma c$.

Definition 2.20 (Salisfied alignment). *An alignment A is satisfied for an equalising function γ by two models m, m' of o, o' if and only if all its correspondences are satisfied for γ by m and m'. This is denoted as $m, m' \models_\gamma A$.*

This is useful for defining the classical notions of validity and satisfiability.

Definition 2.21 (Alignment validity). *An alignment A of two ontologies o and o' is said to be valid if and only if*

$$\forall m \in \mathcal{M}(o), \forall m' \in \mathcal{M}(o'), \forall \gamma \in \Gamma, m, m' \models_\gamma A$$

This is denoted as $\models A$.

From the practical perspective, this is not a very useful definition since unless the ontologies have no models, it will be very difficult to find valid alignments. A relaxed definition could consider the validity that, given an equalising function, describes how domains are related. Valid alignments are direct logical consequences of the two ontologies. Thus they do not provide additional information than what is already in these ontologies. Satisfiable alignments offer more information.

Definition 2.22 (Satisfiable alignment). *An alignment A of two ontologies o and o' is said to be satisfiable if and only if*

$$\exists m \in \mathcal{M}(o), \exists m' \in \mathcal{M}(o'), \exists \gamma \in \Gamma; m, m' \models_\gamma A$$

Thus, an alignment is satisfiable if there are models of the ontologies that can be combined in such a way that this alignment makes sense. The satisfiable set of alignments is far larger than the set of valid ones. Again, one can define γ-satisfiable alignments, i.e., alignments satisfiable for a given equalising function. This is what is computed (in its minimal form) by algorithms like those of T-tree (see Sect. 6.2.1).

Given an alignment between two ontologies, the semantics of the aligned ontologies can be defined as follows.

Definition 2.23 (Models of aligned ontologies). *Given two ontologies o and o' and an alignment A between these ontologies, a model m'' of these ontologies aligned by A is a triple $\langle m, m', \gamma \rangle \in \mathcal{M}(o) \times \mathcal{M}(o') \times \Gamma$, such that $m, m' \models_\gamma A$.*

In that respect, the alignment acts as a model filter for the ontologies. It selects the interpretations of ontologies which are coherent with the alignments. This allows transferring information from one ontology to another since reducing the set of models will entail more consequences in each aligned ontology.

These definitions can be generalised to an arbitrary number of alignments and ontologies captured in the concept of a distributed system [Ghidini and Serafini, 1998, Franconi et al., 2003].

Definition 2.24 (Distributed system of networked ontologies). *A distributed system of networked ontologies $\langle \Omega, \Lambda \rangle$ is made of a set Ω of ontologies and a set Λ of alignments between these ontologies. We denote as $\Lambda(o, o')$ the set of alignments in Λ between o and o'.*

Definition 2.25 (Models of distributed systems). *Given a finite set of n ontologies Ω and a finite set of alignments Λ between pairs of ontologies in Ω, a model of the distributed system $\langle \Omega, \Lambda \rangle$ is a $n + 1$-uple of models $\langle m_1 \ldots m_n, \gamma \rangle \in \mathcal{M}(o_1) \times \ldots \mathcal{M}(o_n) \times \Gamma$, such that for each alignment $A \in \Lambda(o_i, o_j)$, A is satisfied by $\langle m_i, m_j, \gamma \rangle$.*

This definition coincides with a coherent model of the world in which all models satisfy all alignments. This is the standpoint of an omniscient observer and it corresponds to the global knowledge of a distributed system as defined in [Fagin et al., 1995].

However, if one agent has an inconsistent ontology then the distributed system has no model. Therefore, even agents not connected to the inconsistent ontology cannot compute reasonable models. Moreover, an agent knowing an ontology and the related alignments would like to use the system by gathering information from its neighbours and considering only the models of this information. Thereby, it would be able to compute consequence through some complete deduction mechanisms. This is important when asking agents to answer queries and corresponds to local knowledge in [Fagin et al., 1995]. This is the knowledge an agent can achieve by communicating only with the agents it is connected to in a distributed system.

From that standpoint, there can be several ways to select the acceptable models given the distributed system:

$\mathcal{M}^1_{\Omega,\Lambda}(o) =$
$\quad \{m \in \mathcal{M}(o); \exists \gamma \in \Gamma; \forall o' \in \Omega, \forall A \in \Lambda(o,o'), \exists m' \in \mathcal{M}(o'); m,m' \models_\gamma A\}$
$\mathcal{M}^2_{\Omega,\Lambda}(o) =$
$\quad \{m \in \mathcal{M}(o); \exists \gamma \in \Gamma; \forall o' \in \Omega, \forall A \in \Lambda(o,o'), \exists m' \in \mathcal{M}^2_{\Omega,\Lambda}(o'); m,m' \models_\gamma A\}$
$\mathcal{M}^3_{\Omega,\Lambda}(o) =$
$\quad \{m \in \mathcal{M}(o); \exists \gamma \in \Gamma; \forall o' \in \Omega, \exists m' \in \mathcal{M}(o'); \forall A \in \Lambda(o,o'), m,m' \models_\gamma A\}$
$\mathcal{M}^4_{\Omega,\Lambda}(o) =$
$\quad \{m \in \mathcal{M}(o); \exists \gamma \in \Gamma; \forall o' \in \Omega, \exists m' \in \mathcal{M}^4_{\Omega,\Lambda}(o'); \forall A \in \Lambda(o,o'), m,m' \models_\gamma A\}$
$\mathcal{M}^5_{\Omega,\Lambda}(o) =$
$\quad \{m \in \mathcal{M}(o); \exists \gamma \in \Gamma; \forall o' \in \Omega, \forall A \in \Lambda(o,o'), \forall m' \in \mathcal{M}^5_{\Omega,\Lambda}(o'); m,m' \models_\gamma A\}$
$\mathcal{M}^6_{\Omega,\Lambda}(o) =$
$\quad \{m \in \mathcal{M}(o); \exists \gamma \in \Gamma; \forall o' \in \Omega, \forall A \in \Lambda(o,o'), \forall m' \in \mathcal{M}(o'); m,m' \models_\gamma A\}$

The selection of the γ would ideally be made after the models are chosen because they determine the domains on which the equalising function is built. In practice, the satisfiability condition will only select those equalising functions that are compatible with each of the models.

These approaches have been ordered from the more optimistic to the more cautious. $\mathcal{M}^1_{\Omega,\Lambda}$ selects the models that satisfy each alignment in at least one model of the connected ontology. $\mathcal{M}^3_{\Omega,\Lambda}$ is stronger since it requires that the same model of the connected ontology satisfies all the alignments between the two ontologies. $\mathcal{M}^6_{\Omega,\Lambda}$ is very strong since all alignments must be satisfied by all models of the connected ontologies. $\mathcal{M}^2_{\Omega,\Lambda}$, $\mathcal{M}^4_{\Omega,\Lambda}$ and $\mathcal{M}^5_{\Omega,\Lambda}$ are fixed point characterisations that, instead of considering the initial models of the connected agents, consider their selected models by the same function. This contributes propagating the constraints to the whole connected components of the distributed system. While for $\mathcal{M}^2_{\Omega,\Lambda}$ and $\mathcal{M}^4_{\Omega,\Lambda}$ this strengthens the constraints, for $\mathcal{M}^5_{\Omega,\Lambda}$, this relaxes them with regard to $\mathcal{M}^6_{\Omega,\Lambda}$. Here, an inconsistent model is a problem only to related agents and only for versions $\mathcal{M}^1_{\Omega,\Lambda}$, $\mathcal{M}^3_{\Omega,\Lambda}$ and $\mathcal{M}^4_{\Omega,\Lambda}$ which require the existence of a model for each related ontology.

Each of these options allows the definition of a semantics for distributed systems that is different from the model of distributed system considered above. It is also analogous to the distributed knowledge of the system following [Fagin et al., 1995]. $\mathcal{M}^5_{\Omega,\Lambda}$ would correspond to the semantics given in [Franconi et al., 2003, Bouquet et al., 2004a]. A definition of acceptable models for an ontology corresponding to option $\mathcal{M}^4_{\Omega,\Lambda}$ is given as follows.

Definition 2.26 (Models of an ontology modulo alignments). *Given a distributed system $\langle \Omega, \Lambda \rangle$, the models of $o \in \Omega$ modulo Λ are those models of o, such that for each ontology o' there exists a model that satisfies all elements of Λ between o and o':*

$$\mathcal{M}^4_{\Omega,\Lambda}(o) =$$
$$\{m \in \mathcal{M}(o); \exists \gamma \in \Gamma; \forall o' \in \Omega, \exists m' \in \mathcal{M}^4_{\Omega,\Lambda}(o'); \forall A \in \Lambda(o,o'), m, m' \models_\gamma A\}$$

One can be even more restrictive, as in local model semantics [Ghidini and Giunchiglia, 2001], by considering only a subset of the possible models of each ontology.

When dealing with ontology matching between a pair of ontologies, the matter of semantics between ontologies is not related to the alignment but to the interpretation of the full distributed system, for instance, depending if one wants to enforce global consistency or not. In this book we will not take a position on such a matter and will only retain the basic interpretation framework provided above.

Finally, such a formalism contributes to the definition of the meaning of alignments: it describes what are the consequences of ontologies with alignments, i.e., what can be deduced by an agent. However, it does not describe what the correct alignments are: matching is not a deductive task but an inductive one. The framework is nevertheless particularly useful for deciding if delivered alignments are consistent, i.e., if distributed systems have a model or not. Hence, it is useful for specifying what is expected from matching algorithms and how they should be designed or evaluated.

2.6 Summary

In this chapter, we have first described different kinds of data and conceptual models and observed an expressivity hierarchy of them. Formal ontologies turn out to be their most elaborate form. This means that the work that has already been devoted to matching various data and conceptual models can be reused for ontology matching. This also means that by developing and reviewing ontology matching this book will help wider areas as well.

In fact, on the one side, for example, schema matching is usually performed with the help of techniques trying to guess the meaning encoded in the schemas. On the other side, ontology matching systems primarily try to exploit knowledge explicitly encoded in the ontologies. In real-world applications, schemas and ontologies usually have both well defined and obscure terms, and contexts in which they occur, therefore, solutions from both problems would be mutually beneficial. Consequently, we focus our attention on ontology matching as a task that comprises many characteristics of other forms of matching between data and conceptual models.

Then, we focused on identifying what semantic heterogeneity is and why it requires matching. We have presented various reasons why mismatches may occur between ontologies. Their variety and the fact that they often occur together constrain to develop multiple approaches for matching ontologies. These techniques will be classified in the next chapter and further detailed in latter ones.

Finally, we have defined the action of matching ontologies and its result: the alignment. The interpretation of alignment has been provided first generally but informally, before introducing a semantics for it in distributed systems. This semantics enables the reader to understand how the result of ontology matching have to be interpreted and what is expected from ontology matchers.

Part II

Ontology matching techniques

3
Classifications of ontology matching techniques

Having defined what the matching problem is, we attempt at classifying the techniques that can be used for solving this problem. The major contributions of the previous decades are presented in [Larson et al., 1989, Batini et al., 1986, Kashyap and Sheth, 1996, Parent and Spaccapietra, 1998], while the topic through the recent years have been surveyed in [Rahm and Bernstein, 2001, Wache et al., 2001, Kalfoglou and Schorlemmer, 2003b]. These three works address the matching problem from different perspectives (artificial intelligence, information systems, databases) and analyse disjoint sets of systems. [Shvaiko and Euzenat, 2005] have attempted to consider the above mentioned works together, focusing on schema-based matching methods, and aiming to provide a common conceptual basis for their analysis. Here, we follow and extend this work on classifying matching approaches and use it in the following chapters for organising the presentation.

In this chapter we first consider various dimensions along which a classification can be elaborated (§3.1). We then present our classification based on several of these dimensions (§3.2). Finally, we discuss some alternative classifications of matching approaches that have been proposed so far (§3.3).

3.1 Matching dimensions

There are many independent dimensions along which algorithms can be classified. Following the definition of the matching process in Fig. 2.8, we may primarily classify algorithms according to (i) the input of the algorithms, (ii) the characteristics of the matching process, and (iii) the output of the algorithms. The other characteristics, such as parameters, resources, and input alignments, are considered less important. Let us discuss these three main aspects in turn.

3.1.1 Input dimensions

These dimensions concern the kind of input on which algorithms operate. As a first dimension, algorithms can be classified depending on the data or conceptual models in which ontologies are expressed. For example, the Artemis system (§6.1.6) supports the relational, object-oriented, and entity-relationship models; Cupid (§6.1.11) supports XML and relational models; QOM (§6.3.4) supports RDF and OWL models. A second possible dimension depends on the kind of data that the algorithms exploit: different approaches exploit different information in the input ontologies. Some of them rely only on schema-level information, e.g., Cupid (§6.1.11), COMA (§6.1.12); others rely only on instance data, e.g., GLUE (§6.2.5); and others exploit both schema- and instance-level information, e.g., QOM (§6.3.4). Even with the same data models, matching systems do not always use all available constructs, e.g., S-Match (§6.1.19) when dealing with attributes discards information about datatypes, e.g., string or integer, and uses only the attributes names. In general, some algorithms focus on the labels assigned to the entities, some consider their internal structure and the types of their attributes, and others consider their relations with other entities (see next section for details).

3.1.2 Process dimensions

A classification of the matching process could be based on its general properties, as soon as we restrict ourselves to formal algorithms. In particular, it depends on the *approximate* or *exact* nature of its computation. Exact algorithms compute the precise solution to a problem; approximate algorithms sacrifice exactness for performance [Ehrig and Sure, 2004]. All of the techniques discussed in the remainder of the book can be either approximate or exact. Another dimension for analysing the matching algorithms is based on the way they interpret the input data. We identify three large classes based on the intrinsic input, external resources, or some semantic theory of the considered entities. We call these three classes *syntactic*, *external*, and *semantic* respectively, and discuss them in detail in the next section.

3.1.3 Output dimensions

Apart from the information that matching systems exploit and how they manipulate it, the other important class of dimensions concerns the form of the result these systems produce. The form of the alignment might be of importance: is it a one-to-one alignment between the ontology entities? Has it to be a final correspondence? Is any relation suitable?

Some other significant distinctions in the output results have been indicated in [Giunchiglia and Shvaiko, 2003a]. One dimension concerns whether systems deliver a graded answer, e.g., that the correspondence holds with 98% confidence or 4/5 probability; or an all-or-nothing answer, e.g., that the correspondence definitely holds or not. In some approaches, correspondences between ontology entities are determined using distance measures. This is used for providing an alignment expressing equivalence between these entities. Another dimension concerns the kind of relations between entities a system can provide. Most of the systems focus on equivalence (=), while a few others are able to provide a more expressive result, e.g., equivalence, subsumption ($\sqsubseteq$), and incompatibility ($\perp$) [Giunchiglia et al., 2004, Bouquet et al., 2003c].

In the next section we present a classification of elementary techniques that draws simultaneously on these criteria.

3.2 Classification of matching approaches

To ground and ensure a comprehensive coverage for our classification we have analysed state of the art approaches used for ontology matching. Chap. 6 reports a partial list of systems and approaches which have been scrutinised pointing to (some of) the most important contributions. We have used the following guidelines for building our classification:

Exhaustivity: The extension of categories dividing a particular category must cover its extension, i.e., their aggregation should give the complete extension of the category.

Disjointness: In order to have a proper tree, the categories dividing one category should be pairwise disjoint by construction.

Homogeneity: In addition, the criteria used for further dividing one category should be of the same nature, i.e., should come from the same dimension introduced in Sect. 3.1. This usually helps guarantee disjointness.

Saturation: Classes of concrete matching techniques should be as specific and discriminative as possible in order to provide a fine-grained distinction between possible alternatives. These classes have been identified following a *saturation* principle: they have been added and modified until the saturation was reached, i.e., taking into account new techniques did not require introducing new classes or modifying them.

Disjointness and exhaustivity of the categories ensure stability of the classification, namely that new techniques will not occur in between two categories. Classes of matching techniques represent the state of the art. Obviously, with appearance of new techniques, they might be extended and further detailed.

The exact vs. approximate opposition has not been used because each of the methods described below can be implemented as exact or approximate algorithms, depending on the goals of the matching system.

We build on the previous work on classifying automated schema matching approaches of [Rahm and Bernstein, 2001] which distinguishes between *elementary* (individual) matchers and *composition* of matchers. Elementary matchers comprise *instance-* and *schema-based*, *element-* and *structure-level*, *linguistic* and *constraint-based* matching techniques. Also *cardinality* and *auxiliary information*, e.g., thesauri, global schemas, can be taken into account.

For classifying elementary matching techniques, we have introduced two synthetic classifications in [Shvaiko and Euzenat, 2005], based on what we have found to be the most salient properties of the matching dimensions (see Fig. 3.1). These two classifications are presented as two trees sharing their leaves. The leaves represent classes of elementary matching techniques and their concrete examples. Two synthetic classifications are:

- *Granularity/Input Interpretation* classification is based (i) on the matcher granularity, i.e., element- or structure-level, and then (ii) on how the techniques generally interpret the input information.
- *Kind of Input* classification is based on the kind of input which is used by elementary matching techniques.

The overall classification of Fig. 3.1 can be read both in descending (focusing on how the techniques interpret the input information) and ascending (focusing on the kinds of manipulated objects) manner in order to reach the *Basic Techniques* layer.

Elementary matchers are distinguished by the *Granularity/Input interpretation* layer according to the following classification criteria:

- *Element-level vs. structure-level:* Element-level matching techniques compute correspondences by analysing entities or instances of those entities in isolation, ignoring their relations with other entities or their instances. Structure-level techniques compute correspondences by analysing how entities or their instances appear together in a structure. This criterion for schema-based approaches is the same as first introduced in [Rahm and Bernstein, 2001], while element-level vs. structure-level separation for instance-based approaches follows the work in [Kang and Naughton, 2003].
- *Syntactic vs. external vs. semantic:* The key characteristic of the syntactic techniques is that they interpret the input with regard to its sole structure following some clearly stated algorithm. External are the techniques exploiting auxiliary (external) resources of a domain and common knowledge in order to interpret the input. These resources may be human input or some thesaurus expressing the relationships between terms. Semantic techniques use some formal semantics, e.g., model-theoretic semantics, to interpret the input and justify their results. In case of a semantic-based matching system, exact algorithms are complete with regard to the semantics, i.e., they guarantee a discovery of all the possible alignments, while approximate algorithms tend to be incomplete.

To emphasise the differences with the initial classification of [Rahm and Bernstein, 2001], the new categories or classes are marked in bold face. In particular, in the Granularity/Input Interpretation layer we detail further the element- and structure-level matching by introducing the syntactic vs. semantic vs. external criteria. The reasons for having these three categories are as follows. Our initial criterion was to distinguish between internal and external techniques. By *internal* we mean techniques exploiting information which comes only with the input ontologies. *External* techniques are as defined above. Internal techniques can be further detailed by distinguishing between syntactic and semantic interpretation

3.2 Classification of matching approaches

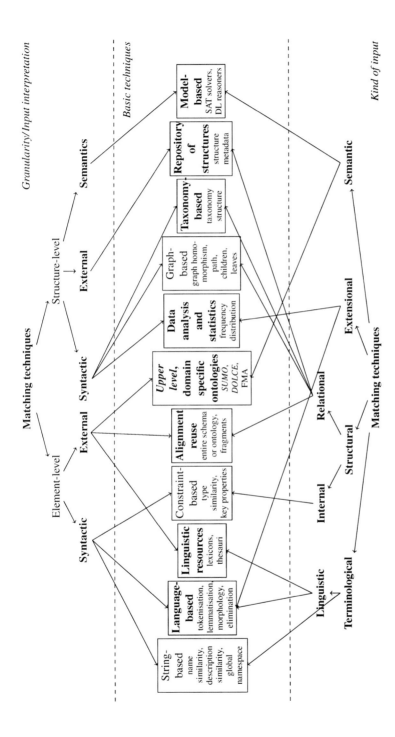

Fig. 3.1. The retained classifications of elementary matching approaches. The upper classification is based on granularity and input interpretation; the lower classification is based on the kind of input. The middle layer features classes of basic techniques. The novelty of this classification in comparison with our previous work in [Shvaiko and Euzenat, 2005] includes *extensional* category of techniques as well as *data analysis and statistics* class of methods.

of input, also as defined above. The same distinction can be introduced, to some extent, for the external techniques. In fact, we can qualify some oracles, e.g., Cyc [Lenat and Guha, 1990], WordNet [Miller, 1995], SUMO [Niles and Pease, 2001], DOLCE [Gangemi et al., 2003], as syntactic or semantic, but not user input. Thus, we do not detail *external* techniques any further and we omit in Fig. 3.1 the theoretical category of *internal* techniques, as opposed to *external*. We also omit in further discussions element-level semantic techniques, since semantics is usually given in a structure, and, hence, there are no element-level semantic techniques.

Distinctions between classes of elementary matching techniques in the *Basic Techniques* layer of our classification are motivated by the way a technique interprets the input information in each concrete case. In particular, a label can be interpreted as a string (a sequence of letters from an alphabet) or as a word or a phrase in some natural language, a hierarchy can be considered as a graph (a set of nodes related by edges) or a taxonomy (a set of concepts having a set-theoretic interpretation organised by a relation which preserves inclusion). Thus, we introduce the following classes of elementary ontology matching techniques at the element-level: string-based, language-based, based on linguistic resources, constraint-based, alignment reuse, and based on upper level and domain specific formal ontologies. At the structure-level we distinguish between graph-based, taxonomy-based, based on repositories of structures, model-based, and data analysis and statistics techniques.

The *Kind of Input* layer classification is concerned with the type of input considered by a particular technique:

- The first level is categorised depending on which kind of data the algorithms work on: strings (*terminological*), structure (*structural*), models (*semantics*) or data instances (*extensional*). The two first ones are found in the ontology descriptions. The third one requires some semantic interpretation of the ontology and usually uses some semantically compliant reasoner to deduce the correspondences. The last one constitutes the actual population of an ontology.
- The second level of this classification decomposes further these categories if necessary: terminological methods can be string-based (considering the terms as sequences of characters) or based on the interpretation of these terms as linguistic objects (*linguistic*). The structural methods category is split into two types of methods: those which consider the internal structure of entities, e.g., attributes and their types (*internal*), and those which consider the relation of entities with other entities (*relational*).

Following the above mentioned guidelines for building a classification, the terminological category should be divided into linguistic and non linguistic techniques. However, since non linguistic techniques are all string-based, this category has been discarded. This presentation is the one followed in the presentation of basic techniques (Chap. 4).

We discuss below the main classes of the *Basic Techniques* layer according to the above classification in more detail. The order follows that of the *Granularity/Input Interpretation* classification and these techniques are divided in two sections concerning element-level techniques (§3.2.1) and structure-level techniques (§3.2.2). Finally, techniques which are marked in italic in Fig. 3.1 (techniques based on upper level ontologies) have not been implemented in any matching system yet. However, we argue that their appearance seems reasonable in the near future.

3.2.1 Element-level techniques

Element-level techniques consider ontology entities or their instances in isolation from their relations with other entities or their instances.

String-based techniques

String-based techniques are often used in order to match names and name descriptions of ontology entities. These techniques consider strings as sequences of letters in an alphabet. They are typically based on the following intuition: the more similar the strings, the more likely they are to denote the same concepts. Usually, distance functions map a pair of strings to a real number, where a smaller value of the real number indicates a greater similarity between the strings. Some examples of string-based techniques which are extensively used in matching systems are prefix, suffix, edit, and n-gram distances. Various such string comparison techniques are presented in Sect. 4.2.1.

Language-based techniques

Language-based techniques consider names as words in some natural language, e.g., English. They are based on natural language processing techniques exploiting morphological properties of the input words. Several of these techniques are presented in Sect. 4.2.2 (intrinsic techniques).

Usually, they are applied to names of entities before running string-based or lexicon-based techniques in order to improve their results. However, we consider these language-based techniques as a separate class of matching techniques, since they can be naturally extended, for example, in a distance computation (by comparing the resulting strings or sets of strings).

Constraint-based techniques

Constraint-based techniques are algorithms which deal with the internal constraints being applied to the definitions of entities, such as types, cardinality (or multiplicity) of attributes, and keys. These techniques are presented in Sect. 4.3.1.

Linguistic resources

Linguistic resources such as lexicons or domain specific thesauri are used in order to match words (in this case names of ontology entities are considered as words of a natural language) based on linguistic relations between them, e.g., synonyms, hyponyms. Several such methods are presented in Sect. 4.2.2 (extrinsic techniques).

Alignment reuse

Alignment reuse techniques represent an alternative way of exploiting external resources, which record alignments of previously matched ontologies. For instance, when we need to match ontology o' and o'', given the alignments between o and o', and between o and o'' available from the external resource. Alignment reuse is motivated by the intuition that many ontologies to be matched are similar to already matched ontologies, especially if they are describing the same application domain. These techniques are particularly promising when dealing with large ontologies consisting of hundreds and thousands of entities. In these cases, first, large match problems are decomposed into smaller subproblems, thus generating a set of ontology fragments matching problems. Then, reuse of previous match results can be more effectively applied at the level of ontology fragments rather than at the level of entire ontologies. The approach was first introduced in [Rahm and Bernstein, 2001] and was later implemented as two matchers, i.e., (i) reuse alignments of entire ontologies, or (ii) their fragments [Do and Rahm, 2002, Aumüller et al., 2005, Rahm et al., 2004].

Upper level and domain specific formal ontologies

Upper level ontologies can also be used as external sources of common knowledge. Examples are the upper Cyc ontology [Lenat and Guha, 1990], the Suggested Upper Merged Ontology (SUMO) [Niles and Pease, 2001] and Descriptive Ontology for Linguistic and Cognitive Engineering (DOLCE) [Gangemi et al., 2003]. The key characteristic of these ontologies is that they are logic-based systems, and therefore, matching techniques exploiting them are based on semantics. For the moment, we are not aware of any matching system which uses this kind of techniques. However, it is quite reasonable to assume that this will happen in the near future. In fact, for example, the DOLCE ontology aims at providing a formal specification (axiomatic theory) for the top level part of WordNet. Therefore, systems exploiting WordNet in their matching process may also consider using DOLCE as a potential semantic extension.

Domain specific formal ontologies can also be used as external sources of background knowledge. Such ontologies are focusing on a particular domain and use terms in a sense that is relevant only to this domain and which is not related to similar concepts in other domains. For example, in the anatomy domain, an ontology such as The Foundational Model of Anatomy (FMA) can be used as the context for the other medical ontologies to be matched (as long as it is known that the reference ontology covers the ontologies to be matched). This can be used for providing the missing structure when matching poorly structured resources [Aleksovski et al., 2006]. These methods are discussed in Sect. 4.5.1.

3.2.2 Structure-level techniques

Contrary to element-level techniques, structure-level techniques consider the ontology entities or their instances to compare their relations with other entities or their instances.

Graph-based techniques

Graph-based techniques are graph algorithms which consider the input ontologies as labelled graphs. The ontologies (including database schemas, and taxonomies) are viewed as labelled graph structures. Usually, the similarity comparison between a pair of nodes from the two ontologies is based on the analysis of their positions within the graphs. The intuition behind this is that, if two nodes from two ontologies are similar, their neighbours must also be somehow similar. Different graph-based techniques are described in Sect. 4.3.2.

Along with purely graph-based techniques, there are other more specific structure-based techniques, for instance, involving trees.

Taxonomy-based techniques

Taxonomy-based techniques are also graph algorithms which consider only the specialisation relation. The intuition behind taxonomic techniques is that *is-a* links connect terms that are already similar (being interpreted as a subset or superset of each other), therefore their neighbours may be also somehow similar. This intuition can be exploited in several different ways presented in Sect. 4.3.2.

Repository of structures

Repositories of structures store ontologies and their fragments together with pairwise similarity measures, e.g., coefficients in the [0 1] range between them. Unlike alignment reuse, repositories of structures store only similarities between ontologies, not alignments. In the following, to simplify the presentation, we call ontologies, or their fragments, as structures. When new structures are to be matched, they are first checked for similarity against the structures which are already available in the repository. The goal is to identify structures which are sufficiently similar to be worth matching in more detail, or reusing already existing alignments, thus, avoiding the match operation over the dissimilar structures. Obviously, the determination of similarity between structures should be computationally cheaper than matching them in full detail. The approach of [Rahm *et al.*, 2004] to matching two structures proposes to use some metadata describing these structures, such as structure name, root name, number of nodes, maximal path length, etc. These indicators are then analysed and aggregated into a single coefficient, which estimates similarity between them. For example, two structures may be found as an appropriate match if they both have the same number of nodes.

Model-based techniques

Model-based (or semantically grounded) algorithms handle the input based on its semantic interpretation, e.g., model-theoretic semantics. The intuition is that if two entities are the same, then they share the same interpretations. Thus, they are well grounded deductive methods. Examples are propositional satisfiability and description logics reasoning techniques. They are further reviewed in Sect. 4.5.

Data analysis and statistics techniques

Data analysis and statistical techniques are those which take advantage of a (hopefully large) representative sample of a population in order to find regularities and discrepancies. This helps in grouping together items or computing distances between them. Among data analysis techniques we discuss distance-based classification, formal concepts analysis (§4.4.1) and correspondence analysis; among statistical analysis methods we consider frequency distributions.

We exclude from this category learning techniques which require a sample of the result. These techniques are considered specifically in Sect. 5.4 as strategies.

3.3 Other classifications

Let us now consider some other available classifications of matching techniques.

[Ehrig, 2007] introduced a classification based on two orthogonal dimensions. These can be viewed as horizontal and vertical dimensions. The horizontal dimension includes three layers that are built one on top of another:

Data layer: This is the first layer. Matching between entities is performed here by comparing only data values of simple or complex datatypes.

Ontology layer: This is the second layer which, in turn, is further divided into four levels, following the 'layer cake' of [Berners-Lee et al., 2001]. These are semantic nets, description logics, restrictions and rules. For example, at the level of semantic nets, ontologies are viewed as graphs with concepts and relations, and, therefore, matching is performed by comparing only these. The description logics level brings a formal semantics account to ontologies. Matching at this level includes, for example, determining taxonomic similarity based on the number of subsumption relations separating two concepts. This level also takes into account instances of entities, therefore, for example, assessing concepts to be the same, if their instances are similar. Matching at the levels of restrictions and rules is typically based on the idea that if similar rules between entities exist, these entities can be regarded as similar. This typically requires processing higher order relations.

Context layer: Finally, this layer is concerned with the practical usage of entities in the context of an application. Matching is performed here by comparing the usages of entities in ontology-based applications. One of the intuitions behind such matching methods is that similar entities are often used in similar contexts.

The vertical dimension represents specific *domain knowledge* which can be situated at any layer of the horizontal dimension. Here, the advantage of external resources of domain specific knowledge, e.g., Dublin Core for the bibliographic domain, is considered for assessing the similarity between entities of ontologies.

[Doan and Halevy, 2005] classifies matching techniques into (i) rule-based and (ii) learning-based. Typically, rule-based techniques work with schema-level information, such as entity names, datatypes and structures. Some examples of rules are that two entities match if their names are similar or if they have the same number of neighbour entities. Learning-based approaches often work with instance-level information, thereby performing matching, for example, by comparing value formats and distributions of data instances underlying the entities under consideration. However, learning can also be done at the schema-level and from the previous matches, e.g., as proposed in the LSD approach (§6.2.4).

[Zanobini, 2006] classifies matching methods into three categories following the cognitive theory of meaning and communication between agents:

Syntactic: This category represents methods that use purely syntactic matching methods. Some examples of such methods include string-based techniques, e.g., edit distance between strings (§4.2.1) and graph matching techniques, e.g., tree edit distance (§4.3.2).

Pragmatic: This category represents methods that rely on comparison of data instances underlying the entities under consideration in order to compute alignments. Some examples of such methods include automatic classifiers, e.g., Bayesian classifier (§4.4), and formal concepts analysis (§4.4.1).

Conceptual: This category represents methods that work with concepts and compare their meanings in order to compute alignments. Some examples of such methods include techniques exploiting external thesauri, such as WordNet (§4.2.2), in order to compare senses among the concepts under consideration.

There were also some classifications mixing the process dimension of matching together with either input dimension or output dimension. For example, [Do, 2005] extends the work of [Rahm and Bernstein, 2001] by adding a *reuse-oriented* category of techniques on top of schema-based vs. instance-based separation, meaning that reuse-oriented techniques can be applied at schema and instance level. However, these techniques can also include some input information, such as user input or alignments obtained from previous match operations.

[Giunchiglia and Shvaiko, 2003a] classified matching approaches into *syntactic* and *semantic*. At the matching process dimension these correspond to syntactic and conceptual categories of [Zanobini, 2006], respectively. However, these have been also constrained by a second condition dealing with the output dimension: syntactic techniques return coefficients in the $[0\ 1]$ range, while semantic techniques return logical relations, such as equivalence, subsumption. Combining methods that work with concepts as well as return logical relations has been defined as a semantic matching problem in [Giunchiglia and Shvaiko, 2003a].

Finally, we notice that the more the ontology matching field progresses, the wider the variety of techniques that come into use at different levels of granularity. For example, machine learning methods, which where often applied only to the instance level information, also started being applied more widely to schema level information. We believe that such a cross-fertilisation will gain more evidence in future. Therefore, ultimately, it could be the case that any mathematical method will find appropriate uses for ontology matching.

3.4 Summary

Following the complexity of ontology definition, there is a variety of techniques that can be used. The classifications discussed in this chapter provide a common conceptual basis for organising them, and, hence, can be used for comparing (analytically) different existing ontology matching systems as well as for designing new ones, taking advantages of state of the art solutions. The classifications of matching methods also provide some guidelines which help in identifying families of matching approaches.

This chapter has shown the difficulty of having a clear cut classification of algorithms. In Sect. 3.2 we provided two such classifications based on granularity and input interpretation on the one side and the kind of input on the other side. They will be used for organising the presentation of basic techniques in the next chapter.

4
Basic techniques

The goal of ontology matching is to find the relations between entities expressed in different ontologies. Very often, these relations are equivalence relations that are discovered through the measure of the similarity between the entities of ontologies.

We present here some of the basic methods for assessing the similarity or the relations between ontology entities. By basic, we mean that these methods base their judgment on one particular kind of features of these entities. Chap. 5, in turn, shows how the results of these methods can be combined.

In this chapter, we first introduce basic concepts related to similarity (§4.1). Then, we consider basic methods following the 'kind of input' layer of the classification of Chap. 3: entity names (§4.2), structure (§4.3), extension (§4.4) and semantics (§4.5).

4.1 Similarity, distances and other measures

There are many ways to assess the similarity between two entities. The most common way amounts to defining a measure of this similarity. We present some characteristics of these measures.

Definition 4.1 (Similarity). *A similarity* $\sigma : o \times o \to \mathbb{R}$ *is a function from a pair of entities to a real number expressing the similarity between two objects such that:*

$$\forall x, y \in o, \sigma(x,y) \geq 0 \qquad \textit{(positiveness)}$$
$$\forall x \in o, \forall y, z \in o, \sigma(x,x) \geq \sigma(y,z) \qquad \textit{(maximality)}$$
$$\forall x, y \in o, \sigma(x,y) = \sigma(y,x) \qquad \textit{(symmetry)}$$

The dissimilarity is a dual operation. It is defined as follows.

Definition 4.2 (Dissimilarity). *Given a set o of entities, a dissimilarity* $\delta : o \times o \to \mathbb{R}$ *is a function from a pair of entities to a real number such that:*

$$\forall x, y \in o, \delta(x,y) \geq 0 \qquad \text{(positiveness)}$$
$$\forall x \in o, \delta(x,x) = 0 \qquad \text{(minimality)}$$
$$\forall x, y \in o, \delta(x,y) = \delta(y,x) \qquad \text{(symmetry)}$$

Some authors consider a 'non symmetric (dis)similarity', [Tverski, 1977]; we then use the term non symmetric measure or pre-similarity. There are more constraining notions of dissimilarity, such as distances and ultrametrics.

Definition 4.3 (Distance). *A distance (or metric) $\delta : o \times o \to \mathbb{R}$ is a dissimilarity function satisfying the definiteness and triangular inequality:*

$$\forall x, y \in o, \delta(x,y) = 0 \text{ if and only if } x = y \qquad \text{(definiteness)}$$
$$\forall x, y, z \in o, \delta(x,y) + \delta(y,z) \geq \delta(x,z) \qquad \text{(triangular inequality)}$$

Definition 4.4 (Ultrametric). *Given a set o of entities, an ultrametric is a metric such that:*

$$\forall x, y, z \in o, \delta(x,y) \leq \max(\delta(x,z), \delta(y,z)) \qquad \text{(ultrametric inequality)}$$

Very often, the measures are normalised, especially if the similarity of different kinds of entities must be compared. Reducing each value to the same scale in proportion to the size of the considered space is the common way to normalise.

Definition 4.5 (Normalised (dis)similarity). *A (dis)similarity is said to be normalised if it ranges over the unit interval of real numbers $[0\ 1]$. A normalised version of a (dis)similarity σ (respectively, δ) is denoted as $\overline{\sigma}$ (respectively, $\overline{\delta}$).*

It is easy to see that to any normalised similarity $\overline{\sigma}$ corresponds a normalised dissimilarity $\overline{\delta} = 1 - \overline{\sigma}$ and vice versa. In the remainder, we will consider mostly normalised measures and assume that a dissimilarity function between two entities returns a real number between 0. and 1.

From the above definitions, the similarity and dissimilarity are complete functions that map pairs of entities to real numbers. An alternative representation for such a function on a finite set of entities is a *matrix* (see Example 4.14). The matrix has the advantage of being a finite data structure that can be exchanged between programs.

4.2 Name-based techniques

Some terminological methods compare strings. They can be applied to the name, the label or the comments of entities in order to find those which are similar. This can be used for comparing class names and/or URIs.

Throughout this section, the set $\mathbb{S}$ will represent the set of strings, i.e., the sequences of letters of any length over an alphabet $\mathbb{L}$: $\mathbb{S} = \mathbb{L}*$. The empty string is denoted as ϵ, and $\forall s, t \in \mathbb{S}$, $s + t$ is the concatenation of the strings s and t. $|s|$ denotes the length of the string s, i.e., the numbers of characters it contains. $s[i]$ for $i \in [1\ |s|]$ stands for the letter in position i of s.

Example 4.6 (Strings). The string 'article' is made of the letters a, r, t, i, c, l and e. Its length is 7 characters. 'peer-reviewed' and ' ' are two other strings (so '-' and ' ' are letters in the alphabet) and their concatenation 'peer-reviewed'+' '+'article' provides the string 'peer-reviewed article' whose length is 21.

A string s is the substring of another string t, if there exist two strings s' and s'', such that $s' + s + s'' = t$ (denoted as $s \in t$). Two strings are equal ($s = t$) if and only if $s \in t$ and $t \in s$. The number of occurrences of s in t (denoted as $s\#t$) is the number of distinct pairs s', s'', such that $s' + s + s'' = t$.

Example 4.7 (Substrings). The string 'peer-reviewed article' has the string 'review' as a substring because 'peer-'+'review'+'ed article'='peer-reviewed article'. The string 'homonymous' has three occurences of the string 'o', two occurences of the string 'mo' and only one occurence of the string 'nym'.

The main problem in comparing ontology entities on the basis of their labels occurs due to the existence of synonyms and homonyms:

Synonyms are different words used to name the same entity. For instance, Article and Paper are synonyms in some contexts;

Homonyms are words used to name different entities. For instance, peer as a noun has a sense 'equal' as well as another sense 'member of the nobility'. The fact that a word can have multiple senses is also known as *polysemy*.

Consequently, it is not possible to deduce with certainty that two entities are the same if they have the same name or that they are different because they have different names. There are more reasons than synonymy and homonymy why this could happen. In particular:

- Words from different languages, such as English, French, Italian, Spanish, German, Greek, are used to name the same entities. For instance, the word Book in English is Livre in French and книга in Russian.
- Syntactic variations of the same word often occur according to different acceptable spellings, abbreviations, use of optional prefixes or suffixes, etc. For instance, Compact disc, CD, C.D. and CD-ROM can be considered equivalent in some contexts. However, in some other contexts, CD may mean Corps diplomatique and in some others change directory.

These kinds of variations can occur within one ontology but can be even more frequent across ontologies. However, the way in which things are named remains very important in every day communication and names remain a good index of similarity or dissimilarity. Moreover, many different techniques have been designed for assessing the similarity of two terms notwithstanding the similarity or dissimilarity of the strings which denote them.

There are two main categories of methods for comparing terms depending on their consideration of character strings only (§4.2.1) or using some linguistic knowledge to interpret these strings (§4.2.2).

4.2.1 String-based methods

String-based methods take advantage of the structure of the string (as a sequence of letters). String-based methods will typically find classes Book and Textbook to be similar, but not classes Book and Volume.

There are many ways to compare strings depending on the way the string is viewed: for example, as an exact sequence of letters, an erroneous sequence of letters, a set of letters, a set of words. [Cohen *et al.*, 2003b] compares various string-matching techniques, from distance like functions to token-based distance functions. We discuss the most frequently used methods.

We distinguish between (i) normalisation techniques which are used for reducing strings to be compared to a common format, (ii) substring or subsequence techniques that base similarity on the common letters between strings, (iii) edit distances that further evaluate how one string can be an erroneous version of another, (iv) statistical measures that establish the importance of a word in a string by weighting the relation between two strings and (v) path comparisons.

Normalisation

Before comparing actual strings which have a meaning in natural language, there are normalisation procedures that can help improve the results of subsequent comparisons. In particular:

Case normalisation consists of converting each alphabetic character in the strings into their lower case counterpart. For example, CD becomes cd and SciFi becomes scifi.

Diacritics suppression consists of replacing characters with diacritic signs with their most frequent replacements. For example, replacing Montréal with Montreal.

Blank normalisation consists of normalising all blank characters, such as blank, tabulation, carriage return, or sequences of these, into a single blank character.

Link stripping consists of normalising some links between words, such as replacing apostrophes and blank underline into dashes or blanks. For example, peer-reviewed becomes peer reviewed.

Digit suppression consists of suppressing digits. For example, book24545-18 becomes book.

Punctuation elimination suppresses punctuation signs. For example, C.D. becomes CD.

These normalisation operations must be used with care for several reasons. In particular:

- they are often language-dependent, e.g., they work for occidental languages;
- they are order dependent: they do not guarantee to bring the same results when applied in any order;
- they can result in loosing some meaningful information; for example, carbon-14 becomes carbon or sentence separation, which is very useful for parsing, is lost;
- they may reduce variations, but increase synonyms. For example, in French livre and livré are different words respectively meaning book and shipped.

String equality

String equality returns 0 if the strings under consideration are not identical and 1 if they are identical. This can be taken as a similarity measure.

Definition 4.8 (String equality). *String equality is a similarity* $\sigma : \mathbb{S} \times \mathbb{S} \to [0\ 1]$ *such that* $\forall x, y \in \mathbb{S}$, $\sigma(x,x) = 1$ *and if* $x \neq y, \sigma(x,y) = 0$.

It can be performed after some syntactic normalisation of the string, e.g., downcasing, encoding conversion, accent normalisation.

This measure does not explain how strings are different. A more immediate way of comparing two strings is the Hamming distance which counts the number of positions in which the two strings differ [Hamming, 1950]. We present here the version normalised by the length of the longest string.

Definition 4.9 (Hamming distance). *The Hamming distance is a dissimilarity* $\delta : \mathbb{S} \times \mathbb{S} \to [0\ 1]$ *such that:*

$$\delta(s,t) = \frac{\left(\sum_{i=1}^{\min(|s|,|t|)} s[i] \neq t[i]\right) + ||s| - |t||}{\max(|s|,|t|)}$$

Substring test

Different variations can be obtained from the string equality, such as considering that strings are very similar when one is a substring of another:

Definition 4.10 (Substring test). *Substring test is a similarity* $\sigma : \mathbb{S} \times \mathbb{S} \to [0\ 1]$ *such that* $\forall x, y \in \mathbb{S}$, *if there exist* $p, s \in \mathbb{S}$ *where* $x = p + y + s$ *or* $y = p + x + s$, *then* $\sigma(x,y) = 1$, *otherwise* $\sigma(x,y) = 0$.

This is obviously a similarity. This measure can be refined in a substring similarity which measures the ratio of the common subpart between two strings.

Definition 4.11 (Substring similarity). *Substring similarity is a similarity* $\sigma : \mathbb{S} \times \mathbb{S} \to [0\ 1]$ *such that* $\forall x, y \in \mathbb{S}$, *and let* t *be the longest common substring of* x *and* y:

$$\sigma(x,y) = \frac{2|t|}{|x| + |y|}$$

It is easy to see that this measure is indeed a similarity. One could also consider a subsequence similarity as well. This definition can be used for building functions based on the longest common prefix or longest common suffix.

Thus, for example, the similarity between article and aricle would be $4/7 = .57$, while between article and paper would be $1/7 = .14$, and, finally, between article and particle would be $6/7 = .86$.

A prefix or suffix pre-similarity can be defined on this model from the prefix and suffix tests, which test whether one string is the prefix or suffix of another. These measures would not be symmetric. Prefix and suffix pre-similarity can be useful as a test for strings denoting a more general concept than another (in many languages, adding clauses to a term would restrict its range). For instance, reviewed article is more specific than article. It can also be used for comparing strings and similar abbreviations, e.g., ord and order.

The n-gram similarity is also often used in comparing strings. It computes the number of common n-grams, i.e., sequences of n characters, between them. For instance, trigrams for the string article are: art, rti, tic, icl, cle.

Definition 4.12 (n-gram similarity). *Let $ngram(s, n)$ be the set of substrings of s of length n. The n-gram similarity is a similarity $\sigma : \mathbb{S} \times \mathbb{S} \to \mathbb{R}$ such that:*

$$\sigma(s,t) = |ngram(s,n) \cap ngram(t,n)|$$

The normalised version of this function is as follows.

$$\overline{\sigma}(s,t) = \frac{|ngram(s,n) \cap ngram(t,n)|}{\min(|s|,|t|) - n + 1}$$

This function is quite efficient when only some characters are missing.

Thus, for example, the similarity between article and aricle would be $2/4 = .5$, while between article and paper would be 0, and, finally, between article and particle would be $5/6 = .83$.

It is possible, to add extra characters at the beginning and end of strings for dealing with too small strings.

Edit distance

Intuitively, an edit distance between two objects is the minimal cost of operations to be applied to one of the objects in order to obtain the other one. Edit distances were designed for measuring similarity between strings that may contain spelling mistakes.

Definition 4.13 (Edit distance). *Given a set Op of string operations ($op : \mathbb{S} \to \mathbb{S}$), and a cost function $w : Op \to \mathbb{R}$, such that for any pair of strings there exists a sequence of operations which transforms the first one into the second one (and vice versa), the edit distance is a dissimilarity $\delta : \mathbb{S} \times \mathbb{S} \to [0\ 1]$ where $\delta(s,t)$, is the cost of the less costly sequence of operations which transforms s into t.*

$$\delta(s,t) = \min_{(op_i)_I; op_n(...op_1(s))=t} (\sum_{i \in I} w_{op_i})$$

In string edit distance, the operations that are usually considered include insertion of a character $ins(c, i)$, replacement of a character by another $sub(c, c', i)$ and deletion of a character $del(c, i)$. It can be easily checked that these operations are such that $ins(c, i) = del(c, i)^{-1}$ and $sub(c, c', i) = sub(c', c, i)^{-1}$. Each operation is assigned a cost and the distance between two strings is the sum of the cost of each operation on the less costly set of operations.

The Levenshtein distance [Levenshtein, 1965] is the minimum number of *insertions*, *deletions*, and *substitutions* of characters required to transform one string into the other. It is the edit distance with all costs equal to 1. The Needleman–Wunch distance [Needleman and Wunsch, 1970], in turn, is the edit distance with a higher costs for ins and del.

It can be proved that the edit distance is indeed a distance if $\forall op \in Op, w_{op} = w_{op^{-1}}$.

Example 4.14. The (rounded) Levenshtein distance table between the class labels of ontologies in Fig. 2.7 (p. 37):

	Science	Children	Book	Person	DVD	Textbook	Product	Pocket	Publisher	Popular	CD
Politics	0.75	1.00	0.88	0.88	1.00	1.00	0.75	0.75	0.67	0.75	1.00
Thing	0.71	0.75	1.00	1.00	1.00	0.88	1.00	1.00	0.89	1.00	1.00
Autobiography	0.92	0.85	0.85	0.92	1.00	0.85	0.92	0.92	0.85	0.85	1.00
Novel	0.86	0.88	0.80	1.00	1.00	1.00	0.86	0.67	0.89	0.71	1.00
Biography	1.00	0.89	0.78	0.89	1.00	1.00	0.89	0.89	1.00	0.89	1.00
Writer	0.86	0.75	1.00	1.00	1.00	0.88	0.86	0.83	0.67	0.86	1.00
Essay	1.00	1.00	1.00	0.83	1.00	1.00	1.00	1.00	0.89	0.86	1.00
Volume	0.86	0.75	0.83	1.00	1.00	1.00	0.71	0.83	0.78	0.71	1.00
LiteraryCritic	0.93	0.93	1.00	0.86	1.00	0.93	0.86	0.93	0.93	0.86	0.93
Poetry	0.86	0.88	0.83	0.83	1.00	0.88	0.71	0.67	0.89	0.71	1.00
Literature	0.80	0.90	1.00	0.80	1.00	0.90	0.80	0.90	0.90	0.80	1.00
Human	0.86	0.88	1.00	0.83	1.00	1.00	1.00	1.00	0.89	0.71	1.00

The closest names are Pocket and Novel, Pocket and Poetry, as well as Writer and Publisher and Politics and Publisher. These names are relatively far from each others (.67). So, in this case no correspondence can be found from such measures alone. However, the same measure on properties will obviously find the correspondence between author and author, for instance.

Other measures compute the cost of an edition operation as a function of the characters or substrings on which the operation applies. For that purpose, they use a cost matrix for each operation. A well known example of such a measure is the Smith–Waterman measure [Smith and Waterman, 1981] which was adapted to compute the distance between biological sequences based on the molecules that were manipulated. Other such measures are the Gotoh [Gotoh, 1981] and Monge–Elkan [Monge and Elkan, 1997] distance functions.

The Jaro measure has been defined for matching proper names that may contain similar spelling mistakes [Jaro, 1976, Jaro, 1989]. It is not based on an edit distance model, but on the number and proximity of the common characters between two strings. This measure is not a similarity because it is not symmetric.

Definition 4.15 (Jaro measure). *The Jaro measure is a non symmetric measure* $\sigma : \mathbb{S} \times \mathbb{S} \to [0\ 1]$ *such that*

$$\sigma(s,t) = \frac{1}{3} \times \left(\frac{|com(s,t)|}{|s|} + \frac{|com(t,s)|}{|t|} + \frac{|com(s,t)| - |transp(s,t)|}{|com(s,t)|} \right),$$

with

$s[i] \in com(s,t)$ *if and only if* $\exists j \in [i - (\min(|s|,|t|)/2\ i + (\min(|s|,|t|)/2]$

and $transp(s,t)$ *are the elements of* $com(s,t)$ *which occur in a different order in s and t.*

For instance, if we again compare article with aricle, aritcle and paper, the number of common letters will respectively be 6, 7 and 1 (because in the last case, the 'e' in paper is too far away from that in article). The number of transposed common letters will be 0, 1 and 0 respectively. As a consequence, the similarities between these strings are: .95, .90 and .45.

This measure has been improved by favouring matches between strings with longer common prefixes [Winkler, 1999].

Definition 4.16 (Jaro–Winkler measure). *The Jaro–Winkler measure* $\sigma : \mathbb{S} \times \mathbb{S} \to [0\ 1]$ *is as follows:*

$$\sigma(s,t) = \sigma_{Jaro}(s,t) + P \times Q \times \frac{(1 - \sigma_{Jaro}(s,t))}{10},$$

such that P is the length of the common prefix and Q is a constant.

In this case, the similarity for the three strings compared to article with $Q = 4$ are: .99, .98 and .45. These measures only improve on the previous ones by explicitly providing a model of mistakes that penalises less the comparison.

Another similar measure is Smoa [Stoilos et al., 2005] which is adapted to the way computer users define identifiers. It depends on common substring lengths and non common substring lengths, the second part being substracted from the first one. This measure has a value between -1 and 1.

Token-based distances

The following techniques come from information retrieval and consider a string as a (multi)set of words (also called bag of words), i.e., a set in which a particular item can appear several times. These approaches usually work well on long texts (comprising many words). For that reason, it is helpful to take advantage of other strings that are attached to ontology entities. This can be adapted to ontology entities as follows:

– By aggregating different sources of strings: identifiers, labels, comments, documentation, etc. Some systems go further by aggregating the tokens that correspond to connected entities [Qu et al., 2006].
– By splitting strings into independent tokens. For example, InProceedings becomes In and Proceedings, peer-reviewed article becomes peer, reviewed and article.

Ontology entities are then identified with *bags of words* (or multisets) suitable for manipulation by using information retrieval techniques. Many different similarities or dissimilarities being applied to sets of entities can thus be applied to these bags of words. For example, the matching coefficient is the complement of the Hamming distance on sets (§4.4.1) and the Dice coefficient is the complement of the Hamming distance on multisets, i.e., using the union, intersection and cardinality of multisets instead of sets.

Original measures are those based on the corpus of such strings, i.e., the set of all such strings found in one of the ontologies or in both of them. These measures are no longer intrinsic to the strings to be compared but depend on the corpus.

They usually consider a bag of words s as a vector $\vec{s}$ belonging to a metric space V in which each dimension is a term (or token) and each position in the vector is the number of occurrences of the token in the corresponding bag of words. This is one way to represent multisets. Each document can be considered as a point in this space identified by its coordinate vector [Salton, 1971, Salton and McGill, 1983].

Once the entities have been transformed into vectors, usual metric space distances can be used: Euclidean distance, Manhattan distance (also known as city blocks) and any instance of the Minkowski distance (see also p. 123). We present here the cosine similarity which measures the cosine of the angles made by two vectors. It is very often used in information retrieval.

Definition 4.17 (Cosine similarity). *Given $\vec{s}$ and $\vec{t}$, the vectors corresponding to two strings s and t in a vector space V, the cosine similarity is the function σ_V : $V \times V \to [0\ 1]$ such that:*

$$\sigma_V(s,t) = \frac{\sum_{i \in |V|} \vec{s}_i \times \vec{t}_i}{\sqrt{\sum_{i \in |V|} \vec{s}_i^{\,2} \times \sum_{i \in |V|} \vec{t}_i^{\,2}}}$$

Some more elaborate techniques use reduced spaces, like those obtained by correspondence analysis, in order to deal with a smaller dimension as well as to automatically map words of similar meanings to the same dimension. A famous example of such a technique, which is by using singular value decomposition, is known as latent semantic indexing [Deerwester et al., 1990].

A very common measure is TFIDF (Term frequency-Inverse document frequency) [Robertson and Jones, 1976] which is used for scoring the relevance of a document, i.e., a bag of words, to a term by taking into account the frequency of appearance of the term in the corpus. It is usually not a measure of similarity: it assesses the relevance of a term to a document. It is used here to assess the relevance of a substring to a string by comparing the frequency of appearance of the string in the document with regard to its frequency in the whole corpus.

Definition 4.18 (Term frequency-Inverse document frequency). *Given a corpus C of multisets, we define the following measures:*

$$\forall t \in \mathbb{S}, \forall s \in C, tf(t,s) = t\#s \qquad (term\ frequency)$$

$$\forall t \in \mathbb{S}, idf(t) = log\left(\frac{|C|}{|\{s \in C; t \in s\}|}\right) \qquad (inverse\ document\ frequency)$$

$$TFIDF(s,t) = tf(t,s) \times idf(t) \qquad (TFIDF)$$

Many systems use measures based on TFIDF. These measures compute, for each term in the strings, their relevance with regard to the corpus based on TFIDF. Then, they use vector space techniques for computing a distance between the two strings. There are several options for doing so depending on the selected space: this can be the whole corpus, the union of terms covered by the two strings or only the intersection of the terms involved in both strings. The most often used aggregation measure is the cosine similarity.

Path comparison

Path difference consists of comparing not only the labels of objects but the sequence of labels of entities to which those bearing the label are related. For instance, in the left-hand ontology of Fig. 2.7, the Science class can be identified by the path Product:Book:Science. In a first approximation, these can be considered as a particular way to aggregate tokens in an ordered fashion. A simple (and only) example is the one which concatenates all the names of the superclasses of classes before comparing them. So the result is dependent on the individual string comparison aggregated in some way.

Definition 4.19 (Path distance). *Given two sequences of strings, $\langle s_i \rangle_{i=1}^{n}$ and $\langle s'_j \rangle_{j=1}^{m}$, their path distance is defined as follows:*

$$\delta(\langle s_i \rangle_{i=1}^{n}, \langle s'_j \rangle_{j=1}^{m}) = \lambda \times \delta'(s_n, s'_m) + (1-\lambda) \times \delta(\langle s_i \rangle_{i=1}^{n-1}, \langle s'_j \rangle_{i=1}^{m-1})$$

such that

$$\delta(\langle\rangle, \langle s'_j \rangle_{j=1}^{k}) = \delta(\langle s_i \rangle_{i=1}^{k}, \langle\rangle) = k$$

with δ' being one of the other string or language-based distance and $\lambda \in [0\ 1]$.

For instance, we can take the string equality distance as δ', scoring 0 when the strings are equal, and .7 as λ. Then if we have to compare Product:Book:Science with Book:Essay:Science and Product:Cultural:Book:Science, the distances will respectively be: .273 and .09.

This measure is dependent on the similarity between the last element of each path: this similarity is affected by a λ penalty but every subsequent step is affected by a $\lambda \times (1-\lambda)^n$ penalty. So this measure takes into account the prefix, but the prefix can only influence the result to an extent which decreases as its distance from the end of the sequence increases. As can be seen, this measure is dependent on the rank of the elements to compare in the path. A more accurate, but expensive, measure, would choose the best match between both paths and penalise the items remote from the end of the path. Another way to take these paths into account is simply to apply them as a distance on sequences, such as described in [Valtchev, 1999].

Summary on string-based methods

The results given so far for these string comparisons are useful if people use very similar strings to denote the same concepts. If synonyms with different structures are used, this will yield a low similarity. Selecting pairs of strings with low similarity, in turn, yields many false positives since two strings can be very similar, e.g., Inproceedings and proceedings, and denote relatively different concepts. These measures are most often used in order to detect if two very similar strings are used. Otherwise, matching must use more reliable sources of information.

There are several software packages for computing string distances. Table 4.1 provides a brief comparison of distances available in four Java packages: Simetrics[1], SecondString[2], the Alignment API[3] and SimPack[4]. A comparison of the metrics of the second package has been provided in [Cohen *et al.*, 2003b].

4.2.2 Language-based methods

So far we have considered strings as sequences of characters. When considering language phenomenon, these strings become texts (theoretical peer-reviewed journal article). Texts can be segmented into words: easily identified sequence of letters that are derived from an entry in a dictionary (theoretical, peer, reviewed, journal, article). These words do not occur in a bag (as used in information retrieval) but in a sequence which has a grammatical structure. Very often words, like peer, bear a meaning and correspond to some concepts, but the more useful concepts to be properly handled in a text are terms, such as peer-review, or peer-reviewed journal.

Terms are phrases that identify concepts; they are thus often used for labelling concepts in ontologies. As a consequence, ontology matching could take great advantage of recognising and identifying them in strings. This amounts to recognise the term Peer-reviewed journal in the labels scientific periodicals reviewed by peers (and not in journal review paper).

Language-based methods rely on using Natural Language Processing (NLP) techniques to help extract the meaningful terms from a text. Comparing these terms and their relations should help assess the similarity of the ontology entities they name

---

[1] http://www.dcs.shef.ac.uk/~sam/stringmetrics.html
[2] http://secondstring.sourceforge.net
[3] http://alignapi.gforge.inria.fr
[4] http://www.ifi.unizh.ch/ddis/simpack.html

Table 4.1. String measures available in Simetrics, SecondString, Alignment API and SimPack Java packages.

Simetrics	SecondString	AlignAPI	SimPack
	n-grams	n-grams	
Levenshtein	Levenshtein	Levenshtein	Levenshtein
Jaro	Jaro	Jaro	
Jaro–Winkler	Jaro–Winkler	Jaro–Winkler	
Needleman–Wunch	Needleman–Wunch	Needleman–Wunch	
		Smoa	
Smith–Waterman			
Monge–Elkan	Monge–Elkan		
Gotoh			
Matching coefficient			
Jaccard	Jaccard		Jaccard
Dice coefficient			Dice coefficient
	TFIDF		TFIDF
Cityblocks			Cityblocks
Euclidean			Euclidean
Cosine			Cosine
Overlap			Overlap
Soundex			

and comment. Although these are based on some linguistic knowledge, we distinguish methods which rely on algorithms only and those which make use of external resources such as dictionaries.

Intrinsic methods: Linguistic normalisation

Linguistic normalisation aims at reducing each form of a term to some standardised form that can be easily recognised. Table 4.2 shows that the same term (theory paper) can appear under many different forms. The work in [Maynard and Ananiadou, 2001] distinguishes three main kinds of term variation: morphological (variation on the form and function of a word based on the same root), syntactic (variation on the grammatical structure of a term) and semantic (variation on one aspect of the term, usually using a hypernym or hyponym). Various subtypes of these broad categories are exemplified in Table 4.2. Multilingual variation, i.e., where the term variant is expressed in a different language, can be naturally added to these. Moreover, these types of variations can be combined in various ways.

Complete linguistic software chains have been developed for quickly obtaining a normal form of strings denoting terms. This is available through shallow parsers or part-of-speech taggers [Brill, 1992]. These usually perform the following functions:

Tokenisation: Tokenisation is the operation described in Sect. 4.2.1. It consists of segmenting strings into sequences of tokens by a tokeniser which recognises

Table 4.2. Variants of the term *theory paper* (adapted from [Maynard, 1999] and [Euzenat *et al.*, 2004a]).

Type	Subtype	Example
Morphological	Inflection	theory papers
	Derivation	theoretical paper
	Inflectional-Derivational	theoretical papers
Syntactic	Insertion	theory review paper
	Permutation	paper on theory
	Coordination	philosophy and theory paper
Morphosyntactic	Derivation-Coordination	philosophical and theoretical paper
	Inflection-Permutation	papers on theory
Semantic		foundational paper
Multilingual	French	article théorique

punctuation, cases, blank characters, digits, etc. For example, peer-reviewed periodic publication becomes ⟨peer, reviewed, periodic, publication⟩.

Lemmatisation: The strings underlying tokens are morphologically analysed in order to reduce them to normalised basic forms. Morphological analysis makes it possible to find flexion and derivations of a root. This involves suppressing tense, gender or number marks. Retrieving the root is called lemmatisation. Currently, systems can use some approximate lemmatisation techniques called stemming [Lovins, 1968, Porter, 1980] which strip suffixes from terms. For example, reviewed becomes review.

Term extraction: More elaborate technologies enable the extraction of terms from a text [Jacquemin and Tzoukermann, 1999, Bourigault and Jacquemin, 1999, Maynard and Ananiadou, 2001, Cerbah and Euzenat, 2001]. It is generally related to what is called corpus linguistics and requires a relatively large amount of text. Terminology extractors identify terms from the repetition of morphologically similar phrases in the texts and the use of patterns, e.g., noun^1 noun^2 → noun^2 on noun^1. This would recognise that the term theory paper is the same term as paper on theory.

Stopword elimination: The tokens that are recognised as articles, prepositions, conjunctions, etc. (usually words, such as to or a), are marked to be discarded because they are considered as non meaningful (empty) words for matching. For example, collection of article becomes collection article.

Once these techniques have been applied, ontology entities are represented as sets of terms, not words, that can be compared with the same techniques as presented before.

86   4 Basic techniques

Extrinsic methods

Extrinsic linguistic methods use external resources, such as dictionaries and lexicons. Several kinds of linguistic resources can be exploited in order to find similarities between terms.

Lexicons. A lexicon, or dictionary, is a set of words together with a natural language definition of these words (see for instance those of Example 4.21). Of course, for a particular word, e.g., Article, there can be several such definitions. Dictionaries can be used with gloss-based distances (see below).

Multi-lingual lexicons. Multi-lingual lexicons are lexicons in which the definition is replaced by the equivalent terms in another language, e.g., Paper in English corresponds to Article in French. Such dictionaries can be very useful if ontology labels are expressed in different languages. They can be used for matching as well as for disambiguating terms, i.e., identifying their intended sense, before matching.

Semantico-syntactic lexicons. Semantico-syntactic lexicons and semantic lexicons are resources used in natural language analysers. They very often not only record names but their categories, e.g., non animate, liquid, and record the types of arguments taken by verbs and adjectives, e.g., to flow takes a liquid as subject and has no object. These are difficult to create and are not much used in ontology matching.

Thesauri. A thesaurus is a kind of lexicon to which some relational information has been added. It usually contains relations, named hypernym, e.g., Biography is a more general term than Autobiography, which is hyponym, synonym, e.g., Paper means the same as Article, antonym, e.g., practice is the opposite of theory. WordNet [Miller, 1995] is such a thesaurus which distinguishes clearly between word senses by grouping words into sets of synonyms (synsets).

Terminologies. A terminology is a thesaurus for terms, which very often contains phrases rather than single words. They are usually domain specific and tend to be less equivocal than dictionaries.

This is not an exhaustive nor an authorised description of linguistic resources but it provides a typology of the kinds of properties on which a similarity between terms can be assessed on a linguistic basis.

These resources can be defined for one language or be specific to some domain. In the latter case, they tend to be more adapted when texts or ontologies concern this domain because they retain specialised senses, or senses that do not exist in the everyday language. They may also contain proper names and common abbreviations that are used in the domain. For instance, a company could expand CD as Compact Disc, PO as Purchase Order instead of Post Office or Project Officer.

It is worth noting that linguistic resources are introduced in order to deal with synonyms (the fact that matching entities are named differently). By increasing the interpretation (sense) of words, they increase the chances of finding the matching terms (true positives). On the other side this also increases homonyms (the fact that more words are available for naming the matching entities) and the chances to match non matching terms (false positives). Dealing with this problem is known as word sense disambiguation [Lesk, 1986, Ide and Véronis, 1998]. Word sense disambiguation tries to restrict the candidate senses (and the candidate matches) from the context, especially by selecting the senses in relation to the other associated words and their senses.

We illustrate the use of external resources with the help of WordNet[5] [Miller, 1995, Fellbaum, 1998]. WordNet is an electronic lexical database for English (it has been adapted to other languages, see for instance EuroWordNet[6]), based on the notion of *synsets* or sets of synonyms. A synset denotes a concept or a sense of a group of terms. WordNet also provides an hypernym (superconcept/subconcept) structure as well as other relations such as meronym (*part of* relations). It also provides textual descriptions of the concepts (*gloss*) containing definitions and examples. We will denote WordNet as a partially ordered synonym resource.

Definition 4.20 (Partially ordered synonym resource). *A partially ordered synonym resource Σ over a set of words W, is a triple $\langle E, \leq, \lambda \rangle$, such that $E \subseteq 2^W$ is a set of synsets, $\leq$ is the hypernym relation between synsets and λ is a function from synsets to their definition (a text that is considered here as a bag of words in W). For a term t, $\Sigma(t)$ denotes the set of synsets associated with t.*

Example 4.21 (WordNet entry). We reproduce here the WordNet (version 2.0) entry for the word author. Each sense is numbered in superscript:

> author[1] *noun*: Someone who originates or causes or initiates something; *Example* 'he was the generator of several complaints'. *Synonym* generator, source. *Hypernym* maker. *Hyponym* coiner.
> author[2] *noun*: Writes (books or stories or articles or the like) professionally (for pay). *Synonym* writer[2]. *Hypernym* communicator. *Hyponym* abstractor, alliterator, authoress, biographer, coauthor, commentator, contributor, cyberpunk, drafter, dramatist, encyclopedist, essayist, folk writer, framer, gagman, ghostwriter, Gothic romancer, hack, journalist, libretist, lyricist, novelist, pamphleteer, paragrapher, poet, polemist, rhymer, scriptwriter, space writer, speechwriter, tragedian, wordmonger, word-painter, wordsmith, Andersen, Assimov...
> author[3] *verb.*: Be the author of; *Example* 'She authored this play'. *Hypernym* write. *Hyponym* co-author, ghost.

This resembles a traditional dictionary entry apart from the *Hypernym* and *Hyponym* features and the explicit mention of the considered sense. The hypernym relations for the senses of the words creator, writer, author, illustrator, and person are presented in Fig. 4.1.

There are at least three families of methods for using WordNet as a resource for matching terms used in ontology entities:

---

[5] http://wordnet.princeton.edu
[6] http://www.illc.uva.nl/EuroWordNet/

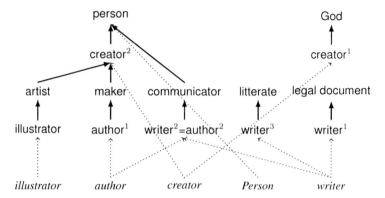

Fig. 4.1. The fragment of the WordNet hierarchy (limited to nouns) dealing with author, writer, creator, illustrator and person.

- considering that two terms are similar because they belong to some common synset;
- taking advantage of the hypernym structure for measuring the distances between synsets corresponding to two terms;
- taking advantage of the definitions of concepts provided by WordNet in order to evaluate the distance between the synsets associated with two terms.

A matcher based on WordNet can be designed by translating the (lexical) relations provided by WordNet to logical relations according to the following rules [Giunchiglia et al., 2004]:

- $t \sqsubseteq t'$, if t is a hyponym or meronym of t'. For example, author is a hyponym of creator, therefore we can conclude that author $\sqsubseteq$ creator.
- $t \sqsupseteq t'$, if t is a hypernym or holonym of t'. For example, Europe is a holonym of France, therefore we can conclude that Europe $\sqsupseteq$ France.
- $t = t'$, if they are connected by synonymy relation or they belong to one synset. For example, writer and author are synonyms, therefore we can conclude that writer = author.
- $t \perp t'$, if they are connected by antonymy relation or they are the siblings in the *part of* hierarchy. For example, Italy and France are siblings in the WordNet *part of* hierarchy, therefore we can conclude that Italy $\perp$ France.

Simple measures can be defined here (we only consider synonyms because they are the basis of WordNet synsets but other relationships can be used as well). The simplest use of synonyms is as follows:

Definition 4.22 (Synonymy similarity). *Given two terms s and t and a synonym resource Σ, the synonymy is a similarity $\sigma : \mathbb{S} \times \mathbb{S} \to [0\ 1]$ such that:*

$$\sigma(s,t) = \begin{cases} 1 & \text{if } \Sigma(s) \cap \Sigma(t) \neq \emptyset \\ 0 & \text{otherwise} \end{cases}$$

This would consider that the similarity between author and writer is maximal (1.) and that between author and creator is minimal (0.).

Example 4.23 (Synonymy). The synonymy similarity between illustrator, author, creator, Person, and writer is given by the following table:

	illustrator	author	creator	Person	writer
illustrator	1.	0.	0.	0.	0.
author	0.	1.	0.	0.	1.
creator	0.	0.	1.	0.	0.
Person	0.	0.	0.	1.	0.
writer	0.	1.	0.	0.	1.

This strict exploitation of synonyms does not allow analysis of how far non synonymous objects are nor how close synonymous objects are. Since synonymy is a relation, all the measures on the graph of relations can be used on WordNet synonyms. Another measure computes the cosynonymy similarity.

Definition 4.24 (Cosynonymy similarity). *Given two terms s and t and a synonym resource Σ, the cosynonymy is a similarity $\sigma : \mathbb{S} \times \mathbb{S} \to [0\ 1]$ such that:*

$$\sigma(s,t) = \frac{|\Sigma(s) \cap \Sigma(t)|}{|\Sigma(s) \cup \Sigma(t)|}$$

Example 4.25 (Cosynonymy similarity). The synonymy similarity between illustrator, author, creator, Person, and writer is given by the following table:

	illustrator	author	creator	Person	writer
illustrator	1.	0.	0.	0.	0.
author	0.	1.	0.	0.	.25
creator	0.	0.	1.	0.	0.
Person	0.	0.	0.	1.	0.
writer	0.	.25	0.	0.	1.

Some elaborate measures take into account that the terms can be part of several synsets and use a measure in the hyponym/hypernym hierarchy between synsets. A simple measure, known as edge-count, counts the number of edges separating two synsets in Σ (or the structural topological dissimilarity, see Sect. 4.3.2). More elaborate measures weight edge count with the position of synsets in the hierarchy. In particular, a measure developed specifically for WordNet is the one proposed by Wu and Palmer. It is presented in Sect. 4.3.2 because the hierarchy is, in this respect, similar to a class hierarchy. All measures defined in Sect. 4.3.2 can be used on the WordNet hypernym graph.

Other measures rely on an information theoretic perspective. They are based on the assumption that the most probable a concept, the less information it carries. So the information content of a concept is inverse to its probability of occurence. In the similarity proposed in [Resnik, 1995, Resnik, 1999], each synset (c) is associated with a probability of occurrence ($\pi(c)$) of an instance of the concept in a particular corpus. Usually, $\pi(c)$ is the the sum of the synset word occurrences divided by the total number of concepts. This probability is obtained from a corpus study. It is such that the more specific the concept, the lower its probability. The Resnik semantic similarity between two terms is a function of the more general synset common to both terms. It considers the maximum information content (or entropy), of the possible such synsets, taken as the negation of the logarithm of the probability of occurence.

Definition 4.26 (Resnik semantic similarity). *Given two terms s and t and a partially ordered synonym resource $\Sigma = \langle E, \leq, \lambda \rangle$ provided with a probability measure π, Resnik semantic similarity is a similarity $\sigma : \mathbb{S} \times \mathbb{S} \to [0\ 1]$ such that:*

$$\sigma(s,t) = \max_{k; \exists c, c' \in E; s \in c \wedge t \in c' \wedge c \leq k \wedge c' \leq k} (-log(\pi(k)))$$

We do not provide examples of corpus-based similarity because the results are dependent on the corpus on which it is based (here for defining π). Examples of such measures based on the Brown corpus[7] are given in [Budanitsky and Hirst, 2006].

This measure uses the maximum, but one could have chosen instead an average or a sum of all the pairs of synsets associated with the two terms.

Other information-theoretic similarities depend on the increase of the information content measure from the terms to their common hypernyms instead of the shared information content. This is the case in the Lin information-theoretic similarity [Lin, 1998]. This method specifies the probabilistic degree of overlap between two synsets:

Definition 4.27 (Information-theoretic similarity). *Given two terms s and t and a partially ordered synonym resource $\Sigma = \langle E, \leq, \lambda \rangle$ provided with a probability π, Lin information theoretic similarity is a similarity $\sigma : \mathbb{S} \times \mathbb{S} \to [0\ 1]$ such that:*

$$\sigma(s,t) = \max_{k; \exists c, c' \in \Sigma; s \in c \wedge t \in c' \wedge c \leq k \wedge c' \leq k} \frac{2 \times log(\pi(k))}{log(\pi(s)) + log(\pi(t))}$$

These similarities are not normalised.

A final way to compare terms found in strings through a thesaurus, like WordNet, is to use the definition (gloss) given to these terms in WordNet. In this case, any dictionary entry $s \in \Sigma$ is identified by the set of words corresponding to $\lambda(s)$. Then any measure defined in Sect. 4.2.1 can be used for comparing the strings [Lesk, 1986].

Definition 4.28 (Gloss overlap). *Given a partially ordered synonym resource $\Sigma = \langle R, \leq, \lambda \rangle$, the gloss overlap between two strings s and t is defined by the Jaccard similarity between their glosses:*

$$\sigma(s,t) = \frac{|\lambda(s) \cap \lambda(t)|}{|\lambda(s) \cup \lambda(t)|}$$

---

[7] http://nora.hd.uib.no/icame/

Example 4.29 (Gloss overlap). For computing the gloss overlap similarity between illustrator, author, creator, Person, and writer, we used the following treatments: take gloss for all senses and add the term name; suppress quotations ('...'); suppress empty words (or, and, the, a, an, for, of, etc.); suppress technical vocabulary, e.g., 'term'; suppress empty phrases, e.g., 'usually including'; keep categories, e.g., law; stem words. The gloss of author is given in Example 4.21.

The results have been taken as sets (not bags, so there is no repetition) of words and syntactically compared, yielding the following table:

	illustrator	author	creator	Person	writer
illustrator	1.	0.05	0.07	0.	0.02
author	0.05	1.	0.	0.	0.19
creator	0.07	0.	1.	0.06	0.02
Person	0.	0.	0.06	1.	0.04
writer	0.02	0.19	0.02	0.04	1.

This result is consistent with the previous measures since the only previously matching pair (author-writer) is still the highest scorer. This measure introduces new relations such as creator-illustrator, but still does not find the (possible) relation between creator and author. This is entirely related to the quality of glosses in WordNet.

Another example of building a matcher by using (WordNet) glosses includes counting the number of occurrences of the label of the source input sense in the gloss of the target input sense. If this number is equal to a threshold, e.g., 1, the less general relation can be returned. The reason for returning the less general relation is due to a common pattern of defining terms in glosses through a more general term. For example, in WordNet creator is defined as 'a person who grows or makes or invents things'. Thus, following this strategy we could find that creator ⊑ person. Some other variations of gloss-based matchers include considering glosses of the parent (children) nodes of the input senses in the WordNet *is a (part of)* hierarchy [Giunchiglia and Yatskevich, 2004]. The relations produced by these matchers depend heavily on the context of the matching task, and therefore, these matchers cannot be applied in all the cases [Giunchiglia *et al.*, 2006c].

Summary on linguistic methods

Many methods presented in this section have been implemented in the Perl package[8] WordNet::similarity [Pedersen *et al.*, 2004] and the Java package SimPack[9] (see Table 4.3). They have been thoroughly compared in [Budanitsky and Hirst, 2006].

Linguistic resources, such as stemmers, part-of-speech taggers, lexicons, and thesauri are invaluable resources since they allow the interpretation of the terms used in the expressions of ontologies. They provide a more accurate apprehension of these labels.

---
[8] http://wn-similarity.sourceforge.net/
[9] http://www.ifi.unizh.ch/ddis/simpack.html

Table 4.3. List of language measures based on WordNet and available in the wn-similarity Perl package and the SimPack Java package (some measures have not been presented yet).

WordNet::similarity	SimPack
Resnik	Resnik
Jiang–Conrath (1997)	
Lin	Lin
Leacock–Chodorow	Leacock–Chodorow
Hirst–St.Onge ([Saint-Onge, 1995])	
Edge count	Edge count
Wu–Palmer	Wu–Palmer
Extended Gloss Overlap	
Vector on gloss	

However, whenever the adequate resources are available for some language, they mainly open new possible matches between entities because they recognise that two terms can denote the same concept. Unfortunately, since they also recognise that the same term may denote several concepts at once, these techniques provide many possible matches from which to choose.

One way to choose among these representations is to take into account the structure of ontology entities in order to select the most coherent matches.

4.3 Structure-based techniques

The structure of entities that can be found in ontologies can be compared, instead of or in addition to comparing their names or identifiers.

This comparison can be subdivided into a comparison of the internal structure of an entity, i.e., besides its name and annotations, its properties or, in the case of OWL ontologies, the properties which take their values in a datatype, or the comparison of the entity with other entities to which it is related. The former is called internal (§4.3.1) and the latter is called relational structure (§4.3.2). The internal structure is the definition of entities without reference to other entities; the relational structure is the set of relations that an entity has with other entities. As expected, the internal structure is primarily exploited in database schema matching, while the relational structure is more important in matching formal ontologies and semantic networks.

4.3.1 Internal structure

Internal structure based methods are sometimes referred to as constraint-based approaches in the literature [Rahm and Bernstein, 2001]. These methods are based on the internal structure of entities and use such criteria as the set of their properties, the range of their properties (attributes and relations), their cardinality or multiplicity, and the transitivity or symmetry of their properties to calculate the similarity between them.

Entities with comparable internal structures or properties with similar domains and ranges in two ontologies can be numerous. For that reason, these kinds of methods are commonly used to create correspondence clusters rather than to discover accurate correspondences between entities. They are usually combined with other element-level techniques, such as terminological methods, and are responsible for reducing the number of candidate correspondences. They can be used with other approaches as a preprocessing step to eliminate most of the properties that are clearly incompatible.

For illustrating these methods we consider the properties associated with the Product and Volume entities in the example of Fig. 4.2 (the expected correspondences are given in Fig. 2.9, p. 48).

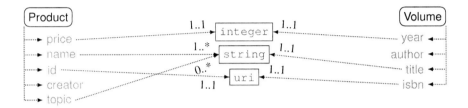

Fig. 4.2. Two sets of properties to be compared.

If we start from the elements of Fig. 4.2, there is no chance that pure terminological similarity methods find them very similar, though year and creator may appear the same to some edit distance methods. A linguistic method may be better able to find a relationship between creator and author.

Comparing the internal structure of ontology entities amounts to comparing their properties and composing the obtained result: the system can evaluate the similarity between all components considered next (names, keys, datatypes, domains, cardinalities) or multiplicities and combine the results. The combination operation is considered in Sect. 5.2, we focus here on the elementary comparison.

Property comparison and keys

In database schemas, unlike in formal ontologies, tables are provided with keys: a combination of properties whose values uniquely identify an object. For a Book, it would typically be the international standard book number (isbn), for a Person it can be his or her name, birth place and date.

This information is primarily very useful for recognising that two individuals are the same. Thus, keys are mostly used in extensional methods as a means to identify individuals and then apply methods on common set of instances (§4.4).

However keys can also be used for identifying classes: two classes identified in the same way are likely to represent the same set of objects. Moreover, even if two schemas use different keys for the same class, e.g., identifying Person with a social security number, there can be secondary keys that perform the same functions, e.g., that the social security number is also considered a key in the other class. So, when provided with keys, if they are highly compatible (similar names and types), it is plausible that the classes are equivalent.

For instance, if Product has id as a key and Volume has isbn as a key, it can be considered that these properties should correspond in case where the classes are the same. This can be considered possible because both properties have the same type (uri).

Datatype comparison

Property comparison involves comparing the property datatype (in OWL, this can be the range of the relation or a Restriction applied to the property in the class). Contrary to objects that require interpretations, datatypes can be considered objectively and it is possible to determine how close a datatype is to another (ideally this can be based on the interpretation of datatypes as sets of values and the set-theoretic comparison of these datatypes [Valtchev, 1999, Valtchev and Euzenat, 1997]).

We distinguish here between a datatype, which corresponds to the way the values are stored in a computer (like integer, float, string or uri), and a domain, which characterises a subset of a particular datatype (like [10 12] or '*book'). Datatypes are considered here and domains are addressed in the next section.

Datatypes are not fully disjoint, though there are rules by which an object of one type can be thought of as an object of another type and rules by which a value of some type can be converted in the memory representation of another type (known as *casting* in programming languages).

Ideally, the proximity between datatypes should be maximal when these are the same types, lower when the types are compatible (for instance, integer and float are compatible since they can be cast one into the other) and the lowest when they are non compatible. In addition, domain comparison should ideally be based on datatype comparison and the comparison of the sets of values covered by these domains. The compatibility between property datatypes can be assessed by using an underlying table lookup. An example of a part of such a table is given in Table 4.4.

Table 4.4. Part of a datatype compatibility table.

	char	fixed	enumeration	int	number	string
string	0.7	0.4	0.7	0.4	0.5	1.0
number	0.6	0.9	0.0	0.9	1.0	0.5

Such a table can be extracted, for languages like OWL, from the type hierarchy of XML Schema datatypes (see Fig. 4.3). In the example of Fig. 4.2, it can be considered that since a uri is a subclass of string, the isbn may be related to name.

Example 4.30 (Datatype comparison). In the example of Fig. 4.2, data type comparison would let us match price with year, both name and topics with title, and id with

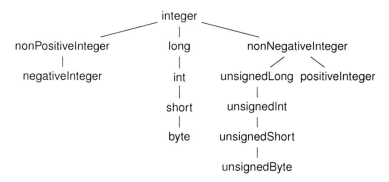

Fig. 4.3. Fragment of the XML Schema datatype hierarchy [Biron and Malhotra (ed.), 2004].

isbn. creator and author are left aside because they are object-valued properties. This comparison yields interesting results since it finds the expected matches. However, it also finds incorrect ones (price-year and topics-title) so these methods cannot be used in isolation.

Domain comparison

Depending on the entities to be considered, what can be reached from a property can be different: in classes these are domains while in individuals these are values. Moreover, they can be structured in sets or sequences. It is thus important to consider this fact in the comparison.

[Valtchev, 1999] proposes a framework in which the types or domains of properties must be compared on the basis of their interpretations: sets of values. Type comparison is based on their respective size, in which the size of a type is the cardinality or multiplicity of the set of values it defines. The distance between two domains is then given by the difference between their size and that of their common generalisation. This measure is usually normalised by the size of the largest possible distance attached to a particular datatype. We give here an instance of this type of measure.

Definition 4.31 (Relative size distance). *Given two domain expressions e and e' over a datatype τ, the relative size distance $\delta : 2^\tau \times 2^\tau \to [0\ 1]$, is as follows:*

$$\delta(e, e') = \frac{|gen_\tau(e \vee e')| - |gen_\tau(e \wedge e')|}{|\tau|},$$

such that $gen_\tau(.)$ provides the generalisation of a type expression and $\vee$ and $\wedge$ correspond to the union and intersection of the types.

Example 4.32 (Relative size distance). Consider a property age in one class to be compared with the property age of three other classes (schoolchild, teenager and grown-up). The first property has a domain of [6 12], while the others have respective domains expressed by: [7 14], [14 22] and ≥ 10. All these properties have datatype integer. The generalisation of these four domains are the domains themselves, the union with [6 12] is respectively [6 14], [6 22], [6 +∞[, and the intersection is respectively [7 12], ∅, and [10 12]. As a consequence, the distance will be respectively $3/|\tau|$, $17/|\tau|$ and $|\tau| - 3/|\tau|$. This corresponds to some intuition that the distance between domains depends on the difference between the values they cover in isolation and in common.

There are three advantages of this measure. The most obvious one is that it is normalised. The second one is that it is totally general (it is not expressed in terms of integers). The third one is that it can easily be mapped to the usual measures that are often used.

Usually, a common generalisation depends on the type: it is a set for enumerated types and an interval for ordered types (it can also be a set of intervals). In the case of dense types, the size of a domain is the usual measure of its size (Euclidean distance can be used for real or floating point numbers). The case of infinite types has to be taken adequately (by evaluating the largest possible domain in a computer or by normalising with regard to the actual corpus) [Valtchev, 1999]. Normalising over the largest distance in the corpus, if possible, is often a good idea. Indeed, it is not reasonable, for example, to normalise the age of people with that of planets or their size even if they use the same unit. Another advantage of this framework is that it encompasses value comparisons which can be considered as singletons and compared with domains if necessary.

Comparing multiplicities and properties

Properties can be constrained by multiplicities (as they are called in UML). Multiplicities are the acceptable cardinalities of the set of values of a property (for a given object). Similar to compatibilities between datatypes, compatibility between cardinalities can be established based on a table look-up. An example of such a table for DTDs is given in Table 4.5, following the work in [Lee et al., 2002].

Table 4.5. A cardinality compatibility table.

	*	+	?	none
*	1.0	0.9	0.7	0.7
+	0.9	1.0	0.7	0.7
?	0.7	0.7	1.0	0.8
none	0.7	0.7	0.8	1.0

In OWL, cardinalities or multiplicities are expressed through the minCardinality, maxCardinality and cardinality restrictions. Multiplicities can be expressed as an interval of the set of positive integers [0 +∞[. As such they are domains of the integer type. Two multiplicities are compatible if the intersection of the corresponding intervals is non empty. Any measure on the integer datatype can be used for assessing the similarity between multiplicities (see previous paragraph). However, in this case we choose a simpler distance inspired from the Jaccard similarity.

Values can be collected by a particular construction (set, list, multiset) on which cardinality constraints are applied. Again, it is possible to compare these constructed datatypes by comparing (i) the datatypes on which they are constructed and (ii) the cardinalities that are applied to them. For instance, sets of 2 and 3 children are closer to a set of 3 people than to a set of 10–12 flowers (if children are people). This technique is used in [Euzenat and Valtchev, 2004].

Definition 4.33 (Multiplicity similarity). *Given two multiplicity expressions $[b\ e]$ and $[b'\ e']$, the multiplicity similarity is a similarity between non negative integer intervals $\sigma : 2^\tau \times 2^\tau \to [0\ 1]$, such that:*

$$\sigma([b\ e], [b'\ e']) = \begin{cases} 0 & \text{if } b' > e \text{ or } b > e' \\ \dfrac{\min(e, e') - \max(b, b')}{\max(e, e') - \min(b, b')} & \text{otherwise} \end{cases}$$

For instance, if we have to compare multiplicity $[0\ 6]$ with $[2\ 8]$, $[8\ 12]$ and $[0\ +\infty]$, the comparison will respectively yield .5, 0. and $6/MAXINT$ (the latter is very low but remains non null because it is compatible with the initial multiplicity).

Example 4.34 (Multiplicity comparison). In the example of Fig. 4.2, multiplicity comparison can be used to further match id with isbn because they will both have a cardinality of $[1\ 1]$ and, unfortunately, will match price with year as well. However, is can also be used to prefer matching name rather than topic to title because they have the same multiplicities ($[1\ +\infty]$ instead of $[0\ +\infty]$).

Other features

Other internal structural factors have been considered in database schema matching. Since these are internal features, they can be very dependent on the knowledge model. For example, the work in [Navathe and Buneman, 1986] discusses such additional property characteristics as uniqueness, static semantic integrity constraints, dynamic semantic integrity constraints, security constraints, allowable operations and scale.

It is also possible in some languages to consider collection constructors, e.g., Set, List, Bag or multiset, Array, and their compatibility. It is then necessary to compare sets or lists of objects, e.g., the sequence of topics or the set of authors of a Book. In this case, general techniques can be used for assessing the similarity or distance between these sets depending on the similarity applying to the type of their elements. Concerning sets, these methods will be presented in Sect. 4.4.1 in the context of extension comparison. Concerning sequences, they can be adapted from some of the measures that have been presented in Sect. 4.2.1 which have considered strings as sequences of characters and paths as sequences of strings. In addition, Sect. 5.3.2 explains how to compare sets of objects with similarities.

In [Ehrig and Sure, 2004], it is proposed that the definition of a set of rules can be used for determining similarity between ontology entities. They point out that some features from OWL related to internal structure, such as symmetry and restrictions of values, could be used, but are discarded at the moment, as they do not have any wide distribution.

Summary on internal structure

Internal structure, including the names of entities, is very important for matching because it provides a basis on which algorithms can rely. The techniques for comparing them are efficient and easy to implement.

However, the internal structure does not provide much information on the entities to compare: many very different types of objects can have properties with the same datatypes. On the one hand, they can be used for eliminating incompatible correspondences and promoting compatible ones. On the other hand, it is always possible that different models of a concept use different, and incompatible, types. For these reasons, internal structure comparisons must always be used jointly with other techniques.

4.3.2 Relational structure

An ontology can be considered to be a graph whose edges are labelled by relation names (mathematically speaking, this is the graph of the multiple relations of the ontology: $\leq, \in, \perp, :, =$). Finding the correspondences between elements of such graphs corresponds to solving a form of the graph homomorphism problem [Garey and Johnson, 1979]. Namely it can be related to finding a maximum common directed subgraph.

Definition 4.35 (Maximum common directed subgraph problem). *Given two directed graphs $G = \langle V, E \rangle$ and $G' = \langle V', E' \rangle$, does there exist $F \subseteq E$ and $F' \subseteq E'$ and a pair of functions $f : V \to V'$ and $f^{-1} : V' \to V$ such that:*

- $\forall \langle u, v \rangle \in E|_F, \langle f(u), f(v) \rangle \in E'|_{F'}$;
- $\forall \langle u', v' \rangle \in E'|_{F'}, \langle f^{-1}(u'), f^{-1}(v') \rangle \in E|_F$;
- $\forall u \in V|_F, f^{-1}(f(u)) = u$;
- $\forall u' \in V'|_{F'}, f(f^{-1}(u')) = u'$;
- *there is no other $F \subseteq H \subseteq E$ and $F' \subseteq H' \subseteq E'$ satisfying these properties.*

Note that graph matching is another type of problem which is presented in Sect. 5.7.3.

In ontology matching, the problem is encoded as an optimisation problem (finding the isomorphic subgraphs minimising some distance like the dissimilarity between matched objects or maximising similarity). These subgraphs do not have to be maximal. Moreover, the problem is very often adapted for multipartite graphs separating classes from properties.

The similarity comparison between two entities from two ontologies can be based on the relations of these entities with the other entities in the ontologies: the more two entities are similar, the more their related entities should be alike. This remark can be exploited in several ways depending on the kind of relations considered. Moreover, given the transitive nature of some relations, it is natural to extend this remark through transitivity. Roughly, for each pair of relations, we can come up with 5 different ways of comparing the relations [Euzenat et al., 2004a]:

r comparing the entities in direct relation through r;
r^- comparing the entities in the transitive reduction of relation r;
r^+ comparing the entities in the transitive closure of relation r;
r^{-1} comparing the entities coming through a relation r;
$r \uparrow$ comparing entities which are ultimately in r^+ (the maximal elements of the closure).

These relations are exemplified as follows:

Example 4.36 (Exploiting relations in an ontology). Given the left-hand ontology of Fig. 2.7, the relations based on subClass from Book are as follows:

$$subclass(\text{Book}) = subclass^-(\text{Book}) = \{\text{Science, Pocket, Children}\}$$
$$subclass^+(\text{Book}) = \{\text{Science, Pocket, Textbook, Popular, Children}\}$$
$$subclass^{-1}(\text{Book}) = \{\text{Product}\}$$
$$subclass \uparrow (\text{Book}) = \{\text{Textbook, Popular, Pocket, Children}\}$$

Table 4.6 displays the different ways of comparing two ontology entities based on their relations with other entities. Of course, an approach can combine several of the above criteria [Mädche and Staab, 2002, Euzenat and Valtchev, 2004, Bach et al., 2004].

As can be observed from Table 4.6, some features have type String and can be compared with the techniques proposed in Sect. 4.2.1. However, those with type Class or Property really induce a graph structure. Moreover, the values which are labelled by Set(·) are more difficult to deal with because this means that many edges labelled by the feature will appear in the graph. The last part of the table is, in fact, relevant to the extensional methods that will be presented in Sect. 4.4.

There are three types of relations that have been considered so far in relational structure techniques: taxonomic relations, mereologic relations and all the involved relations. These are considered below.

Taxonomic structure

The taxonomic structure, i.e., the graph made with the subClassOf relation, is the backbone of ontologies. For this reason, it has been studied in detail by researchers and is very often used as a comparison source for matching classes.

Table 4.6. Features on which comparison of ontology entities can be made. The table reads: Two *Entities* are similar if their *Features* are similar. This table is an adapted version of tables reported in [Ehrig, 2007], [Euzenat *et al.*, 2004a] and [Euzenat and Valtchev, 2004].

Entity	Feature	OWL	Type
Class	name	rdf:label	String
	id	rdf:ID	String
	comments	rdf:comment	String
	same classes	owl:sameClassAs	Set(Class)
	properties	*property*	Set(Property)
	ultimate properties	*property*$\uparrow$	Set(Property)
	direct superclasses	owl:subClassOf$^-$	Set(Class)
	direct subclasses	owl:subClassOf^{-1-}	Set(Class)
	superclasses	owl:subClassOf*	Set(Class)
	subclasses	owl:subClassOf^{-1*}	Set(Class)
	ultimate subclasses	owl:subClassOf$^{-1}\uparrow$	Set(Class)
	direct instances	rdf:type^{-1*}	Set(Individual)
	instances	rdf:type^{-1-}	Set(Individual)
Property	name	rdf:label	String
	id	rdf:ID	String
	comments	rdf:comment	String
	same properties	owl:samePropertyAs	Set(Property)
	domain/range	rdfs:domain/rdfs:range	Class
	direct superproperties	rdfs:subProperty$^-$	Set(Property)
	direct subproperties	rdfs:subProperty^{-1-}	Set(Property)
	superproperties	rdfs:subProperty*	Set(Property)
	subproperties	rdfs:subProperty^{-1*}	Set(Property)
Individual	name	rdf:label	String
	id	rdf:ID	String
	comments	rdf:comment	String
	same individuals	owl:sameAs	Set(Instance)
	direct classes	rdf:type$^-$	Set(Class)
	classes	rdf:type*	Set(Class)
	properties	*property*	Set(Property)

There have been several measures proposed for comparing classes based on the taxonomic structure. The most common ones are based on counting the number of edges in the taxonomy between two classes. The structural topological dissimilarity on a hierarchy [Valtchev and Euzenat, 1997] follows the graph distance, i.e., the shortest path distance in a graph taken here as the transitive reduction of the hierarchy.

Definition 4.37 (Structural topological dissimilarity on hierarchies). *The structural topological dissimilarity $\delta : o \times o \rightarrow \mathbb{R}$ is a dissimilarity over a hierarchy $H = \langle o, \leq \rangle$, such that:*

$$\forall e, e' \in o, \delta(e, e') = \min_{c \in o}[\delta(e, c) + \delta(e', c)]$$

4.3 Structure-based techniques   101

where $\delta(e,c)$ is the number of intermediate edges between an element e and another element c.

This corresponds to the unit tree distance of [Barthélemy and Guénoche, 1992], i.e., with weight 1 on each edge. This function can be normalised by the maximal length of a path between two classes in the taxonomy:

$$\bar{\delta}(e,e') = \frac{\delta(e,e')}{\max_{c,c' \in o} \delta(c,c')}$$

Example 4.38 (Structural topological dissimilarity). We provide the examples of this section based on the taxonomy in Fig. 4.1. We consider that each term corresponds to a class (all senses are considered together) and there exists a top of the hierarchy (on top of Person, litterate, legal document and God).

	illustrator	author	creator	Person	writer
illustrator	0.	.8	.4	.6	1.
author	.8	0.	.4	.6	0.
creator	.4	.4	0.	.2	.6
Person	.6	.6	.2	0.	.4
writer	1.	0.	.6	.4	0.

Again, this corroborates the WordNet data that the closest classes are writer and author.

The results given by such a measure are not always semantically relevant since a long path in a class hierarchy can often be summarised as an alternative short one.

A similar measure is the one of Leacock–Chodorow [Leacock et al., 1998] which is function of the length of the shortest path. It has been introduced for lexicographic taxonomies (§4.2.2). A more elaborate distance of this kind is known as the Wu–Palmer similarity [Wu and Palmer, 1994]. This distance takes into account the fact that two classes near the root of a hierarchy are close to each other in terms of edges but can be very different conceptually, while two classes under one of them which are separated by a larger number of edges should be closer conceptually.

Definition 4.39 (Wu–Palmer similarity). *The Wu–Palmer similarity $\sigma : o \times o \to \mathbb{R}$ is a similarity over a hierarchy $H = \langle o, \leq \rangle$, such that:*

$$\sigma(c,c') = \frac{2 \times \delta(c \wedge c', \rho)}{\delta(c, c \wedge c') + \delta(c', c \wedge c') + 2 \times \delta(c \wedge c', \rho)}$$

where ρ is the root of the hierarchy, $\delta(c,c')$ is the number of intermediate edges between a class c and another class c' and $c \wedge c' = \{c'' \in o; c \leq c'' \wedge c' \leq c''\}$.

Example 4.40 (Wu–Palmer similarity). The Wu–Palmer similarity also provides a figure in coherence with WordNet structure.

	illustrator	author	creator	Person	writer
illustrator	1.	.5	.67	.4	.29
author	.5	1.	.67	.4	1.
creator	.67	.67	1.	.67	.4
Person	.4	0.4	.67	1.	.5
writer	.29	1.	.4	.5	1.

The upward cotopic similarity applies the Jaccard similarity to cotopies. It has been described in [Mädche and Zacharias, 2002] and is as follows:

Definition 4.41 (Upward cotopic similarity). *The upward cotopic similarity* $\sigma : o \times o \to \mathbb{R}$ *is a similarity over a hierarchy* $H = \langle o, \leq \rangle$, *such that:*

$$\sigma(c, c') = \frac{|UC(c, H) \cap UC(c', H)|}{|UC(c, H) \cup UC(c', H)|}$$

where $UC(c, H) = \{c' \in H; c \leq c'\}$ *is the set of superclasses of c.*

Example 4.42 (Upward cotopic similarity). In this case, because all senses count in the cotopy (and not the closest one in terms of path), the result is different from other measures: creator benefits from its position as a superclass of author and illustrator for scoring better than the usual writer-creator pair because they have too many unrelated senses.

	illustrator	author	creator	Person	writer
illustrator	1.	.37	.43	.4	.18
author	.37	1.	.43	.29	.36
creator	.43	.43	1.	.4	.18
Person	.4	.29	.4	1.	.25
writer	.18	.36	.18	.25	1.

These measures cannot be applied as they are in the context of ontology matching since the ontologies are not supposed to share the same taxonomy H, but this can be used in conjunction with a resource of common knowledge, such as WordNet. For that purpose, it is necessary to develop these kinds of measures over a pair of ontologies. In [Valtchev, 1999, Euzenat and Valtchev, 2004], this amounts to using a (local) matching between the elements to be compared (for instance, the hierarchies).

Beside these global measures that take into account the whole taxonomy for assessing the similarity between classes, there are non global measures that have been used in the ontology matching contexts. These measures usually take advantage of the 'direct' part of Table 4.6. Below are some of these measures:

Super or subclass rules: These matchers are based on rules capturing the intuition that classes are similar if their super or subclasses are similar. For example, if superclasses are the same, the actual classes are similar to each other. If subclasses are the same, the compared classes are also similar [Dieng and Hug, 1998, Ehrig and Sure, 2004]. This technique has at least two drawbacks: (i) when there are several sub or superclasses, then, without care, they would all be mapped into the same one, so it is necessary to have some other discriminating features, and (ii) the similarity between the sub or super classes will rely in turn on that of their super or subclasses. This turns this problem into yet another global similarity problem.

Bounded path matching: Bounded path matchers take two paths with links between classes defined by the hierarchical relations, compare terms and their positions along these paths, and identify similar terms. This technique has been introduced in Anchor-Prompt (§6.1.9). For example, in Fig. 2.9, if Book corresponds to Volume and Popular corresponds to Autobiography, then the elements along the paths (Science on one side and Biography and Essay on the other side) must be carefully considered for correspondence. For instance, for deciding that Essay is more general than Science. This technique is primarily guided by two anchors of paths and uses alternative techniques for choosing the best match.

Mereologic structure

The second well known structure after the taxonomic structure is the mereologic structure, i.e., the structure corresponding to a *part-of* relationship. The difficulty for dealing with this kind of structure is that it is not easy to find the properties which carry a mereologic structure. For example, a class Proceedings can have some whole-part relations with a class InProceedings, but it will be expressed through a property communications. These InProceedings objects will in turn have a mereologic structure which is expressed through sections property.

However, if it is possible to detect the relations that support the part-of structure, this can be then used for computing similarity between classes: they will be more similar if they share similar parts. This is even more useful when comparing extensions of classes because it can be inferred that objects sharing the same set of parts will be the same.

Relations

Beside two previous kinds of relations, one can consider the general problem of matching entities based on all their relations. Classes are also related through the definitions of their properties (like author and creator in Fig. 4.2). These properties are also edges of a graph and if they are found similar, they can be used for finding that classes are similar. However, contrary to taxonomic and mereologic structures, the relation graph can contain circuits. How to handle these will be considered in Sect. 5.3. We consider here similarities.

The similarity between nodes can also be based on their relations. For example, in one of the possible ontology representations of schemas of Fig. 2.7, if the Book class is related to the Human class by the author relation in one ontology, and if the

Volume class is related to the Writer class by the author relation in the other ontology, then knowing that classes Book and Volumes are similar, and that relations author and author are similar, we can infer that Human and Writer may be similar too. The similarity among relations in [Mädche and Staab, 2002] is computed according to this principle.

This can be applied to a set of classes and a set of relations. It means that if we have a set of relations $r_1 \ldots r_n$ in the first ontology which are similar to another set of relations $r'_1 \ldots r'_n$ in the second ontology, it is possible that two classes, which are the domains of relations in those two sets, are similar too.

This principle can also be extended to the composition of relations, i.e., instead of considering only the relations asserted at a class, one can consider their composition with relations starting at the domain of this relation. For instance, instead of considering the author relation, one will consider the author·firstname, the author·lastname, or the author·nationality relations.

One of the problems of this approach is that it is based on the use of similarity of relations to infer the similarity of their domain classes or their range classes. This introduces circularity in the computation of similarity. There are several ways to overcome this circularity. As a first alternative, the similarity on relations can be based on their labels using techniques developed in Sect. 4.2.1. As a second alternative, if relations are organised in a taxonomy, then methods considered in the previous subsection can be used as well.

Finally, two extreme solutions, that use the relations for reaching nodes but not for actually matching, are considered by the following approaches:

Children. The similarity between nodes of the graph is computed based on similarity of their children nodes, that is, two non leaf entities are structurally similar if their immediate children sets are highly similar. A more complex version of this matcher is implemented in [Do and Rahm, 2002].

Leaves. The similarity between nodes of the graphs is computed based on similarity of leaf nodes, that is, two non leaf schema elements are structurally similar if their leaf sets are highly similar, even if their immediate children are not [Madhavan et al., 2001, Do and Rahm, 2002]. This is very well adapted to comparing document schemas.

Summary on relational structure

Matching ontologies from their relational (or external) structure is very powerful because it allows all the relations between entities to be taken into account. This must be grounded on other tangible properties, which is why it is often used in combination with internal structural methods and terminological methods.

It is worth considering what are the important relations before using such techniques. The most commonly used structure is the taxonomy because it is the backbone of ontologies and has usually received a lot of attention from designers. In some fields, the mereology relations are as important as taxonomic ones. However, they are difficult to identify because contrary to the subClass relation, they can bear any other name.

The relational structure raises the problem of which part influences what: there is usually a mutual influence between each of the related parts. This is the reason why, beside the similarity equations used for comparing the entities, it is necessary to have an iterative algorithm. This is considered in Sect. 5.3.

4.4 Extensional techniques

When individual representations (or instances) are available, there is a very good opportunity for matching systems. When two ontologies share the same set of individuals, matching is highly facilitated. For example, if two classes share exactly the same set of individuals, then there can be a strong presumption that these classes represent a correct match.

Even when classes do not share the same set of individuals, these allow the grounding of the matching process on tangible indices which do not change easily. For instance, titles of Books do not have any reason to change. So if titles of Books are different, then these are most certainly not the same books. Then, matching can be again based on individual comparisons.

We thus divide extensional methods into three categories: those which apply to ontologies with common instance sets, those which propose individual identification techniques, before using the previous ones, and those which do not require identification, i.e., which work on heterogeneous sets of instances.

4.4.1 Common extension comparison

The easiest way to compare classes when they share instances is to test the intersection of their instance set A and B and to consider that these classes are very similar when $A \cap B = A = B$, more general when $A \cap B = B$ or $A \cap B = A$. The work in [Larson et al., 1989, Sheth et al., 1988] discussed how relationships and entity sets can be integrated primarily based on the set relations: equal ($A \cap B = A = B$), contains ($A \cap B = A$), contained-in ($A \cap B = B$), disjoint ($A \cap B = \emptyset$) and overlap. The problem is the ability to handle faults: small amounts of incorrect data may lead the system to draw a wrong conclusion on domain relationships. Moreover, the dissimilarity has to be 1 when none of these cases apply: for instance, if the classes have some instances in common but not all.

A way to refine this is to use the Hamming distance between two extensions: it corresponds to the size of the symmetric difference normalised by the size of the union.

Definition 4.43 (Hamming distance). *The Hamming distance between two sets is a disimilarity function* $\delta : 2^E \times 2^E \to \mathbb{R}$ *such that* $\forall x, y \subseteq E$:

$$\delta(x, y) = \frac{|x \cup y - x \cap y|}{|x \cup y|}$$

This version of the symmetric difference is normalised. Using such a distance in comparing sets is more robust than using equality: it tolerates some individuals being misclassified and can still produce a short distance.

It is also possible to compute a similarity based on the probabilistic interpretation of the set of instances. This is the case of the Jaccard similarity [Jaccard, 1901].

Definition 4.44 (Jaccard similarity). *Given two sets A and B, let $P(X)$ be the probability of a random instance to be in the set X. The Jaccard similarity is defined as follows:*

$$\sigma(A, B) = \frac{P(A \cap B)}{P(A \cup B)}$$

This measure is normalised and reaches 0 when $A \cap B = \emptyset$ and 1 when $A = B$. It can be used with two classes of different ontologies sharing the same set of instances.

Formal concept analysis

One of the tools of formal concept analysis (FCA) [Ganter and Wille, 1999] is the computation of the concept lattice. The idea behind formal concept analysis is the duality between a set of objects (here the individuals) and their properties: the more properties are constrained, the fewer objects satisfy the constraints. So a set of objects with properties can be organised in a lattice of concepts covering these objects. Each concept can be identified by its properties (the intent) and covers the individual satisfying these properties (the extent).

In ontology matching, the properties can simply be the classes to which the individuals are known to belong and the technique is independent from the origin of the entities, i.e., whether they come from the same ontology or not. From this data set, formal concept analysis computes the concept lattice (or Galois lattice). This is performed by computing the closure of the instances×properties Galois connection. This operation starts with the complete lattice of the power set of extent (respectively, intent) and keeps only the nodes which are closed under the connection, i.e., starting with a set of properties, it determines the corresponding set of individuals, which itself provides a corresponding set of properties; if this set is the initial one, then it is closed and is preserved, otherwise, the node is discarded. The result is a concept lattice, like the one computed in Fig. 4.4 from the table.

For instance, let us start with the table of Fig. 4.4. The table displays a small set of instances and the classes they belong to (from both ontologies). The right-hand side of Fig. 4.4 displays the corresponding concept lattice. From this lattice the following correspondences can be extracted:

Science = Essay	Science $\geq$ Biography	Essay $\geq$ Popular
Science $\geq$ Autobiography	Popular = Biography	Popular = Autobiography
Literature $\geq$ Pocket	Novel = Pocket	

The result is not accurate. However, it is possible to weight these results by first eliminating the redundant correspondences and by providing a confidence according to the size of the extent covered by the correspondence.

4.4 Extensional techniques

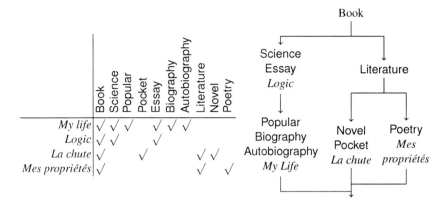

Fig. 4.4. A 'formal context' and the corresponding concept lattice.

4.4.2 Instance identification techniques

If a common set of instances does not exist, it is possible to try to identify which instance from one set corresponds to which other instance from the other set. This method is usable when one knows that the instances are the same. This works, for example, when integrating, two human resource databases of the same company, but does not apply for those of different companies or for databases of events which have no relations.

A first natural technique for identifying instances is to take advantage of keys in databases. Keys can be either internal to the database, i.e., generated unique surrogates, in which case they are not very useful for identification, or external identification, in which case there is high probability that these identification keys are present in both data sets (even if they are not present as keys). In such a case, if they are used as keys, we can be sure that they uniquely identify an individual (like isbn).

When keys are not available, or they are different, other approaches to determine property correspondences use instance data to compare property values. In databases, this technique has been known as record linkage [Fellegi and Sunter, 1969, Elfeky et al., 2002] or object identification [Lim et al., 1993]. They aim at identifying multiple representations of the same object within a set of objects. They are usually based on string-based and internal structure-based techniques (§4.2 and §4.3.1).

If values are not precisely the same but their distributions can be compared, it is possible to apply global techniques. This case is covered in the next section.

4.4.3 Disjoint extension comparison

When it is not possible to directly infer a dataset common to both ontologies, it is easier to use approximate techniques for comparing class extensions. These methods can be based on statistical measures about the features of class members, on the similarities computed between instances of classes or based on a matching between entity sets.

Statistical approach

The instance data can be used to compute some statistics about the property values found in instances, such as maximum, minimum, mean, variance, existence of null values, existence of decimals, scale, precision, grouping, and number of segments. This allows the characterising of the domains of class properties (§4.3.1) from the data. In practice, if dealing with statistically representative samples, these measures should be the same for two equivalent classes of different ontologies.

Example 4.45 (Statistical matching). Consider two ontologies with instances. The analysis of numerical properties size and weight in one ontology and hauteur and poids in the other reveals that they have different average values but the same coefficient of variation, i.e., standard deviation divided by mean, which, in turn, reveals comparable variability of size and hauteur on the one hand and weight and poids on the other hand. This is typically what happens when values are expressed in different units. The ratio of average values of size/hauteur is 2.54 and that of weight/poids is 28.35.

These values have been established based on the whole population. They can be used for comparing the statistical characteristics of these properties in the classes of the ontologies. For instance, the average value of the size property for the Pocket class significantly differs from that of the global population and, once divided by 28.35, is very close to that of the Livredepoche class (also differing from the whole population in the same manner). Hence, these two classes could be considered as similar.

Other approaches, like [Li and Clifton, 1994], propose methods that utilise data patterns and distributions instead of data values and domains. The result is a better fault tolerance and a lower time-consumption since only a small portion of data values are needed due to the employment of data sampling techniques. In general, applying internal structure methods to instances allows a more precise characterisation of the actual contents of schema elements, thus, more accurately determining corresponding datatypes based, for example, on the discovered value ranges and character patterns.

These methods have, however, one prerequisite: they work better if the correspondences between properties are known (otherwise they could match different properties on the basis of their domain). This is already a matching problem to be solved.

Similarity-based extension comparison

Similarity-based techniques do not require the classes to share the same set of instances, though they can still be applied in that case. In particular, the methods based on common extensions always return 0 when the two classes do not share any instances, disregarding the distance between the elements of the sets. In some cases, it is preferable to compare the sets of instances. This requires a (dis)similarity measure between the instances that can be obtained with the other basic methods.

4.4 Extensional techniques

In data analysis, the linkage aggregation methods allow the assessment of the distance between two sets whose objects are only similar. They thus allow us to compare two classes on the basis of their instances.

Definition 4.46 (Single linkage). *Given a dissimilarity function $\delta : E \times E \to \mathbb{R}$, the single linkage measure between two sets is a disimilarity function $\Delta : 2^E \times 2^E \to \mathbb{R}$ such that $\forall x, y \subseteq E$, $\Delta(x, y) = \min_{(e,e') \in x \times y} \delta(e, e')$.*

Definition 4.47 (Full linkage). *Given a dissimilarity function $\delta : E \times E \to \mathbb{R}$, the complete linkage measure between two sets is a disimilarity function $\Delta : 2^E \times 2^E \to \mathbb{R}$ such that $\forall x, y \subseteq E$, $\Delta(x, y) = \max_{(e,e') \in x \times y} \delta(e, e')$.*

Definition 4.48 (Average linkage). *Given a dissimilarity function $\delta : E \times E \to \mathbb{R}$, the average linkage measure between two sets is a disimilarity function $\Delta : 2^E \times 2^E \to \mathbb{R}$ such that $\forall x, y \subseteq E$, $\Delta(x, y) = \frac{\sum_{(e,e') \in x \times y} \delta(e,e')}{|x| \times |y|}$.*

Other linkage measures have been defined. Each of these methods has its own benefits, e.g., maximising shortest distance, minimising longest distance, minimising average distance. Another method from the same family is the Hausdorff distance measuring the maximal distance of a set to the nearest point in the other set [Hausdorff, 1914]:

Definition 4.49 (Hausdorff distance). *Given a dissimilarity function $\delta : E \times E \to \mathbb{R}$, the Hausdorff distance between two sets is a disimilarity function $\Delta : 2^E \times 2^E \to \mathbb{R}$ such that $\forall x, y \subseteq E$,*

$$\Delta(x, y) = \max(\max_{e \in x} \min_{e' \in y} \delta(e, e'), \max_{e' \in y} \min_{e \in x} \delta(e, e'))$$

Matching-based comparison

The problem with the former distances, but average, is that their value is a function of the distance between one pair of members of the sets. The average linkage, on the other hand, has its value function of the distance between all the possible comparisons.

Matching-based comparisons [Valtchev, 1999] consider that the elements to be compared are those which correspond to each other, i.e., the most similar one.

To that extent, the distance between two sets is considered as a value to be minimised and its computation is an optimisation problem: that of finding the elements of both sets which correspond to each others. In particular, it corresponds to solving a bipartite graph matching problem (§5.7.3).

Definition 4.50 (Match-based similarity). *Given a similarity function $\sigma : E \times E \to \mathbb{R}$, the match-based similarity between two subsets of E is a similarity function $MSim : 2^E \times 2^E \to \mathbb{R}$ such that $\forall x, y \subseteq E$,*

$$MSim(x, y) = \frac{\max_{p \in Pairings(x,y)} \left(\sum_{\langle n,n' \rangle \in p} \sigma(n, n') \right)}{\max(|x|, |y|)},$$

with $Pairings(x, y)$ being the set of mapping of elements of x to elements of y.

This match-based similarity already requires an alignment of entities to be computed. It also depends on the kind of alignment that is required. Indeed, the result will be different depending on whether the alignment is required to be injective or not. The match-based comparison can also be used when comparing sequences [Valtchev, 1999].

Summary on extensional techniques

Knowing extension information is invaluable for ontology matching because this provides information that is independent from the conceptual part of the ontology. Indeed, ontologies are views of the world and this is the reason why there can be numerous different ontologies on the same topic (and the reason why they have to be matched). Extension information is supposed to be less prone to variability and can be used to accurately match classes.

This extension information is even more useful when a set of individuals characterised in both ontologies is available. This provides an easy way to compare the overlap between two classes.

There are situations, however, in which data instance information is not available. This can be caused by the unavailability of data (connection data to a web service is not available) or for confidentiality reasons. In such a situation, the other techniques are the only possible ones.

4.5 Semantic-based techniques

The key characteristics of semantic methods is that model-theoretic semantics is used to justify their results. Hence they are deductive methods. Of course, pure deductive methods do not perform very well alone for an essentially inductive task like ontology matching. They hence need a preprocessing phase which provides 'anchors', i.e., entities which are declared, for example, to be equivalent (based on the identity of their names or user input for instance). The semantic methods act as amplifiers of these seeding alignments.

We thus include in semantic techniques particular methods for anchoring the ontologies (§4.5.1). They are based on the use of existing formal resources for initiating an alignment that can be further considered by deductive methods (§4.5.2).

4.5.1 Techniques based on external ontologies

When two ontologies have to be matched, they often lack a common ground on which comparisons can be based. In this section we focus on using intermediate formal ontologies for that purpose. These intermediate ontologies can define the common context or background knowledge [Giunchiglia et al., 2006c] for the two ontologies to be matched. The intuition is that a background ontology with a comprehensive coverage of the domain of interest of the ontologies to be matched helps in the disambiguation of multiple possible meanings of terms.

This common ground can often be found by relating the ontologies to external resources. These resources can differ on three specific dimensions:

Breadth: whether they are general purpose resources or domain specific resources. By using specialised resources, e.g., the Formal Model of Anatomy in medicine, one can be sure that the concepts in the contextualised resources can be matched accurately to their corresponding concepts in the ontology. However, by using more general resources there is more probability that an alignment already exists and can be exploited right away.

Formality: whether they are pure ontologies with semantic descriptions or informal resources such as WordNet. By using formal resources, e.g., DOLCE or the Formal Model of Anatomy, it is possible to reason within or across these formal models in order to deduce the relation between two terms. By using informal resources, e.g., WordNet, it is possible to extend the set of senses that are covered by a term and to increase the number of terms which can express these concepts. There is thus more opportunity to match terms.

Status: whether these resources are considered as references such as ontologies, thesauri or they are sets of instances or annotated documents that are shared.

Since non pure ontological resources such as WordNet have been considered in Sect. 4.2.2 and extensional resources have been dealt with in Sect. 4.4.1, we concentrate here on using external formal ontologies.

Contextualising ontologies can typically be achieved by matching these ontologies with a common upper-level ontology that is used as external source of common knowledge, e.g., Cyc [Lenat and Guha, 1990], Suggested Upper Merged Ontology (SUMO) [Niles and Pease, 2001] or Descriptive Ontology for Linguistic and Cognitive Engineering (DOLCE) [Gangemi et al., 2003].

Example 4.51 (Using upper-level ontologies as background knowledge). An experiment has been carried out by expressing fishery resources (such as databases and thesauri) within the DOLCE upper level ontology [Gangemi, 2004]. The goal was to merge these resources into a common Core Ontology of Fisheries. It has involved transforming manually the resources into lightweight ontologies expressed with respect to DOLCE and then using reasoning facilities for detecting relations and inconsistencies between entities of this ontology.

An approach proposed in [Aleksovski et al., 2006] works in two steps:

Anchoring (also known as contextualising) is matching ontologies o' and o'' to the background ontology o. This can be done by using any available methods presented in this book, usually non sophisticated ones.

Deriving relations is the (indirect) matching of ontologies o' and o'' by using the correspondences discovered during the anchoring step. Since concepts of ontologies o' and o'' become a part of the background ontology o via anchors, checking if these concepts are related, can be therefore performed by using a reasoning service (§4.5.2) in the background ontology. Intuitively, combining the anchor relations with the relations between the concepts of the reference ontology is used to derive the relations between concepts of o' and o''.

Example 4.52 (Using domain specific formal ontologies as background knowledge).
Suppose we want to match the anatomy part of the CRISP[10] directory to the anatomy part of the MeSH[11] meta-thesaurus. In this case the FMA ontology[12] can be used as background knowledge which gives the context to the matching task. The result of anchoring is a set of matches with three different kinds of relations: $\equiv, \preceq, \succeq$ between concepts from FMA, and CRISP or MeSH.

For example, the concept of brain from CRISP, denoted by Brain_{CRISP}, could be easily anchored to the concept brain of FMA, denoted by Brain_{FMA}. Similarly, the concept of head from MeSH, denoted by Head_{MeSH}, could be anchored to a background knowledge concept Head_{FMA}. In the reference ontology FMA there is a *part of* relation between Brain_{FMA} and Head_{FMA}. Therefore, we can derive that Brain_{CRISP} is a part of Head_{MeSH}.

Since the domain specific ontology provides the context for the matching task, the concept of Head was correctly interpreted as meaning the upper part of the human body, instead of, for example, meaning a chief person. This is not so straightforward as can be shown by replacing FMA with WordNet: in WordNet the concept of Head has 33 senses (as a noun). Finally, once the context of the matching task has been established, as our example shows, various heuristics, such as string-based techniques, can improve the anchoring step.

There are some other techniques which attempt at using not one context ontology but as many as possible. These ontologies are typically taken from the web, selected for relevance, i.e., that they contain enough matches with the initial ontologies, and the result is a consensus between the results provided with these ontologies [Sabou et al., 2006a].

Once these initial alignments have been obtained, they can be exploited further by deductive techniques.

4.5.2 Deductive techniques

The basis of the semantic techniques are the merging of two ontologies and the search for correspondences A such that $o, o' \models A$. Of course, this can apply only if A can be considered as a formula of the language. For instance, this can apply if it is a subsumption relation between two entities e and e': $e \sqsubseteq e'$. These semantic techniques can also be used for testing the satisfiability of alignments (§2.5.4), in particular, for discarding alignments which lead to an inconsistent merge of both ontologies.

Examples of semantic techniques are propositional satisfiability, modal satisfiability techniques, or description logic based techniques.

---

[10] http://crisp.cit.nih.gov/
[11] http://www.nlm.nih.gov/mesh/
[12] http://sig.biostr.washington.edu/projects/fm/

Propositional techniques

An approach for applying propositional satisfiability (SAT) techniques to ontology matching includes the following steps [Giunchiglia and Shvaiko, 2003a, Bouquet and Serafini, 2003, Giunchiglia *et al.*, 2004, Shvaiko, 2006]:

1. Build a theory or domain knowledge ($Axioms$) for the given input two ontologies as a conjunction of the available axioms. The theory is constructed by using matchers discussed in the previous sections, e.g., those based on WordNet, or those using external ontologies (§4.5.1).
2. Build a matching formula for each pair of classes c and c' from two ontologies. The criterion for determining whether a relation holds between two classes is the fact that it is entailed by the premises (theory). Therefore, a matching query is created as a formula of the following form:

$$Axioms \rightarrow r(c, c')$$

   for each pair of classes c and c' for which we want to test the relation r (within $=, \sqsubseteq, \sqsupseteq, \perp$). c and c' are also sometimes called contexts.
3. Check for validity of the formula, namely that it is true for all the truth assignments of all the propositional variables occurring in it. A propositional formula is valid if and only if its negation is unsatisfiable, which is checked by using a SAT solver.

SAT solvers are correct and complete decision procedures for propositional satisfiability, and therefore, they can be used for an exhaustive check of all the possible correspondences. In some sense, these techniques compute the deductive closure of some initial alignment [Euzenat, 2007].

Example 4.53 (Propositional logic relation inference).
Step 1. Suppose that classes images and Europe belong to one ontology, while another ontology has classes pictures and Europe (as well). A matcher which uses WordNet can determine that images = pictures. Also many other matchers can find that classes of Europe in both ontologies are identical, i.e., Europe = Europe. Then translating the relations between classes under consideration into propositional connectives in the obvious way results in the following $Axioms$:

$$(\text{images} \leftrightarrow \text{pictures}) \wedge (\text{Europe} \leftrightarrow \text{Europe})$$

Step 2. Suppose c is defined as Europe ⊓ images which intuitively stands for the concept of European images, while c' is defined as pictures ⊓ Europe which intuitively stands for the concept of pictures of Europe. Let us also suppose that we want to know if c is equivalent ($\leftrightarrow$) to c'. Thus, this matching task requires constructing the following formula:

$$((\text{images} \leftrightarrow \text{pictures}) \wedge (\text{Europe} \leftrightarrow \text{Europe})) \rightarrow$$
$$((\text{Europe} \wedge \text{images}) \leftrightarrow (\text{Europe} \wedge \text{pictures}))$$

Step 3. Negation of this formula turns out to be unsatisfiable, and therefore, the equivalence relation holds. See also Chap. 9 for a detailed discussion of this example.

Notice that this technique, beside pruning the incorrect correspondences, also discovers the new ones between complex concepts. In the example above c is defined by combining (taking intersection of) such atomic concepts as Europe and images. And, similarly for c'. These are simple examples of complex concepts being bounded by the expressive power of a propositional language. The relation between such complex concepts as (Europe ∧ images) and (Europe ∧ pictures) was not available after the first step, and has being discovered as a result of deduction.

This technique can only be used for matching tree-like structures, such as classifications, taxonomies, without taking properties or roles into account. Modal SAT can be used, as proposed in [Shvaiko, 2004], for extending the methods related to propositional SAT to binary predicates.

Description logic techniques

In description logics, the relations, e.g., $=, \sqsubseteq, \sqsupseteq, \bot$, can be expressed with respect to subsumption. The subsumption test, can be used to establish the relations between classes in a purely semantic manner. In fact, first merging two ontologies (after renaming) and then testing each pair of concepts and roles for subsumption is enough for matching terms with the same interpretation (or with a subset of the interpretations of the others) [Bouquet *et al.*, 2006].

Example 4.54 (Description logic relation inference). Consider two minimal description logic ontologies:

$$\text{Micro-company} = \text{Company} \sqcap \leq_5 \text{employee}$$

meaning that a Micro-company is a Company with at most 5 employees and

$$\text{SME} = \text{Firm} \sqcap \leq_{10} \text{associate}$$

meaning that a SME is a Firm with at most 10 associates. The following initial alignment (expressed in description logic syntax) includes:

$$\text{Company} = \text{Firm}$$
$$\text{associate} \sqsubseteq \text{employee}$$

It expresses that Company is equivalent to Firm and associate is a subclass of employee. This obviously entails:

$$\text{Micro-company} \sqsubseteq \text{SME}$$

i.e., Micro-company is a subclass of SME.

There are other uses of description logic techniques which are relevant to ontology matching. For example, in a spatio-temporal database integration scenario, as first motivated in [Parent and Spaccapietra, 2000] and later developed in [Sotnykova *et al.*, 2005], the inter-schema correspondences are initially proposed by the integrated schema designer and are encoded together with input schemas in the $\mathcal{ALCRP}(\mathcal{S}_2 \oplus \mathcal{T})$ language. Then, description logic reasoning services are used to check the satisfiability of the two source schemas and the set of inter-schema correspondences. If some objects are found unsatisfiable, the inter-schema correspondences should be reconsidered. A similar approach in the context of alignment debugging has also been investigated in [Meilicke *et al.*, 2006].

Summary on semantic techniques

As it was mentioned in the beginning, semantics techniques cannot find the correspondences alone. However, they are invaluable when correspondences are generated in order to ensure the completeness, i.e., find all the correspondences that must hold, and the consistency, i.e., find correspondences that lead to inconsistency, of the alignment.

Only a few of these techniques have been developed so far (usually, databases had only simple semantic theories so these techniques were not developed in this field). However, with the improvement of deductive tools for dealing with semantic web languages, we believe that we will see more systems using semantic-based techniques.

An important challenge of these techniques is their integration with inductive techniques. Indeed, completing alignments and finding inconsistencies is a crucial step. However, once deductive techniques have been applied, their results might be considered as an input to inductive techniques. For example, for finding more correspondences from the completion or for selecting alternative correspondences instead of inconsistent ones. This theme deserves to be further investigated.

4.6 Summary

We have discussed basic techniques that can be used for building correspondences based on terminological (§4.2), conceptual (§4.3), extensional (§4.4) and semantic (§4.5) arguments. This classification of techniques is a natural one since each of these deals with a partial view of ontologies.

There are many such techniques and our goal was not to present them all. It was rather to propose a panorama of the most used ones so far and to show the direction they take. There is still much work going on in finding better methods in each of these directions.

We have also observed that all these techniques cannot be used in isolation, but that each of them can take advantage of the results provided by the others. Another part of the art of ontology matching relies on selecting and combining these methods in the most adequate way. Combinations of basic matchers is the topic of the next chapter.

5

Matching strategies

The basic techniques presented in Chap. 4 are the building blocks on which a matching solution is built. Once the similarity or (dis)similarity between ontology entities are available, the alignment remains to be computed. This involves more global treatments. In particular, the following aspects of building a working matching system are considered in this chapter:

- aggregating the results of the basic methods in order to compute the compound similarity between entities (§5.2) and organising the combination of various similarities or matching algorithms (§5.1);
- developing a strategy for computing these similarities in spite of cycles and non linearity in the constraints governing similarities (§5.3);
- learning from data the best method and the best parameters for matching (§5.4);
- using probabilistic methods to combine matchers or to derive missing correspondences (§5.5);
- involving users in the loop (§5.6);
- extracting the alignments from the resulting (dis)similarity: indeed, different alignments with different characteristics can be extracted from the same (dis)similarity (§5.7).

5.1 Matcher composition

All the steps mentioned above are considered here under the name of global methods. The goal of a global method is to combine local methods (or basic matchers) in order to define a new matching algorithm. We present here, at the strategic level, some natural ways to combine matchers. For that purpose, we progressively introduce new graphical elements. These are summarised in Fig. A.3 (Appendix A).

So far, we have only presented the outside of the matching process by producing an alignment from two ontologies such as in Fig. 2.8. A natural way of composing the basic matchers consists of improving the matching through the use of sequential composition (see Fig. 5.1). For instance, one would like to first use a matcher based

118   5 Matching strategies

on labels (§4.2) before running another one based on the structure of entities (§4.3) or a semantic matcher (§4.5).

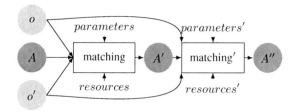

Fig. 5.1. The sequential composition of matchers.

This sequential process can be used, for instance, in on-line data integration. Ontology matching and integration consists of merging data (and sometimes data streams, d and d') expressed in different ontologies (o and o'). For that purpose, the ontologies have to be matched beforehand and the data integration can use this alignment. This is an example of combined off-line and on-line matching.

It can be thought of as:

1. a first matching phase (f), possibly with an instance training set;
2. a data matching phase (f') using the first alignment (A').

This is presented in Fig. 5.2.

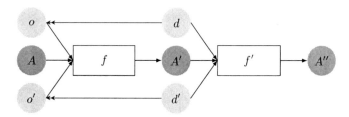

Fig. 5.2. Data integration as another matching process taking advantage of a prior matching of ontologies (o and o') for integrating data flows (d and d').

In this setting, the second phase benefits from the precompiling of the first alignment. Indeed, the second matcher f' can be thought of as a compilation of the first alignment.

However, the sequential combination of matchers is more classically used to improve an alignment. For that purpose, when using similarities or distances, the matchers can be sequentially composed through their similarity matrix. We introduce, in Fig. 5.3, new symbols for matrices as well as a new component for extracting an initial matrix from either an initial alignment or a pair of ontologies (first triangle) and another one for extracting an alignment from a similarity or dissimilarity matrix (second triangle, detailed in Sect. 5.7).

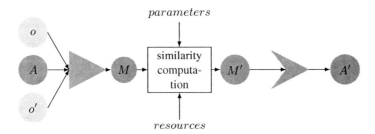

Fig. 5.3. The introduction of a (virtual) matrix which represents a similarity or distance measure between entities to be matched. The first operator builds an initial matrix M from the two ontologies o and o'. The core of the matching algorithm produces a similarity or distance matrix M' from this initial matrix and the description of the ontologies. Finally, alignment A' is extracted from matrix M'.

The sequential composition through a distance or similarity matrix is illustrated in Fig. 5.4.

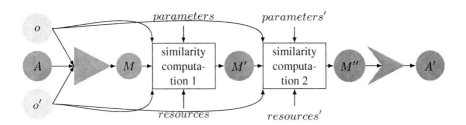

Fig. 5.4. Sequential composition of matchers through similarity.

Another way to combine algorithms consists of running several different algorithms independently and aggregating their results (see Fig. 5.5): this is called parallel composition. Such aggregation techniques can be very different: it can correspond to choosing one of the results on some criterion or merging their results through some operator. For instance, it can consist of running several matching algorithms in parallel and selecting the correspondences which are in all of them (intersection is then used as an aggregation operator) or selecting all the correspondences with their highest confidence.

In the latter case, it is very often more convenient to define the aggregation operators on the similarity or distance matrix (see Fig. 5.6) because there are many

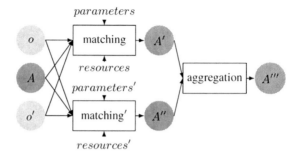

Fig. 5.5. The parallel composition of matchers.

mathematical techniques available for that purpose. These techniques are presented hereafter.

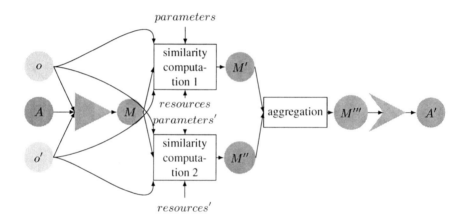

Fig. 5.6. Parallel composition of matchers through similarity.

In fact there are two main kinds of parallel composition:

Heterogeneous parallel composition in which the input is fragmented into different kinds of data (graphs, strings, sets of documents, etc.) and the aggregation takes advantage of all of them (by aggregating their results) or the most promising only. This is the topic of Sect. 5.2.

Homogeneous parallel composition in which the input goes into several competing systems and the aggregation selects the best of these or some consensus between them.

Of course, it is possible to combine these two classes even further.

All these composition techniques are usually implemented within particular matching algorithms (which are presented in Chap. 6). However, there are some systems that offer the opportunity to combine other systems such as FOAM (§8.2.5), Rondo (§8.2.1) or the Alignment API (§8.2.4).

5.2 Similarity aggregation

Compound similarity is concerned with the aggregation of heterogeneous similarities. As explained in Sect. 4.3, structured objects (classes, individuals) are very often involved in many different relations. If it is possible to compute the similarity between each of the ontology entities two objects are related with, these similarities have to be aggregated in order to provide a similarity assessment between the entities themselves. For instance, computing the similarity between two classes requires the aggregation, in a single similarity measure, of the similarity obtained from their names, the similarity of their superclasses, the similarity of their instances and that of their properties.

5.2.1 Triangular norms

Triangular norms are used as conjunction operators in uncertain calculi.

Definition 5.1 (Triangular norm). *A triangular norm T is a function from $D \times D \to D$ (where D is a set ordered by $\leq$ and provided with an upper bound $\top$) satisfying the following conditions:*

$$T(x, \top) = x \qquad \text{(boundary condition)}$$
$$x \leq y \Rightarrow T(x, z) \leq T(y, z) \qquad \text{(monotonicity)}$$
$$T(x, y) = T(y, x) \qquad \text{(commutativity)}$$
$$T(x, T(y, z)) = T(T(x, y), z) \qquad \text{(associativity)}$$

Typical examples of triangular norms are $\min(x, y)$, $x \times y$ and $\max(x+y-1, 0)$. All are normalised if the measures provided to them are normalised; min is the only idempotent norm ($\forall x, \min(x, x) = x$). Triangular norms are the obvious candidates for a combination that requires the highest score from all aggregated values. Due to associativity, triangular norms can be extended to n-ary measures. Any triangular norm over the unit interval can be expressed as a combination of these three functions [Hájek, 1998].

Another triangular norm for aggregating several dimensions is the weighted product.

Definition 5.2 (Weighted product). *Let o be a set of objects which can be analysed in n dimensions. The weighted product between two such objects is as follows:*

$$\forall x, x' \in o, \delta(x, x') = \prod_{i=1}^{n} \delta(x_i, x'_i)^{w_i}$$

such that $\delta(x_i, x'_i)$ is the dissimilarity of the pair of objects along the ith dimension and w_i is the weight of dimension i.

These operators have the drawback that if one of the dimensions has a measure of 0, then the result is also 0.

Example 5.3 (Triangular norms). We consider in this section two ontologies comprising the concepts Product, Provider, Creator for the first one and Book, Translator, Publisher and Writer for the second one.

The two tables below display the result of applying an edit distance and a WordNet-based distance on these labels.

	Book	Translator	Publisher	Writer
Product	.86	.8	.89	.86
Provider	.88	.8	.56	.5
Creator	.86	.5	.89	.57

Normalised Levenshtein distance.

	Book	Translator	Publisher	Writer
Product	.82	.88	.88	.85
Provider	.83	.89	.76	.71
Creator	.82	.53	.88	.85

Alignment API WordNet-based distance.

The following tables display the aggregations of these distances with triangular norms, namely, the min operation and a weighted product.

	Book	Translator	Publisher	Writer
Product	.82	.8	.88	.85
Provider	.83	.8	.56	.5
Creator	.82	.5	.88	.57

Minimum of the distances.

	Book	Translator	Publisher	Writer
Product	.84	.84	.88	.85
Provider	.85	.84	.65	.60
Creator	.84	.51	.88	.70

Weighted product with $w_1 = w_2 = \frac{1}{2}$.

Since the two first similarities where not very dissimilar from each other, the results of the two operators are very similar as well.

Contrary to the multidimensional aggregators, triangular norms tend to imply dependencies between the values of the different dimensions, so that the value given on one dimension can override a value on another dimension.

5.2.2 Multidimentional distances and weighted sums

In case the difference between some properties must be aggregated, one of the most common family of distances are the Minkowski distances. Contrary to the previous ones, these measures are well suited to independent dimensions and tend to balance the values between dimensions.

5.2 Similarity aggregation

Definition 5.4 (Minkowski distance). *Let o be a set of objects which can be analysed in n dimensions, the Minkowski distance between two such objects is as follows:*

$$\forall x, x' \in o, \delta(x, x') = \sqrt[p]{\sum_{i=1}^{n} \delta(x_i, x'_i)^p}$$

where $\delta(x_i, x'_i)$ is the dissimilarity of the pair of objects along the ith dimension.

Instances of the Minkowski distances are the Euclidean distance (when $p = 2$), the Manhattan (a.k.a. City-blocks) distance (when $p = 1$) and the Chebichev distance (when $p = +\infty$). These should be used when aggregating measures from independent dimensions.

Example 5.5 (Minkowski distances). We start with the distance computed on labels with the min aggregation operator in Example 5.3 and a distance obtained from the Hamming distance on the set of instances of concepts. These distances typically take into account independent dimensions.

	Book	Translator	Publisher	Writer
Product	.82	.8	.88	.85
Provider	.83	.8	.56	.5
Creator	.82	.5	.88	.57

Minimum of the distances.

	Book	Translator	Publisher	Writer
Product	.8	1.	1.	1.
Provider	1.	1.	.15	.98
Creator	1.	.83	.99	.22

Distances obtained by using the Hamming distance on sets of the concept instances. The relatively high distance between Product and Book is due to the large number of Products which are not Books.

The aggregation of these two distances using (normalised) Euclidean and Manhattan distances are as follows:

	Book	Translator	Publisher	Writer
Product	.86	.96	1.	.99
Provider	.98	.96	.44	.83
Creator	.97	.73	1.	.46

Normalised Euclidean distance based on the two above dimentions.

	Book	Translator	Publisher	Writer
Product	.86	.96	1.	.98
Provider	.97	.96	.38	.79
Creator	.97	.71	.99	.42

Normalised Manhattan distance based on the two above dimentions.

The values given by the Euclidean distance are lower than those of Manhattan distance, though they are very close.

These distances can be weighted in order to give more importance to some dimensions. They can be normalised by dividing their results by the maximum possible distance (which is not always possible) but they have the main drawback of not being linear if $p \neq 1$. This is a source of problems when trying to find these distances if they are defined as functions of each others (see Sect. 5.3 and [Valtchev, 1999]).

A simple linear aggregation can be further refined by adding weights to this sum. Weighted linear aggregation considers that some of the values to be aggregated do not have the same importance. For instance, similarity in properties is more important than similarity in comments. The aggregation function will thus use a set of weights $w_1, \ldots w_n$ corresponding to a category of entities, e.g., classes, properties. The aggregation function can be defined as follows:

Definition 5.6 (Weighted sum). *Let o be a set of objects which can be analysed in n dimensions, the weighted sum between two such objects is as follows:*

$$\forall x, x' \in o, \delta(x, x') = \sum_{i=1}^{n} w_i \times \delta(x_i, x'_i)$$

where $\delta(x_i, x'_i)$ is the dissimilarity of the pair of objects along the ith dimension and w_i is the weight of dimension i.

The weighted sum can be thought of as a generalisation of the Manhattan distance in which each dimension is weighted. It also corresponds to weighted average with normalised weights. In fact, the weights can be different depending on the categories of the objects aggregated (§5.3.2). Then, the function can use a set of weights w_C^P depending on the category of object C and the kind of value computed P.

This kind of measures can be normalised, if all values are normalised, by having: $\sum_{i=1}^{n} w_i = 1$.

Example 5.7 (Weighted sum). From Example 5.5, it appears that the measure on the instances is more accurate than those on the labels. This can be inferred from the fact that there are no common names in both sets of labels or that there are lower distances in the latter case. Thus, weighting these dimensions could be promising. Let us consider the same input set as in Example 5.5. The computed weighted sums are as follows:

	Book	Translator	Publisher	Writer
Product	.81	.93	.96	.95
Provider	.94	.93	.29	.82
Creator	.94	.72	.95	.34

Normalised weighted sum with $w_{label} = 1/3$ and $w_{inst} = 2/3$.

	Book	Translator	Publisher	Writer
Product	.81	.95	.97	.96
Provider	.96	.95	.25	.86
Creator	.96	.75	.96	.31

Normalised weighted sum with $w_{label} = 1/4$ and $w_{inst} = 3/4$.

The results clearly identify ⟨Provider, Publisher⟩ and ⟨Creator, Writer⟩ as candidate matches. The low similarity between Product and Book prevents from choosing them as a match candidate.

5.2.3 Fuzzy aggregation and weighted average

Fuzzy aggregation operators are used for assimilating homogeneous quantities in a way that preserves the structure of the aggregated domains.

Definition 5.8 (Fuzzy aggregation operator). *A fuzzy aggregation operator f is a function from $D^n \to D$ (with D being a set ordered by $\leq$ and provided with an upper bound $\top$) satisfying $\forall x, x_1, \ldots x_n, y_1, \ldots y_n \in D$ the following conditions:*

$$f(x, \ldots x) = x \quad (idempotency)$$
$$\forall x_i, y_i, x_i \leq y_i \Rightarrow f(x_1, \ldots x_n) \leq f(y_1, \ldots y_n) \quad (increasing\ monotonicity)$$
$$f\ is\ a\ continuous\ function \quad (continuity)$$

min is also a fuzzy aggregation function. A general result about these measures is that for any fuzzy aggregation function f, the aggregation is ordered by $f(x,y) \geq \min(x,y) \geq x \times y \geq \max(x+y-1, 0)$. A typical example of a fuzzy aggregation operator is the weighted average [Gal et al., 2005a].

Definition 5.9 (Weighted average). *Let o be a set of objects which can be analysed in n dimensions. The weighted average between two such objects is as follows:*

$$\forall x, x' \in o, \delta(x, x') = \frac{\sum_{i=1}^{n} w_i \times \delta(x_i, x'_i)}{\sum_{i=1}^{n} w_i}$$

such that $\delta(x_i, x'_i)$ is the dissimilarity of the pair of objects along the ith dimension and w_i is the weight of dimension i.

A simple average function is a function such that all weights are equal. If the values are normalised, the weighted average is normalised. In fact, the normalised weighted sum is also a weighted average (we refer the reader to Example 5.7 for examples).

Fuzzy aggregation functions have to be used when aggregating the results of competing algorithms (which are efficient with respect to some aspects and not with respect to others) and trying to take advantage of all of them. They are very useful if one wants to use a learning algorithm for learning the weights of the measure (see Sect. 5.4). [Gal et al., 2005a] argues that these measures are always preferable to triangular norms for aggregating confidence measures.

5.2.4 Ordered weighted average

Another aggregation operator in this context is the ordered weighted average [Yager, 1988]. It associates weights to the respective positions of the dimension values instead of the dimensions themselves. This allows, in particular, to give more importance to the highest (or the lowest) values. This is important when aggregating matcher results, because this allows retaining only the results of the highest matches disregarding the dimension they come from.

Definition 5.10 (Ordered weighted average). *An ordered weighted average operator f is a function from $D^n \to D$ (with D being a set ordered by $\leq$ and provided with an upper bound $\top$) satisfying $\forall x, x_1, \ldots x_n \in D$, such that:*

$$f(x_1, \ldots x_n) = \sum_{i=1}^{n} w_i \times x'_i$$

where

- *$w_1, \ldots w_n$ is a set of weigths in $[0\ 1]$ such that $\sum_{i=1}^{n} w_i = 1$;*
- *x'_i is the i-th largest element of $(x_1, \ldots x_n)$.*

The ordered weighted average has the properties of an average operator (commutative, monotone and idempotent). The max, min and average functions are special cases of ordered weighted average.

5.3 Global similarity computation

The computation of compound similarity is still local because it only provides similarities by considering the neighbourhood of a node. However, similarity may involve the ontologies as a whole and the final similarity values may ultimately depend on all the ontology entities. Moreover, the distance defined by local methods can be defined in a circular way when the ontology is not reduced to a directed acyclic graph. This is the most common case. For instance, this occurs if the distance between two classes depends on the distances between their instances which themselves depend on the distance between their classes or if there are circuits in the ontology. This is illustrated in Fig. 5.7, in which the similarity between Product and Book depends on the similarity between hasProvider and hasCreator and author, publisher, and translator. In turn, the similarity between these elements ultimately depends on the similarity between Product and Book. Note that the two graphs are homomorphic in many different ways.

In case of circular dependencies, similarity computation in a local fashion is no longer possible. The classical way of dealing with such a problem involves the iterative computation of the distance or similarity refining at each step the last computed values. This is depicted in Fig. 5.8.

For that purpose, strategies must be defined in order to compute this global similarity. We present two such methods here. The first one is defined as a process of

5.3 Global similarity computation

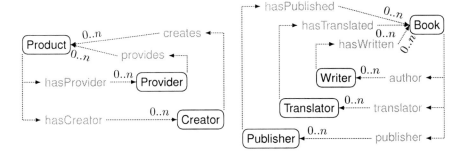

Fig. 5.7. Two typical ontologies containing referential cycles: how do we match them?

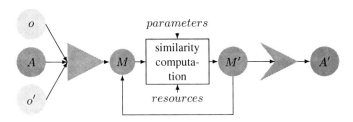

Fig. 5.8. The iterative computation of the fixed point of a similarity or distance function.

propagating the similarity within a graph (§5.3.1) while the second one translates the similarity definitions in a set of equations which is solved by numerical analysis techniques (§5.3.2).

5.3.1 Similarity flooding

Similarity flooding [Melnik *et al.*, 2002] is a generic graph matching algorithm which uses fixed point computation to determine corresponding nodes in the graphs. It is implemented in the Rondo environment (§8.2.1).

The principle of the algorithm is that the similarity between two nodes must depend on the similarity between their adjacent nodes (whatever are the relations that must be taken into account). To implement this, the algorithm proceeds as follows:

1. Transform the ontologies in a directed labelled (multi)graph G in which nodes are pairs of concepts of the ontologies and edges exist between two nodes if there is a relation in both ontologies between the nodes of the two pairs. For instance, in the ontology of Fig. 5.7 ⟨Provider, Writer⟩ is related to ⟨Product, Book⟩ through an edge labelled ⟨hasProvider, hasWritten⟩. In fact, the original Similarity flooding algorithm only connects nodes whose edges have the same label. The graph is closed by symmetry, i.e., there will also be an edge in the reverse direction.

2. Assign weights w to the edges, which are usually $1/n$ in which n is the out degree (the number of outcoming edges) of the source node. The algorithm description does not describe what to do when several edges with different labels link the same pair of concepts or when there is already a reverse edge. One can imagine that the weights are aggregated with a triangular norm (§5.2).
3. Assign initial similarity σ^0 to each node (with some basic method on labels of Sect. 4.2 or with a uniform assignment of 1.0).
4. Compute σ^{i+1} for each node with the chosen formula.
5. Normalise all σ^{i+1} obtained by dividing by the largest value.
6. If no similarity changes more than a particular threshold ϵ, i.e., $\forall e \in o, e' \in o', |\sigma^{i+1}(e, e') - \sigma^{i+1}(e, e')| < \epsilon$, or after a predetermined number of steps, stop; otherwise, go to step 4.

The chosen aggregation function is a weighted linear aggregation in which the weight of an edge is the inverse of the number of other edges with the same label reaching the same pair of entities.

$$\sigma^{i+1}(x, x') = \sigma^0(x, x') + \sum_{\langle\langle y,y'\rangle,p,\langle x,x'\rangle\rangle \in G} \sigma^i(y, y') \times w(\langle\langle y, y'\rangle, p, \langle x, x'\rangle\rangle)$$

Several variations of this formula have been studied, including suppressing the σ^0 term and replacing σ^0 by σ^i, or using $\sigma^0(x, x') + \sigma^i(x, x')$ as the recurrence term. The former accelerates computation, while the latter gives more importance to the initial values.

The convergence of the algorithm is not obvious. [Melnik et al., 2005] provides conditions under which the algorithm converges. This algorithm does provide a similarity measure from which an alignment remains to be extracted (§5.7).

Example 5.11 (Similarity flooding). We start with the ontologies of Fig. 5.7. Since the Similarity flooding algorithm works with the same property names and there is no similar property, we choose to consider that all properties have the same name. From these ontologies is generated the following labelled directed graph (with its weights):

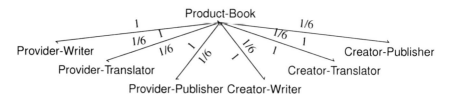

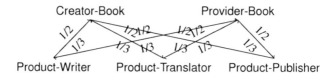

The initial dissimilarity is the one provided in Example 5.7 for the weighted sum with weights of $1/4$ and $3/4$ respectively.

σ^0	Book	Translator	Publisher	Writer
Product	.19	.05	.03	.04
Provider	.04	.05	.75	.14
Creator	.04	.25	.04	.69

The first iteration of the Similarity flooding algorithm computing σ^1 is below (on the left is the σ^i values and on the right is the normalised result):

σ^1	Book	Translator	Publisher	Writer
Product	2.11	0.08	0.06	0.07
Provider	0.10	0.08	0.78	0.17
Creator	0.10	0.28	0.07	0.72

σ^1.

$\bar{\sigma}^1$	Book	Translator	Publisher	Writer
Product	1.00	0.04	0.03	0.03
Provider	0.05	0.04	0.37	0.08
Creator	0.05	0.13	0.03	0.34

Normalised σ^1.

The iterative procedure carries on and, with a value of $\epsilon = .1$, stops at the 17th iteration with the following result:

σ^{17}	Book	Translator	Publisher	Writer
Product	1.95	.09	.07	.08
Provider	.11	.22	.92	.31
Creator	.11	.42	.21	.86

σ^{17}.

$\bar{\sigma}^{17}$	Book	Translator	Publisher	Writer
Product	1,00	.05	.04	.04
Provider	.06	.11	.47	.16
Creator	.06	.21	.11	.44

Normalised σ^{17}.

From these similarity values, it is possible to extract the expected correspondences: Product=Book, Publisher=Provider and Writer=Creator.

5.3.2 Similarity equation fixed point

OLA [Euzenat and Valtchev, 2004] (§6.3.8) provides a method for dealing with circularities and dependencies between similarity definitions.

In this case, the similarity values can only be expressed as a set of equations where each variable corresponds to the similarity between a pair of nodes. There are as many equations as variables. The structure of each equation follows the definition of the respective similarity function for the underlying node category.

Given two classes c and c', the resulting class similarity function is as follows:

$$\sigma_C(c,c') = \pi_L^C \sigma_L(\lambda(c), \lambda(c'))$$
$$+ \pi_O^C MSim_O(\mathcal{I}(c), \mathcal{I}'(c'))$$
$$+ \pi_S^C MSim_C(\mathcal{S}(c), \mathcal{S}'(c'))$$
$$+ \pi_P^C MSim_P(\mathcal{A}(c), \mathcal{A}'(c'))$$

in which $\lambda(\cdot)$ $\mathcal{I}(\cdot)$, $\mathcal{S}(\cdot)$, $\mathcal{A}(\cdot)$ are the functions returning respectively the label, instances, super and subclasses, and properties of a class. $MSim$-measures are similarities between sets of ontology entities which we explain below.

The function is normalised since the sum of weights is equal to one, i.e., $\pi_L^C + \pi_S^C + \pi_O^C + \pi_P^C = 1$, whereas each factor that ranges over collections of nodes or feature values is averaged by the size of the largest collection.

If each of the similarity expressions were a linear aggregation of other similarity variables, this system would be solvable directly, since all variables are of degree one. However, in the case of OWL, and of many other languages, the system is not linear since there could be many candidate pairs for the best match. These correspond to the Set($\cdot$) type in Table 4.6 (p. 100). The similarity may depend on matching the multiple edges with the similar labels outgoing from the nodes under consideration. In this approach, the similarity is computed by a $MSim$ function that first finds an alignment between the set of considered entities and then computes the aggregated similarity with respect to this matching.

In this respect, the OLA algorithm solves a very specific problem, namely a maximal weight graph matching problem (§5.7.3) with weights depending on the matching.

Nevertheless, the resolution of the resulting system can still be carried out as an iterative process that simulates the computation of the greatest fixed point of a vector function, as shown by Bisson [Bisson, 1992]. The point consists of defining an approximation of the $MSim$-measures, solving the system, replacing the approximations by the newly computed solutions and iterating. The first values for these $MSim$-measures are the maximum similarity found for a pair, without considering the dependent part of the equations. The subsequent values are those of the complete similarity formula filled by the solutions of the system. Note that the local matching may change from one step to another depending of the current similarity values.

However, the system is converging because the similarities can only increase (the non dependent part of the equation remains and all dependencies are positive) and, in the case that similarity values are bounded, e.g., to 1, the similarity is bounded. The iterations will stop when no gain above a particular ϵ value is provided by the last iteration. If the algorithm converges, we cannot guarantee that it does not stop at a local optimum (that is, finding another matching in the $MSim$-measures would not increase the similarity values). This could be improved by randomly changing these matchings when the algorithm stops.

Some facts are worth mentioning. First, there is no need for a different expression of the similarity functions in the case where there are no effective circular dependencies between similarity values. In fact, the computation mechanism presented

5.3 Global similarity computation

above establishes the correct similarity values even if there is an appropriate ordering of the variables (the ordering is implicitly followed by the step-wise mechanism). Moreover, in case some similarity values (or some similarity or (dis)similarity assertions) are available beforehand, the corresponding equation can be replaced by the assertion or value.

Example 5.12 (OLA algorithm). The problem to be solved is the same as the one defined in Example 5.11, so the label similarity between classes is the same. The label similarity between properties is set to 1. (all similar) for each pair of properties. Thus, the initial similarities are as follows:

σ_L	Book	Translator	Publisher	Writer
Product	.19	.05	.03	.04
Provider	.04	.05	.75	.14
Creator	.04	.25	.04	.69

σ_L	hasPublished	hasTranslated	hasWritten	author	translator	publisher
creates	1.00	1.00	1.00	1.00	1.00	1.00
provides	1.00	1.00	1.00	1.00	1.00	1.00
hasProvided	1.00	1.00	1.00	1.00	1.00	1.00
hasCreated	1.00	1.00	1.00	1.00	1.00	1.00

The equations are made with equal weights on labels and properties for classes ($\pi_L^C = \pi_P^C = 1/2$) and equal weights on label, range and domain for properties ($\pi_L^P = \pi_R^P = \pi_D^P = 1/3$). The initial similarities (based only on the labels) provide the following values:

σ^0	Book	Translator	Publisher	Writer
Product	.10	.03	.02	.02
Provider	.02	.03	.38	.07
Creator	.02	.13	.02	.35

σ^0	hasPublished	hasTranslated	hasWritten	author	translator	publisher
creates	.33	.33	.33	.33	.33	.33
provides	.33	.33	.33	.33	.33	.33
hasProvided	.33	.33	.33	.33	.33	.33
hasCreated	.33	.33	.33	.33	.33	.33

The first iteration really takes into account the relations between entities and yields the following result:

σ^1	Book	Translator	Publisher	Writer
Product	.26	.19	.18	.19
Provider	.19	.19	.54	.24
Creator	.19	.29	.19	.51

σ^1	hasPublished	hasTranslated	hasWritten	author	translator	publisher
creates	.37	.41	.48	.35	.35	.35
provides	.49	.37	.39	.35	.35	.35
hasProvided	.35	.35	.35	.39	.37	.49
hasCreated	.35	.35	.35	.48	.41	.37

After 3 iterations the values do not change more than $\epsilon = .1$ and after 10 iterations they do not change more than $\epsilon = .01$ yielding the result as follows:

σ^{10}	Book	Translator	Publisher	Writer
Product	.46	.29	.27	.28
Provider	.28	.32	.74	.37
Creator	.28	.44	.31	.70

σ^{10}	hasPublished	hasTranslated	hasWritten	author	translator	publisher
creates	.59	.63	.72	.52	.52	.52
provides	.73	.59	.61	.52	.52	.52
hasProvided	.52	.52	.52	.61	.59	.73
hasCreated	.52	.52	.52	.72	.63	.59

For both values of ϵ the best match is always the same. It is the same as in Example 5.11 for classes and, in addition, for properties it is as follows: creates=hasWritten, provides=hasPublished, hasProvided=publisher and hasCreated=author.

The two presented methods have some similarity: both methods work iteratively on a set of equations extracted from a graphical form of the ontologies. Both methods ultimately depend on the computed proximities between non described language elements, i.e., data type names, values, URIRefs, property type names, etc. These proximities are propagated throughout the graph structure by the similarity dependencies.

Moreover, Similarity flooding is highly dependent on the identity of edge labels, while the OLA algorithm takes similarity between properties into account. Nonetheless it also considers local mappings between alternative matching edges instead of averaging over all the potential matches. That is, the OLA algorithm attempts to identify the subclasses which match and propagate their similarity – which should be high – while Similarity flooding propagates an average similarity between all pairs of subclasses which should be lower than the average similarity between all pairs of matching subclasses. Finally, the convergence of the Similarity flooding is not proved in general.

Another kind of global computation that may be necessary in ontology matching is learning which requires the manipulation of the whole matching process in order to improve its performance.

5.4 Learning methods

In Sect. 4.4, we have discussed techniques used for structurally inducing class relations from data. This section is concerned with algorithms which learn how to sort alignments through the presentation of many correct alignments (positive examples) and incorrect alignments (negative examples). The main difference between both approaches is that the techniques of this section require some sample data to learn from. This can be provided by the algorithm itself and judged by users, for instance, by having only a subset of the correspondences under judgment, or this can be brought from external resources.

Matchers using machine learning usually operate in two phases: (i) the learning or training phase and (ii) the classification or matching phase. During the first phase, training data for the learning process is created, for example, by manually matching two ontologies, and the system learns a matcher from this data. During the second phase, the learnt matcher is used for matching new ontologies. There can be a feedback on the obtained alignment which can be fed into the step (i) again. Learning can be processed on-line, such that the system can continuously learn, or off-line, so its speed is not relevant but its accuracy is.

Usually this process is carried out by dividing a data set, i.e., set of positive and sometime negative examples of alignments into a training set (typically 80% of data) and a control set (typically 20% of data) which is used for evaluating the performances of the learning algorithm.

There are many types of information that a learner can exploit. These include: word frequencies, formats, positions, properties of value distributions. A multistrategy learning approach is useful when several learners are used, each on handling a particular kind of pattern that it learns best. Finally, results produced by various learners can be combined with the help of a meta-learner [Doan *et al.*, 2003].

In this section we consider some of the well-known machine learning methods which have been used for text categorisation, such as Bayes learning (§5.4.1), WHIRL learning (§5.4.2), neural networks (§5.4.3), decision trees (§5.4.4), and stacked generalisation (§5.4.5).

5.4.1 Bayes learning

The naive Bayes learner is one of the simplest and most effective text classifiers [Good, 1965, Domingos and Pazzani, 1996, McCallum and Nigam, 1998]. It represents a probabilistic induction algorithm.

Let us suppose that we want to match attribute x from one ontology to one (y_i) of the attributes ($y_1, \ldots, y_m$, $i = 1, \ldots, m$) from another ontology. The approach views values of attributes as sets of tokens. Suppose V denotes a set of underlying values of attribute x: $V = \{t_1, \ldots, t_n\}$, where t_j is the j-th token, $j = 1, \ldots, n$. Tokens, in turn, are obtained by applying a normalisation technique, such as lemmatisation (§4.2.2), to the words in the data instance. Suppose that $P(y_i)$ is the *a priori* probability that x matches y_i, i.e., without having seen any tokens of x. Then, $P(V)$ stands for the probability of observing values V in x. Finally, $P(V|y_i)$ stands for the conditional probability of observing values V, given that x matches y_i. The Bayes theorem describes how to optimally predict the attribute for a previously unseen data instance, given a training example. The chosen attribute is the one that maximises *a posterior* probability, i.e., after having seen the values V, that x matches y_i. It is denoted as $P(y_i|V)$ and is computed as follows:

$$P(y_i|V) = \frac{P(V|y_i) \times P(y_i)}{P(V)}$$

This is called the Bayes rule. The naive Bayes classifier has a *naive* assumption that the tokens t_j appear in V *independently* of each other given y_i. Based on this assumption the parameters (tokens) of each attribute can be learnt separately; this in turn greatly simplifies learning. Thus, if the attributes are independent given the class, $P(V|y_i)$ can be decomposed into the product of $P(t_1|y_i) \times \ldots P(t_n|y_i)$ and $P(V)$ can be omitted from the Bayes rule for obvious reasons. Henceforth, the Bayes rule can be rewritten as follows:

$$P(y_i|V) = P(y_i) \times \prod_{1 \leq j \leq n} P(t_j|y_i)$$

The independence assumption often does not hold in practice. However, in many applications, the violation of this assumption does not lead to degradation in effectiveness of the approach [Domingos and Pazzani, 1996].

The probabilities of the latter formula can be computed using the training data: $P(y_i)$ can be estimated by the proportion of examples that have been matched to y_i; $P(t_j|y_i)$ can be estimated as $k(t_j, y_i)/k(y_i)$, where $k(y_i)$ is the total number of tokens of all training instances with attribute y_i, and $k(t_j, y_i)$ is the number of occurrences of token t_j in all training instances with attribute y_i. Based on the above formula the corresponding confidence scores can be designed in an obvious way.

Example 5.13 (Naive Bayes learning). Suppose that we have established manually that attributes creator and name of one ontology match respectively attributes author and title of another one. The process works in two steps.

Training phase. Let us also suppose that $\{Bertrand\ Russell\}$ is an instance of the creator attribute and $\{My\ life\}$ is an instance of the name attribute. Thus, based on this information the following training examples can be fed into the classifier: $\langle\{$Bertrand, Russell$\}$, author$\rangle$ and $\langle\{$My, life$\}$, title$\rangle$. The second one declares that $\{$My, life$\}$ is a title and it has two tokens. By inspecting the training instances the learner builds its *internal classification model*. For example, by noticing that if a word such as life occurs frequently in data instances positively related to title and infrequently in those related to other fields, their underlying attribute is therefore likely to match the title attribute on how to classify data instances. If the training set is statistically representative, these frequencies can be transformed into probabilities and the Bayes rule can be used. This can also be applied to classify instances in classes, for instance, using ⟨{title:My title:life}, class:biography⟩.

Matching phase. Let {Life, of, Pi} be an instance of the attribute h1 from the structure of a web site which we want to match against attributes of the second ontology above. The learner uses its internal classification model to predict an attribute for the given instance as well as its confidence score, e.g., ⟨author, 0.2⟩, ⟨title, 0.8⟩. Based on the confidence scores, it can be concluded that h1 is a match for title.

5.4.2 WHIRL learner

WHIRL is an extension of conventional relational databases to perform *soft joins* based on the similarity of textual identifiers (not only based on equivalence of atomic values) [Cohen, 1998]. It has been also used for inductive classification of text and turns out to be competitive with other inductive classification systems, such as C4.5 decision trees [Cohen and Hirsh, 1998]. The WHIRL approach to text classification can be viewed as a kind of nearest neighbour classification algorithm.

WHIRL has been used in matching for learning both schema-level and instance-level information [Doan et al., 2003]. In the case of schema information, training examples could be of the following type ⟨*expanded label′, label*⟩, where *label′* belongs to ontology o' and *label* belongs to ontology o. For example, ⟨location′, address⟩, states that if an ontology entity has the label location, then it matches address. Expansion of *label′* can be done, for instance by including its synonyms, which, in turn, can be obtained from manually created correspondence tables for the domain of interest. WHIRL stores all training examples it has seen. Suppose that we would like to match another ontology o'' to ontology o. Given a *label″* from o'', WHIRL computes the corresponding label in o based on the labels of all examples in its collection that are within a similarity distance from *label″*. The similarity used here is based on TFIDF (§4.2.1) between the expanded labels of the examples. For example, given the label phone from o', WHIRL may generate a prediction as follows: ⟨address, 0.1⟩, ⟨description, 0.2⟩, ⟨agent-phone, 0.7⟩. Based on the confidence scores, it can be concluded that phone is a match for agent-phone.

In the case of instance-level information, this matcher makes use of data content instead of expanded labels. A training example of this case is of the form ⟨*data instances′, label*⟩, where *data instances′* belong to ontology o' and *label* belongs to ontology o. When matching a new ontology o'' to ontology o, the TFIDF distance between any two examples is the distance between data instances of o'' and the WHIRL collection of data instances.

5.4.3 Neural networks

Artificial neural networks are made up of nodes (or neurons) and weighted connections between them. Nodes are grouped into layers, having input, output and either none, one or more hidden layers. Usually each node in a hidden layer is connected to all nodes of the preceding and the following layer. Neural networks have been widely used in practice due to their ability of adaptation. Several types of neural networks exist and have been used for various tasks in ontology matching, such as discovering correspondences among attributes via categorisation and classification [Li and Clifton, 1994] or learning matching parameters, such as matcher weights, to tune matching systems with respect to a particular matching task [Ehrig et al., 2005]. We focus here on the first task mentioned above, while learning matching parameters is addressed in Sect. 5.4.5.

Given schema-level and instance-level information, it is sometimes useful to cluster this input into m categories in order to lower the computational complexity of further manipulations with data. The *self-organising map* network and the corresponding self-organisation learning algorithm can be used for this purpose [Kohonen, 2001]. It categorises n nodes of the input layer into m categories of the output layer. Usually m is predefined based on how detailed the categories should be by setting the radius of clusters. Input patterns or attributes, e.g., field length and datatype, are viewed as dimensions in n-dimensional feature space. The neurons in the network are organising themselves according to the characteristics of given input patterns. This results in a clustered neuron structure, where neurons with similar properties are arranged in related areas on the map. Every node in the output layer represents a cluster centre.

For neural networks, matching is viewed as a classification problem. The *back-propagation* algorithm can be used for this purpose. Back-propagation is a supervised learning algorithm which is used to train a network to recognise input patterns and give corresponding similarity scores. First, the feature weights are loaded into the input nodes. Then, they are propagated forward in order to generate the output. If a misclassification occurs, the error is backpropagated in order to change the weights of connections in the network. Weights are modified until the errors in the output layer are not minimised anymore.

Example 5.14 (Neural networks – adapted from [Li and Clifton, 1994]). Given an ontology, some of its attributes, such as Employee.id, Dept.Employee and Payrol.SSN, can be clustered into one category, since their input characteristics as well as intended meanings are close to each other. The corresponding vector of cluster centre weights can be as follows: $\langle 0, 0.1, 0, \ldots \rangle$, where vector components stand for the features: the first position stands for datatype, the second position stands for length, and so on and so forth. The key feature for grouping the attributes mentioned above was the field length, since its value (0.1) is higher than that of others (0.0). In fact, ID fields typically use the full field all the time, while, e.g., name fields use less and vary.

Fig. 5.9 shows a three-layer network for recognising m categories of patterns, given n features. The number of nodes in the hidden layer can be arbitrary. It is usually chosen based on experiments in order to obtain the shortest training time.

Training phase. The training data for the neural network is composed of vectors of cluster center weights and their target categories. For example, the vector con-

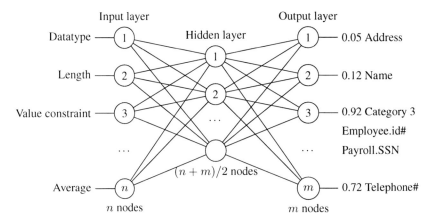

Fig. 5.9. Neural network. Input attribute Health_Plan_Insured# characteristics

sidered previously, i.e., $\langle 0, 0.1, 0, 0, \dots \rangle$, is tagged with its target category, which is number 3. The back propagation algorithm will then adjust the weights so that attributes characteristics corresponding to these attributes will result in a output vector as close as possible to $\langle 0, 0, 1, 0, \dots \rangle$ indicating that the most likely category is 3.

Matching phase. The matching phase includes feeding into the network trained on features of ontology o a new pattern of n characteristics, e.g., of the attribute health_Plan.Insured#, from another ontology o'. Based on the internal classification model of the network, it determines the similarity of this pattern and each of m categories. For instance, this attribute matches category 3 (id numbers) with the similarity score 0.92 and category 1 with the similarity score 0.05.

5.4.4 Decision trees

Decision trees classifiers are made up of a set of rules which are applied in a sequential way and ultimately lead to a decision. Unlike probabilistic methods, e.g., naive Bayes, which are numeric in nature, and, therefore, not easily interpretable by humans, non numeric or symbolic algorithms do not have this drawback. Decision trees are an example of such algorithms. A possible method for learning a decision tree for a category can follow a *divide and conquer* strategy. In a training set T of instances characterised by features and their category, a feature f_1 is selected, which discriminates the population in the best way (with regard to the set of categories). Then, T is partitioned into two subsets, the subset T_1^{yes} corresponding to feature f_1, and the subset T_1^{no} without this feature. This procedure is recursively applied to T_1^{yes} and T_1^{no}. It stops if all instances in a subset are assigned to the same category. It generates a tree of rules with an assignment to actual categories in the leaves. Decision tree learner can be tolerant and accept that some of the instances are misclassified if this produces a large simplification of the tree. This is useful when there can be errors in the training sets.

138   5 Matching strategies

Decision trees have been used in ontology matching for various tasks, such as discovering correspondences among entities [Xu and Embley, 2003] and learning parameters of matching systems, e.g., thresholds, to adapt automatically to a given matching task [Ehrig et al., 2005]. We focus here on the former, while learning matching parameters is addressed in Sect. 5.4.5.

Example 5.15 (Decision tree). Given a large training alignment between instances from the two ontologies of Fig. 2.7, decision tree learning, e.g., C4.5 decision tree induction system [Quinlan, 1993], is applied to generate rules for identifying new instances. Fig. 5.10 shows a fragment of decision tree that can be learnt. The decision first states that it can only match Books from the first ontology into the second one. Then it distinguishes Books having one author, which is a Professor, from those having no Professor as authors. It is then able to consider that if an author is a topic of the Book, then this one must be classified as an Autobiography, otherwise it should be an Essay.

The decision tree has been built with some tolerance: the numbers after the target categories indicate the number of instances in the training set which have been correctly and incorrectly recognised.

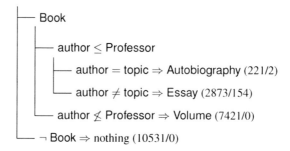

Fig. 5.10. Fragment of a binary decision tree. Each node is labelled by a condition that must be satisfied by the item to classify. When no further classification is possible, the resulting target category is indicated together with the number (in parenthesis) of correctly/incorrectly classified items in the training set.

The decision tree fragment displayed in Fig. 5.10 can be rewritten as mapping from the source ontology to the target ontology. The mapping rule corresponding to the Autobiograpohy branch can be written as:

Book$(e) \land \exists e'$; author$(e, e') \land$ Professor$(e') \land \exists e''$; author$(e, e'') \land$ topic(e, e'')
$\Rightarrow$Autobiography(e)

It is possible to use the same kind of techniques for learning from the structure instead of the instances. [Xu and Embley, 2003] shows how to use decision trees in order to learn rules for matching terms in WordNet.

5.4.5 Stacked generalisation

Stacked generalisation is an approach to combine multiple learning algorithms [Wolpert, 1992]. From the ontology matching perspective, by using this approach we can learn to aggregate several basic learners, e.g., naive Bayes and WHIRL, on a particular label.

The training phase of stacked generalisation works in two steps. The first step deals with collecting the output of each learner, thereby resulting in a new set of data. First, let $D^0 = \{\langle c_i, x_i \rangle\}_{i=1,...m}$ be the training dataset, such that c_i is a category and x_i is an instance represented by the vector of its features. In terms of ontology matching, c_i can be an entity from ontology o and x_i is an entity of ontology o', e.g., an individual represented by its attributes. The c_i is the category to which x_i should be assigned (or, for ontology matching, the entity corresponding to x_i). In order to avoid overfitting, i.e., that the training data does include the query instances, a *cross-validation* is performed. In particular, D^0 is randomly partitioned into p almost equal parts, $D_1^0, \ldots D_p^0$, such that D_k^0 represents a test and $\bar{D}_k^0 = D^0 - D_k^0$, represents a training set for the k-th class of cross-validation. Given q basic learning algorithms (matchers), which are called *level-0 generalisers*, using the l-th matching algorithm on training set $\bar{D}_k$ results in a *model* M_l^k, which has been learnt. Such a model, given a vector of features characterising an object, returns a prediction which is the category it should be assigned. These are the *level-0 models*. Let M_l^i be the prediction of the model M_l^k on $x_i \in D_k$. The final result of cross-validation is the set D^1 which consists of exactly one prediction for each of the training examples:

$$D^1 = \{\langle c_i, \langle M_1^i, \ldots M_q^i \rangle\rangle\}_{i=1,...m}$$

The output of the first step is used as the input data for the second step, where another learning algorithm, called *level-1 generaliser*, is employed. In turn, it derives a *level-1 model* M' for c_i with respect to $M_1^i, \ldots M_q^i$. Thus, while level-0 classifiers deal with the possible assignment of entities to categories with regard to their attributes, level-1 classifiers deal with the possible assignment of the same entities to the same categories with regard to the categories predicted by the classifiers.

During the classification phase, given a new instance x, the models which have been learnt, produce a vector $\langle M_1, \ldots M_q \rangle$, which in turn is taken as input by M', whose output is the final classification for that instance.

[Ting and Witten, 1999] identified that the best results are obtained in stacked generalisation for classification tasks when (i) the higher-level model combines the *confidence* and not the predictions of the lower-level models, and (ii) the *multiresponse linear regression* is used as a level-1 generaliser compared to such learning algorithms as C4.5 decision tree (§5.4.4), or naive Bayes (§5.4.1).

Besides multiresponse linear regression, there are other algorithms that may work equally well on this task, such as neural networks (§5.4.3).

Example 5.16 (Stacked generalisation, adapted from [Doan et al., 2003]). Suppose that two basic learners are used: (i) the WHIRL learner working with labels of entities (§5.4.2), and (ii) the naive Bayes learner working with data instances of entities

(§5.4.1). Names of these matchers are abbreviated in this example as WHIRL and NB respectively.

Training phase. Consider the label address from ontology o. Examples of corresponding training data from ontology o' for basic learners are shown in Table 5.1. The first and the second columns list respectively the labels, e.g., location, and the underlying data instances, e.g., ⟨Miami, FL⟩, of some entities from ontology o'. The fourth and the fifth columns describe confidence scores S as produced by WHIRL and naive Bayes based on input from the first three columns via the cross-validation. For example, $S_{WHIRL}^{address}$(location) $= 0.5$, while $S_{NB}^{address}$(⟨Miami, FL⟩) $= 0.8$. Finally, the last column indicates whether the correspondence under consideration holds or not. For example, location from o' actually matches address from o, and therefore, the corresponding value in the last column is 1, while listed-price does not match address, and therefore, the corresponding value in the last column is 0.

Table 5.1. Training data for basic learners and stacked generalisation.

o' label	o' instance	o label	WHIRL	NB	True predictions
location	Miami, FL	address	0.5	0.8	1
listed-price	250K	address	0.4	0.3	0
phone	(305) 729 0831	address	0.3	0.6	0
comments	Fantastic house	address	0.6	0.1	0
location	Boston, MA	address	0.5	0.9	1
listed-price	320K	address	0.2	0.2	0

The information from three right-most columns is used as input for the linear regression [Breiman, 1996, Birkes and Dodge, 2001]. Results of the WHIRL and naive Bayes learners stand for the confidence scores (S), while the last column represents values of the response variable. As the result of least square error minimisation the weight assigned to the pair of WHIRL and label address is 0.2, i.e., $W_{WHIRL}^{address} = 0.2$; while $W_{NB}^{address} = 0.9$. The interpretation of these weights is that, based on stacked generalisation training, the naive Bayes learner appears to be much more reliable compared to WHIRL in their predictions about address.

Matching phase. Let us suppose that we want to match the entity with label area and instance ⟨Seattle, WA⟩ from yet another ontology o'' to entities of ontology o. Consider the case of the entity with label address from o. WHIRL will analyse the label area and generate its confidence score, e.g., ⟨address, 0.4⟩. The naive Bayes learner, in turn, will analyse the data contents and generate its confidence score, e.g., ⟨address, 0.8⟩. By using the weights obtained during the training phase of stacked generalisation, the weighted average of the confidence scores can be computed as follows: $0.4 \times 0.2 + 0.8 \times 0.9 = 0.8$, thus, yielding the combined prediction, which is ⟨address, 0.8⟩.

Concluding this overview of the stacked generalisation method, we note that there are other techniques with similar goals. Examples include *boosting* and *bag-*

ging. To combine the decisions of the individual models (matchers), boosting uses a weighted majority vote and bagging uses unweighted majority vote. However, they require a considerable number of models because they rely on varying the data distribution to obtain a diverse set of models from a single learning algorithm, while stacking can work with only a few level-0 models [Ting and Witten, 1999].

5.5 Probabilistic methods

Similarly to learning methods, probabilistic methods can also be universally used in ontology matching to enhance some available matching candidates, combine matchers, etc. In this section we discuss an example of the methods based on probabilistic inference, namely Bayesian networks.

5.5.1 Bayesian networks

A Bayesian belief network or simply a Bayesian network is a probabilistic approach for modelling causes and effects. Bayesian networks are made up of (i) a directed acyclic graph, containing nodes (also called variables) and arcs, and (ii) a set of *conditional probability tables*. Arcs between nodes stand for conditional dependencies and indicate the direction of influence. For example, an arc from node X_1 (called parent) to node X_2 (called child) means that X_1 has a direct influence on X_2. How a node influences another node (based on past experience) is defined by conditional probability tables for the nodes. $P(X|parents(X))$ is a conditional probability of variable X, where $parents(X)$ is a set of all and only nodes directly influencing X. Graph and conditional probability tables allow the construction of the joint probability distribution of all variables, namely:

$$P(X_1, \ldots, X_n) = \prod_i P(X_i|parents(X_i)), i = 1, \ldots, n$$

Given values for some nodes, it is possible to infer probability distribution for values of other nodes. In the simplest case, a Bayesian network can be specified by an expert and after some values of nodes are made observable it can be used to perform inference, thus making predictions or diagnosing causes. When not all variables are observable, it is necessary to identify dependencies between variables, which, in turn, can be solved by learning a Bayesian network that fits to the data [Russell and Norvig, 1995].

Bayesian networks have been modelled and used differently in ontology matching. For example, in the work of [Pan *et al.*, 2005], two ontologies are translated into two Bayesian networks and matching is performed as evidential inference between these Bayesian networks. Another work [Mitra *et al.*, 2005] uses Bayesian networks to enhance existing matches, e.g., by deriving missed matches. Let us consider the latter in more detail.

142   5 Matching strategies

Example 5.17 (Bayesian network, adapted from [Mitra et al., 2005]).
The Bayesian network is built with mappings and uses meta-rules based on the semantics of the ontology language that expresses how each mapping affects other related mappings. External matchers are adapted to produce initial probability distributions for mappings, which are in turn used to infer probability distributions for other mappings.

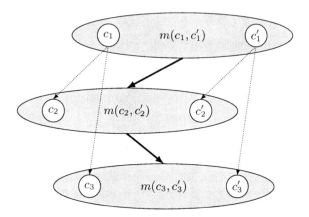

Fig. 5.11. Bayesian network graph.

Nodes in the Bayesian network graph represent matches between pairs of classes or properties from two distinct ontologies, see Fig. 5.11. Solid arrows in the Bayesian network graph represent the influences between its nodes, while dotted arrows stand for the relations in the ontologies under consideration. For example, the mapping between concepts $c_1 \in o$ and $c'_1 \in o'$ affects the mapping between concepts $c_2 \in o$ and $c'_2 \in o'$, which in turn affects the mapping between $c_3 \in o$ and $c'_3 \in o'$. Conditional probability tables are generated by exploiting generic meta-rules. For example, if two concepts c_1 and c'_1 match and there is a relationship q between $c_1 \in o$ and $c_2 \in o$ and a matching relationship q' between $c'_1 \in o'$ and $c'_2 \in o'$, then we can increase the probability of match between $c_2 \in o$ and $c'_2 \in o'$. The probability distribution of the child node is derived from the probability distribution of the parent node using a set of constants. By running a Bayesian network, posterior probabilities for each node are generated.

5.6 User involvement and dynamic composition

Another element that must be taken into account when designing the architecture of a matching system is the availability of users. This was one of the requirements from some applications mentioned in Chap. 1. There are three areas in which users can be involved in a matching solution: (i) by providing initial alignments (and parameters) to the matchers, (ii) by dynamically combining matchers, and (iii) by providing feedback to the matchers in order for them to adapt their results. These three aspects are considered next.

5.6.1 Providing input

This is the user task to provide input to the matching systems. This obviously covers providing the ontologies to be matched.

More difficult is choosing the system parameters, which always depend on the method. Some algorithms may provide advice based on a priori analysis of the data, but these techniques are specific to the matchers themselves.

Another important input that users can provide is an initial alignment. An initial alignment will constrain the matching system to produce an alignment complying with the initial alignment. This is an opportunity for users to control the algorithm behaviour.

5.6.2 Manual matcher composition

Most of the individual methods can be composed through a general purposes programming language, but this is not user-level composition.

We distinguish three ways to compose matchers:

Built-in composition corresponds to most of the algorithms presented in Chap. 6: the methods are composed according to the principles presented in Sect. 5.1. This composition is part of the algorithm and is applied to any data set which is given to the system.

Opportunistic composition would correspond to a system which chooses the next method to run in regard of the input data and/or currently achieved results. Systems like Falcon-AO (§6.3.9) or H-Match (§6.1.7) feature a limited implementation of this (choosing dynamically to use a particular matcher or not).

User-driven composition is used in environments in which users have many different methods that they can apply following their needs. Examples of such environments include Rondo (§8.2.1), COMA++ (§8.2.2), and Protégé (§8.3.2).

For example, a system can involve users in the dynamic assembly of matchers as follows. An interface could provide a library of basic matchers, filters and aggregators as presented in Sect. 5.1 that users can assemble by some graphic interaction. Then the output of these methods could be materialised so that users can inspect them. Inspection can involve actually applying the resulting alignment for existing data in order to see the effects. Users may also dynamically change the parameters and quickly see the effects of these changes on the data. Finally it should be possible to save the designed architecture in order to use it in applications.

5.6.3 Relevance feedback

User feedback for each alignment or each specific correspondence found by a matching system can be used for improving the results by changing the local matcher parameters. Usually, these parameters can be:

- parameters of a matcher;
- thresholds used for filtering the results (§5.7); and
- aggregation parameters (§5.2).

Usually, systems compute a distance between the feedback and the result provided by the system. This is called the error. The computation of the error depends greatly on the kind of feedback which is provided: users discarding unwanted correspondences, adding new correspondences or modifying the proposed correspondences.

Once the error is computed, the system has to select the parameter values that would minimise this error. This can sometimes be done directly: for instance, when the only parameter to set is a threshold, the system can directly compute a threshold that will provide the minimal error. However, very often such a method does not exist and it is necessary to use an indirect method that estimates the error reduction given some parameter value changes and searches the best combination. Most of the methods presented in Sect. 5.4 can be used for this task. This kind of algorithms is used, for instance, in APFEL (§6.4.1).

As users can help the system to work, the system can also help users. This is especially considered in Sect. 8.3 when addressing ontology editors that offer support to ontology matching and in Chap. 9 when detailing strategies for explaining matching results.

Finally, the history of the prior matching actions can be used to bias the ranking computation toward the ontology regions that are likely to be relevant to the selected entity. For example, if the neighbours of an entity have all been matched with entities from the same region of the other ontology, it is likely that this entity will find a match in this region too. Taking the existing matches and the user action history into account makes the process of mapping creation more interactive and personalised [Bernstein *et al.*, 2006].

5.7 Alignment extraction

The matching goal is to identify a satisfactory set of correspondences between ontologies. A (dis)similarity measure between the entities of both ontologies provides a large set of correspondences. Those which will be part of the resulting alignment remain to be extracted on the basis of the similarity. This can be achieved by a specialised extraction method which acts on the similarity matrix or on some pre-alignment already extracted. We distinguish between the extractor itself, which converts a (dis)similarity matrix into an alignment, and a filter, which reduces the candidate correspondences in one of these formats. This is depicted in Fig. 5.12.

Fig. 5.12. Similarity filter, alignment extractor and alignment filter.

Similarity filters transform the (dis)similarity matrix by, for instance, by zeroing all cells under some threshold or by unit-ing those above a threshold. Alignment extractors generate an alignment from a similarity matrix. They are the main topic of this section. Alignment filters can further manipulate with alignments by using the same types of operations as similarity filters.

The alignment filter can be users: an alignment can be obtained by displaying the entity pairs with their similarity scores and ranks, leaving the choice of the appropriate pairs up to users. This user input can be taken as the definitive answer in helper environments, as the definition of an anchor for helping the system (§6.1.9) or as relevance feedback in learning algorithms (§5.4).

One could go a step further and attempt to define algorithms that automate alignment extraction from similarity scores. Various strategies may be applied to the task depending on the properties of the target alignment.

This problem can be defined as follows:

Definition 5.18 (Alignment extraction problem). *Given two sets of entities o and o' and a similarity function $\sigma : o \times o' \to [0\ 1]$, extract an alignment $A \subseteq o \times o'$.*

This problem statement is underconstrained since $o \times o'$ is a solution to this problem. So the goal of this section is to consider how to further constrain the problem of alignment extraction. One guide for doing so has been introduced in Sect. 2.5.2 as the totality and injectivity constraints on alignments.

We present two main strategies based on trimming the (dis)similarity after some threshold (§5.7.1) and on determining an optimal overall (dis)similarity of the extracted alignment (§5.7.3). In between, we present a kind of filter that has been found useful in matching algorithms (§5.7.2).

5.7.1 Thresholds

If neither ontology needs to be completely covered by the alignment, a threshold-based filtering would allow us to retain only the most similar entity pairs. Without the injectivity constraint, the pairs scoring above the threshold represent a sensible alignment.

Thus, applying thresholds requires that the extracted alignment is of sufficient quality. An easier way to proceed consists of selecting correspondences over a particular threshold. Several methods can be found in the literature [Do and Rahm, 2002, Ehrig and Sure, 2004]:

Hard threshold retains all the correspondence above threshold n;

Delta threshold consists of using as a threshold the highest similarity value out of which a particular constant value d is subtracted;
Proportional threshold consists of using as a threshold the percentage of the highest similarity value;
Percentage retains the $n\%$ correspondences above the others.

The Rondo system (§6.1.13) provides an original alignment extraction method (SelectThreshold) which normalises the similarity of each node by the best similarity it has with another node (the result is not symmetric anymore). It then selects for the alignment the pairs of nodes for which the normalised similarity of both nodes is above some defined threshold.

Example 5.19 (Thresholding methods). We start from the weighted sum distance obtained in Example 5.7 with $1/4 - 3/4$ weights. This distance is converted into a similarity as in the following table:

	Book	Translator	Publisher	Writer
Product	.19	.05	.03	.04
Provider	.04	.05	.75	.14
Creator	.05	.25	.04	.69

- An ideal hard threshold of .23 would select the ⟨Provider, Publisher⟩, ⟨Creator, Writer⟩, and ⟨Creator, Translator⟩ as correspondences.
- A delta threshold with the same .23 value would select only the two first ones as the corresponding hard threshold would be $.75 - .23 = .52$.
- On the contrary, the use of a proportional threshold of .23 would result in a $.75 \times .23 = .17$ hard threshold so selecting ⟨Product, Book⟩ in addition to the three ones above.
- The percentage threshold of .23 would select the $12 \times 23\% \approx 3$ initially selected pairs.
- The SelectThreshold method would also yield the set of four correspondences mentioned above for a threshold of .23.

5.7.2 Strengthening and weakening

Some approaches, such as [Ehrig and Sure, 2004], use a sigmoïd function between 0. and 1. ($sig_a(x) = 1/(1 + e^{-a(x-0.5)})$ with a being a parameter for the slope). This allows reinforcing values higher than 0.5 and to weaken those lower than 0.5. This treatment is meant to clearly separate two zones: the positive and negative correspondences (see Fig. 5.13).

Other functions, such as $1 - sin(arccos(x))$, can have an opposite effect: discarding the nonconclusive measures and dispatching the highest ones on the unit interval. This treatment is well justified by considering that very similar entities are indeed similar but loosely similar entities give non conclusive results. Of course, it is possible to shift these functions in order to select threshold other than .5.

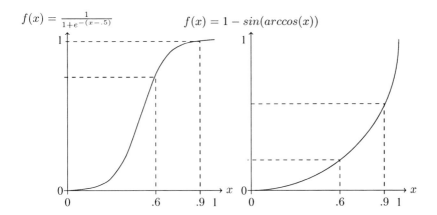

Fig. 5.13. Sigmoïd and trigonometric smoothing functions.

Example 5.20 (Strengthening and weakening). The two tables below display the similarity table of Example 5.19 filtered through the two functions displayed in Fig. 5.13.

	Book	Translator	Publisher	Writer
Product	.02	0.	0.	0.
Provider	0.	0.	.95	.01
Creator	0.	.05	0.	.91

Filtered by $y = \frac{1}{1+e^{-12(x-.5)}}$.

	Book	Translator	Publisher	Writer
Product	.02	0.	0.	0.
Provider	0.	0.	.34	.01
Creator	0.	.03	0.	.28

Filtered by $y = 1 - sin(arccos(x))$.

The sigmoïd function provides high values for the best matches and lower ones for the worse matches while the other proposed function requires higher similarities than .75 to single them out: it reduces all values.

5.7.3 Optimising the result

If an injective mapping is required then some choices need to be made in order to maximise the quality of the alignment. This quality is typically measured on the total similarity of the matched entity pairs. Consequently, the matching algorithm must optimise the global criteria rather than maximising the local similarity at each entity pair.

To sum up, the alignment computation can be viewed as a less constrained version of the basic set similarity functions $MSim$ (§4.4.3). Indeed, its target features are the same: (i) maximal global similarity, (ii) exclusivity, and (iii) maximal cardinality (in correspondences). However, (ii) and (iii) are not mandatory: they depend on injectivity and totality requirements, respectively. Extracting an alignment from a similarity table is typically what is called graph matching [Berge, 1970, Lovász and Plummer, 1986] and more precisely weighted bipartite graph matching (for injective alignments) or covering (for total alignments).

A greedy alignment extraction algorithm could construct the correspondences step-wise, at each step selecting the most similar pair and deleting its members from the table. The algorithm will then stop whenever no pair remains whose similarity is above the threshold. The greedy strategy is not optimal (see Example 5.23); however the ground on which a high similarity is forgotten to the advantage of lower similarities can be questioned and thus the greedy algorithm could be preferred in some situations.

There are two notions of optimal matching of two sets in this context, the first one is a local optimum called *stable marriage* which consists of an assignment (of only one object of the first set to only one object of the second set) such that any permutation of two assignments provides a worse result. An algorithm for computing stable marriages is the Gale–Shapley algorithm [Gale and Shapley, 1962].

Definition 5.21 (Stable marriage problem). *Given two sets of entities o and o' and a similarity function $\sigma : o \times o' \to [0\ 1]$, extract a one-to-one alignment $M \subseteq o \times o'$, such that for any $\langle p, q \rangle \in M$ and $\langle r, s \rangle \in M$, $\sigma(p,q) + \sigma(r,s) \geq \sigma(p,s) + \sigma(r,q)$.*

The second notion is the global optimum, or *maximum weight matching*. It is the assignment for which there does not exist any other assignment with better weighting. It can be computed by the Hungarian method [Munkres, 1957].

Definition 5.22 (Maximum weight graph matching problem). *Given two sets of entities o and o' and a similarity function $\sigma : o \times o' \to [0\ 1]$, extract a one-to-one alignment $M \subseteq o \times o'$, such that for any one-to-one alignment $M' \subseteq o \times o'$,*

$$\sum_{\langle p,q \rangle \in M} \sigma(p,q) \geq \sum_{\langle p,q \rangle \in M'} \sigma(p,q)$$

If weights represent dissimilarities instead of similarities, the problem to solve is the dual minimum weight graph matching.

Example 5.23 (Greedy, stable marriage and maximum weight). Let us consider the following similarity table for the concepts of Fig. 5.7 from which we want to extract a one-to-one alignment.

	Book	Translator	Publisher	Writer
Product	.84	0.	.90	.12
Provider	.12	0.	.84	.60
Creator	.60	.05	.12	.84

The greedy algorithm would select first the highest scoring (.90) correspondence ⟨Product, Publisher⟩ and discard the corresponding line and column. It would then select the next highest scoring (.84) one, ⟨Creator, Writer⟩ and then the remaining best one ⟨Provider, Book⟩. The alignment made of these three correspondences scores 1.96.

However, there are better alignments. For instance, by replacing the last two elements with ⟨Creator, Book⟩ and ⟨Provider, Writer⟩, we obtain an alignment scoring 2.1. This alignment is stable (checking that any single permutation yield a lower score than this one is left as an exercise).

This stable marriage is, however, not the maximum weight matching which is made of ⟨Product, Book⟩, ⟨Provider, Publisher⟩, and ⟨Creator, Writer⟩ and scores 2.52.

5.8 Summary

We have presented the strategic issues involved in creating matching solutions besides using basic matchers presented in Chap. 4. In particular, this involves composing the basic matchers, aggregating their results, dealing with circularities and extracting alignments from the matcher. This also covers the use of learning algorithms and involving users in the matching process.

The craft of ontology matching systems is a delicate art that combines basic matchers in the most advantageous way. This chapter has presented techniques used for assembling the components of a matching system. In most of the cases, the appropriate architecture depends on the problem to solve. Are there any independent basic matchers that can apply to the data? Is the data highly intricated? Are users available to evaluate the result? Must the expected result be injective? These questions meet the requirements of Chap. 1 and their answers lead to different assemblies of components.

With regard to the requirements of Chap. 1, the techniques presented in this chapter are often a matter of trade-off: between completeness and correctness of the alignment for threshold application, between quality and computation time for the choice of global similarity computation.

Fig. 5.14 displays a fictitious example involving several of the methods. It (i) runs several basic matchers in parallel, (ii) aggregates their results, (iii) selects some correspondences on the basis of their (dis)similarity, (iv) extracts an alignment, (v)

uses a semantic algorithm to amplify the selected alignment, and (vi) reiterate this process if necessary.

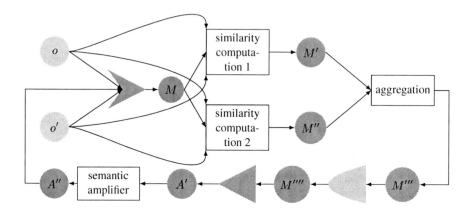

Fig. 5.14. A fictitious matching strategy.

The next part of the book describes how various implemented systems take advantage of the basic matchers discussed and how they compose them in a coherent system (Chap. 6). This will illustrate the diversity of approaches. We will then consider how to evaluate the merit of these systems experimentally (Chap. 7), because the mere theoretical consideration of a system capability and architecture is not a sufficiently convincing ground on which to judge its performances.

Part III

Systems and evaluation

6
Overview of matching systems

This chapter is an overview of matching systems which have emerged during the last decade. There have already been some comparisons of matching systems, in particular in [Parent and Spaccapietra, 2000, Rahm and Bernstein, 2001, Do et al., 2002, Kalfoglou and Schorlemmer, 2003b, Noy, 2004a, Doan and Halevy, 2005, Shvaiko and Euzenat, 2005]. Our purpose here is not to compare them in full detail, though we give some comparisons, but rather to show their variety, in order to demonstrate in how many different ways the methods presented in previous chapters have been practically exploited.

We have followed two principles in deciding whether a matching solution should be included in this chapter: it must have an implementation and an archival publication describing it at the time of writing. We have also excluded from consideration the systems which assume that alignments have already been established, and use this assumption as a prerequisite of running the actual system. These approaches include such information integration systems as: Tsimmis [Chawathe et al., 1994], Observer [Mena et al., 1996], SIMS [Arens et al., 1996], InfoSleuth [Fowler et al., 1999, Nodine et al., 2000], Kraft [Preece et al., 2000], Picsel [Goasdoué et al., 2000], DWQ [Calvanese et al., 2002a], AutoMed [Boyd et al., 2004], and InfoMix [Leone et al., 2005]. Even if we have considered around 50 systems and approaches in this chapter, this overview is not exhaustive. An interested reader can find an updated and complete information on the topic at OntologyMatching.org[1], in particular, links to the web sites of the presented systems can be found there. We only mention address of general purpose resources.

We present the matching systems in light of the classifications discussed in Chap. 3. We also point to concrete basic matchers and matching strategies used in the considered systems by referencing to the corresponding subsections of Chap. 4 and Chap. 5. In order to facilitate the presentation we follow two rules. First, the year of the system appearance is considered. Then, if there are some evolutions of the system or very similar systems, these are discussed close to each other. We illus-

---

[1] http://www.ontologymatching.org

trate systems where the matching process is of a particular interest with the help of figures. We tried to adopt a unified presentation for these systems. However, some of these are specific enough so that we did not enforce the terminology of Sect. 2.4 but kept that one as used by system designers.

The structure of this chapter is as follows. We first describe systems which mostly focus on schema-level information (§6.1). Secondly, we discuss systems which concentrate on instance-level information (§6.2). Then, we present systems which exploit both schema-level and instance-level information (§6.3). Finally, we overview meta-matching systems (§6.4).

6.1 Schema-based systems

Schema-based systems, according to the classification of Chap. 3, are those which rely mostly on schema-level input information for performing ontology matching.

6.1.1 DELTA (The MITRE Corporation)

DELTA (Data Element Tool-based Analysis) is a system that semi-automatically discovers attribute correspondences among database schemas [Clifton et al., 1997]. It handles relational and extended entity–relationship (EER) schemas. The idea of the approach is to use textual similarities between data element definitions in order to find matching candidates. The system converts available information about an attribute, e.g., attribute name, datatype, narrative description, into a simple text string, called *document*. The documents describing each database attribute constitute a *document base*. Then, DELTA feeds the document base from the first schema into a full-text information retrieval tool, such as Personal Librarian. Matching is viewed as a Personal Librarian query based on the information from the second schema. The query can be a string of disconnected phrases, a full boolean query, a few relevant words, or an entire document. The tool estimates the similarity (by using natural language heuristics, such as considering that rare or repeated words are more important) between a search pattern and contents of a document base (§4.2.1). It is thus exclusively based on string-based techniques. It returns a ranked list of documents that it considers to be similar. The selection of the final alignment is to be performed by users.

6.1.2 Hovy (University of Southern California)

[Hovy, 1998] describes heuristics used to match large-scale ontologies, such as Sensus and Mikrokosmos, in order to combine them in a single reference ontology. In particular, three types of matchers were used based on (i) concept names, (ii) concept definitions, and (iii) taxonomy structure. For example, the name matcher splits composite-word names into separate words (§4.2.2) and then compares substrings in order to produce a similarity score. Specifically, the name matcher score is computed as the sum of the square of the number of letters matched, plus 20 points if words are exactly equal or 10 points if end of match coincides. For instance, using this strategy, the comparison between Free World and World results in 35 points, while the comparison between cuisine and vine results in 19 points. The definition matcher compares the English definitions of two concepts (§4.2.2). Here, both definitions are first split into individual words. Then, the number and the ratio of shared words in two definitions is computed in order determine the similarity between them. Finally, results of all the matchers are combined based on experimentally obtained formulas. The combined scores between concepts from two ontologies are sorted in descending order and are presented to users for establishing a cutoff value as well as for approving or discarding operations, results of which are saved for later reuse.

6.1.3 TransScm (Tel Aviv University)

TransScm [Milo and Zohar, 1998] provides data translation and conversion mechanisms between input schemas based on schema matching. First, by using rules, the alignment is produced in a semi-automatic way. Then, this alignment is used to translate data instances of the source schema to instances of the target schema. Input schemas are internally encoded as labelled graphs, where some of the nodes may be ordered. Nodes of the graph represent schema elements, while edges stand for the relations between schema elements or their components. Matching is performed between nodes of the graphs in a top-down and one-to-one fashion. Matchers are viewed as rules. For example, according to the *identical* rule, two nodes match if their labels are found to be synonyms based on the built-in thesaurus (§4.2.2); see [Zohar, 1997] for a list of the available rules. The system combines rules sequentially based on their priorities. It tries to find, for the source node, a unique *best* matching target node, or determines a mismatch. In case there are several matching candidates among which the system cannot choose the best one, or if the system cannot match or mismatch a source node to a target node with the given set of rules, user involvement is required. In particular, users with the help of a graphic user interface can add, disable or modify rules to obtain the desired matching result. Then, instances of the source schema are translated to instances of the target schema according to the match rules. For the example of the *identical* rule, translation includes copying the source node components.

6.1.4 DIKE (Università di Reggio Calabria and Università di Calabria)

DIKE (Database Intensional Knowledge Extractor) is a system supporting the semi-automatic construction of cooperative information systems (CISs) from heterogeneous databases [Palopoli *et al.*, 2003b, Palopoli *et al.*, 2003a, Palopoli *et al.*, 1998, Palopoli *et al.*, 2000]. It takes as input a set of databases belonging to the CIS. It builds a kind of mediated schema (called data repository or global structured dictionary) in order to provide a user-friendly integrated access to the available data sources. DIKE focuses on entity-relationship schemas. The matching step is called the extraction of inter-schema knowledge. It is performed in a semi-automatic way.

Some examples of inter-schema properties that DIKE can find are terminological properties, such as synonyms, homonyms among objects, namely entities and relationships, or type conflicts, e.g., similarities between different types of objects, such as entities, attributes, relationships; structural properties, such as object inclusion; subschema similarities, such as similarities between schema fragments. With each kind of property is associated a plausibility coefficient in the [0 1] range. The properties with a lower plausibility coefficient than a dynamically derived threshold are discarded, whereas others are accepted. DIKE works by computing sequentially the above mentioned properties. For example, synonyms and homonyms are determined based on information from external resources, such as WordNet (§4.2.2), and by analysing the distances of objects in the neighbourhood of the objects under consideration (§4.3.2). Some weights are also used to produce a final coefficient. Then, type conflicts are analysed and resolved by taking as input the results of synonyms and hyponyms analysis.

6.1.5 SKAT and ONION (Stanford University)

SKAT (Semantic Knowledge Articulation Tool) is a rule-based system that semi-automatically discovers mappings between two ontologies [Mitra et al., 1999]. Internally, input ontologies are encoded as graphs. Rules are provided by domain experts and are encoded in first order logic. In particular, experts specify initially desired matches and mismatches. For example, a rule President ≡ Chancellor, indicates that we want President to be an appropriate match for Chancellor. Apart from declarative rules, experts can specify matching procedures that can be used to generate the new matches. Experts have to approve or reject the automatically suggested matches, thereby producing the resulting alignment. Matching procedures are applied sequentially. Some examples of these procedures are: string-based matching, e.g., two terms match if they are spelled similarly (§4.2.1), and structure matching, e.g., structural graph slices matching, such as considering nodes near the root of the first ontology against nodes near the root of the second ontology (§4.3.2).

ONION (ONtology compositION) is a successor system to SKAT that semi-automatically discovers mappings between multiple ontologies, in order to enable a unified query answering over those ontologies [Mitra et al., 2000]. Input ontologies, RDF files, are internally represented as labelled graphs. The alignment is viewed as a set of *articulation rules*. The semi-automated algorithm for resolving the terminological heterogeneity of [Mitra and Wiederhold, 2002] forms the basis of the *articulation generator*, ArtGen, for the ONION system. ArtGen, in turn, can be viewed as an evolution of the SKAT system with some added matchers. Thus, it executes a set of matchers and suggests articulation rules to users. Users can either accept, modify or delete the suggestions. They can also indicate the new matches that the articulation generator may have missed. ArtGen works sequentially, first by performing linguistic matching (§4.2.2) and then structure-based matching (§4.3). During the linguistic matching phase, concept names are represented as sets of words. The linguistic matcher compares all possible pairs of words from any two concepts of both ontologies and assigns a similarity score in [0 1] to each pair. The matcher uses a word similarity table generated by a thesaurus-based or corpus-based matcher called the *word relator* to determine the similarity between pairs of words (§4.2.2). The similarity score between two concepts is the average of the similarity scores (ignoring scores of zero) of all possible pairs of words in their names. If this score is higher than a given threshold, ArtGen generates a match candidate. Structure-based matching is performed based on the results of the linguistic matching. It looks for structural isomorphism between subgraphs of the ontologies, taking into account some linguistic clues (see Sect. 6.1.11 for a similar technique). The structural matcher tries to match only the unmatched pairs from the linguistic matching, thereby complementing its results.

6.1.6 Artemis (Università di Milano and Università di Modena e Reggio Emilia)

Artemis (Analysis of Requirements: Tool Environment for Multiple Information Systems) [Castano et al., 2000] was designed as a module of the MOMIS mediator system [Bergamaschi et al., 1999, Bergamaschi et al., 1998] for creating global views. It performs affinity-based analysis and hierarchical clustering of database schema elements. Affinity-based analysis represents the matching step: in a sequential manner it calculates the name, structural and global affinity coefficients exploiting a common thesaurus. The common thesaurus is built with the help of ODB-Tools [Beneventano et al., 1998], WordNet (§4.2.2) or manual input. It represents a set of intensional and a set of extensional relationships which depict intra- and inter-schema knowledge about classes and attributes of the input schemas. Based on global affinity coefficients, a hierarchical clustering technique categorises classes into groups at different levels of affinity. For each cluster it creates a set of global attributes and the global class. Logical correspondence between the attributes of a global class and source schema attributes is determined through a mapping table.

6.1.7 H-Match (Università degli Studi di Milano)

H-Match [Castano et al., 2006] is an automated ontology matching system. It has been designed to enable knowledge discovery and sharing in the settings of open networked systems, in particular within the Helios peer-to-peer framework [Castano et al., 2005]. The system handles ontologies specified in OWL. Internally, these are encoded as graphs using the H-model representation [Castano et al., 2005]. H-Match inputs two ontologies and outputs (one-to-one or one-to-many) correspondences between concepts of these ontologies with the same or closest intended meaning. The approach is based on a similarity analysis through affinity metrics, e.g., term to term affinity, datatype compatibility (§4.3.1), and thresholds. H-Match computes two types of affinities (in the $[0\ 1]$ range), namely *linguistic* and *contextual* affinity. These are then combined by using weighting schemas, thus yielding a final measure, called *semantic* affinity. Linguistic affinity builds on top of a thesaurus-based approach of the Artemis system (§6.1.6). In particular, it extends the Artemis approach (i) by building a common thesaurus involving relations among WordNet synsets such as meronymy and coordinate terms, and (ii) by providing an automatic handler of compound terms, i.e., those composed by more than one token, that are not available from WordNet. Contextual affinity requires consideration of the neighbour concepts, e.g., linked via taxonomical or mereological relations, of the actual concept (§4.3.2).

One of the major characteristics of H-Match is that it can be dynamically configured for adaptation to a particular matching task, because in dynamic settings, the complexity of a matching task is not known in advance. This is achieved by means of four matching models. These are: *surface*, *shallow*, *deep*, and *intensive*, each of which involves different types of constructs of the ontology, see Fig. 6.1. Computation of a linguistic affinity is a common part of all the matching models. In case of the surface model, linguistic affinity is also the final affinity, since this model considers only names of ontology concepts. All the other three models take into account various contextual features and therefore contribute to the contextual affinity. For example, the shallow model takes into account concept properties, whereas the deep and the intensive models extend previous models by including relations and property values, respectively. Each concept involved in a matching task can be processed according to its own model, independently from the models applied to the other concepts within the same task. Finally, by applying thresholds, correspondences with semantic (final) affinity higher than the cut-off threshold value are returned in the final alignment.

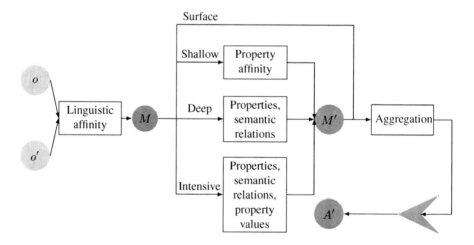

Fig. 6.1. H-Match matching process: H-Match is a conditional system that can use alternatively or in parallel four matching models depending on the resources available.

6.1.8 Tess (University of Massachusetts)

Tess (Type Evolution Software System) is a system to support schema evolution by matching the old and the new versions [Lerner, 2000]. Schemas are viewed as collections of types. It is assumed (since in the given application scenario changes are typically evolutionary, rather than revolutionary) that input schemas are highly similar. Matching is viewed as generation of *derivation rules* to be applied to data. Tess can operate in modes ranging from fully automated to completely manual. Each derivation rule is associated with a similarity metric, which is meant to measure the impact that applying the derivation rule would have on existing data. By defining a threshold for the similarity metric, the user involvement is determined. Matching is performed in three stages. First, the names of the types of old and new versions are compared (§4.2.1). Second, the structural information is taken into account. In particular, type constructors used by the old and new types and the types of components are analysed (§4.3.1). This provides the ability to handle cases in which, for example, component names have been changed, but their types are unchanged. Third, if everything else fails, matching relies upon some ordering information heuristics. Thus, in this case, Tess will try matching components with different names and different types. Finally, based on the derivation rules a transformer is produced which can update data in a database according to a newer version of the schema. In the simplest case, such as the identity derivation rule case, when type names are identical, as in Sect.6.1.3, the derivation function simply copies existing objects. A more complex transformation may include a join operation to combine two related objects into one.

6.1.9 Anchor-Prompt (Stanford Medical Informatics)

Anchor-Prompt [Noy and Musen, 2001] is an extension of Prompt, also formerly known as SMART. It is an ontology merging and alignment tool with a sophisticated prompt mechanism for possible matching terms [Noy and Musen, 1999]. Prompt handles ontologies expressed in such knowledge representation formalisms as OWL and RDF Schema. Anchor-Prompt is a sequential matching algorithm that takes as input two ontologies, internally represented as graphs and a set of anchors-pairs of related terms, which are identified with the help of string-based techniques, such as edit-distance (§4.2.1), , user-defined distance or another matcher computing linguistic similarity. Then the algorithm refines them by analysing the paths of the input ontologies limited by the anchors in order to determine terms frequently appearing in similar positions on similar paths (§4.3.2). Finally, based on the frequencies and user feedback, the algorithm determines matching candidates.

The Prompt and Anchor-Prompt systems have also contributed to the design of other algorithms, such as PromptDiff, which finds differences between two ontologies and provides the editing facility for transforming one ontology into another (see Sect.8.3.2 and [Noy and Musen, 2002b, Noy and Musen, 2003]).

6.1.10 OntoBuilder (Technion Israel Institute of Technology)

OntoBuilder is a system for information seeking on the web [Modica *et al.*, 2001]. A typical situation the system deals with is when users are seeking for a car to be rented. Obviously, they would like to compare prices from multiple providers in order to make an informed decision. OntoBuilder operates in two phases: (i) ontology creation (the *training* phase) and (*ii*) ontology adaptation (the *adaptation* phase). During the training phase an initial ontology (in which user data needs are encoded) is created by extracting it from a visited web site of, e.g., some car rental company. The adaptation phase includes on-the-fly match and interactive merge operations of the related ontologies with the actual (initial) ontology. We concentrate below only on the ontology adaptation phase. During the adaptation phase users suggest the web sites they would like to further explore, e.g., the ones of various car rental companies. Each such site goes through the ontology extraction process. This results in a candidate ontology, which is then merged into the actual ontology. To support this, the best match for each existing term in the actual ontology to terms from the candidate ontology is selected. The selection strategy employs thresholds (§5.7.1). The matching algorithm works in a term to term fashion, sequentially executing various matchers. Some examples of the matchers used are: removing noisy characters and stop terms (§4.2.2), substring matching (§4.2.1). If all else fail, thesaurus look-up is performed (§4.2.2). Finally, mismatched terms are presented to users for manual matching. Some further matchers, such as those for precedence matching, were introduced in later work in [Gal et al., 2005b]. Top-k mappings have been proposed in [Gal, 2006] as an alternative for a single best matching, i.e., top-1 category.

6.1.11 Cupid (University of Washington, Microsoft Corporation and University of Leipzig)

Cupid [Madhavan et al., 2001] implements an algorithm comprising linguistic and structural schema matching techniques, and computing similarity coefficients with the assistance of domain specific thesauri. Input schemas are encoded as graphs. Nodes represent schema elements and are traversed in a combined bottom-up and top-down manner. The matching algorithm consists of three phases (see Fig. 6.2) and operates only with tree-structures, to which non tree cases are reduced.

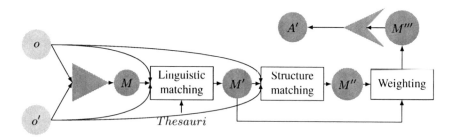

Fig. 6.2. Cupid architecture: it is a very common architecture which mixes parallel and sequential composition. Structure matching takes advantage of the results of linguistic matching, but the results of both of them are taken into consideration for weighting.

The first phase (linguistic matching) computes linguistic similarity coefficients between schema element names (labels) based on morphological normalisation

(§4.2.2), categorisation, string-based techniques, such as common prefix, suffix tests (§4.2.1), and thesauri look-up (§4.2.2). The second phase (structural matching) computes structural similarity coefficients weighted by leaves which measure the similarity between contexts in which elementary schema elements occur (§4.3.2). The third phase (mapping elements generation) aggregates the results of the linguistic and structural matching through a weighted sum (§5.2.2) and generates a final alignment by choosing pairs of schema elements with weighted similarity coefficients, which are higher than a threshold (§5.7.1).

6.1.12 COMA and COMA++ (University of Leipzig)

COMA (COmbination of MAtching algorithms) [Do and Rahm, 2002] is a schema matching tool based on parallel composition of matchers. It provides an extensible library of matching algorithms, a framework for combining obtained results, and a platform for the evaluation of the effectiveness of the different matchers. As from [Do and Rahm, 2002], COMA contains 6 elementary matchers, 5 hybrid matchers, and one reuse-oriented matcher. Most of them implement string-based techniques, such as affix, n-gram, edit distance (§4.2.1); others share techniques with Cupid, e.g., thesauri look-up. An original component, called *reuse-oriented matcher*, tries to reuse previously obtained results for entire new schemas or for their fragments. Schemas are internally encoded as directed acyclic graphs, where elements are the paths. This aims at capturing contexts in which the elements occur. Distinct features of the COMA tool with respect to Cupid are a more flexible architecture and the possibility of performing iterations in the matching process. It presumes interaction with users who approve obtained matches and mismatches to gradually refine and improve the accuracy of match. COMA++ is built on top of COMA by elaborating in more detail the alignment reuse operation. Also it provides a more efficient implementation of the COMA algorithms and a graphical user interface (§8.2.2).

6.1.13 Similarity flooding (Stanford University and University of Leipzig)

The Similarity flooding approach [Melnik *et al.*, 2002] is based on the ideas of similarity propagation. Schemas are presented as directed labelled graphs grounded on the OIM specification [MDC, 1999]. The algorithm manipulates them in an iterative computation to produce an alignment between the nodes of the input graphs. The technique starts from string-based comparison, such as common prefix, suffix tests (§4.2.1), of the vertices labels to obtain an initial alignment which is refined through iterative computation. The basic concept behind the Similarity flooding algorithm is the similarity spreading from similar nodes to the adjacent neighbours through propagation coefficients. From iteration to iteration the similarity measure is spread to the graph until a fixed point is reached or the computation is stopped. The full process is described in Sect. 5.3.1. The result of this step is a refined alignment which is further filtered to produce the final alignment.

6.1.14 XClust (National University of Singapore)

XClust is a tool for integrating multiple DTDs [Lee et al., 2002]. Its integration strategy is based on clustering. Given multiple DTDs, it clusters them according to their similarity. This aims at facilitating the work of system integrators by allowing them to focus on already similar DTDs of single clusters. Clustering is applied recursively until a manageable number of DTDs is obtained. XClust works in two phases: (i) DTD similarity computation, and (ii) DTD clustering. During the first phase, given a set of DTDs, pairwise similarities between their underlying labelled trees are computed. This is done by using several matchers which exploit schema names as well as some structural information. For example, the *basic similarity* is computed as a weighted sum of a WordNet-based matcher that looks for synonyms among names of schema elements (§4.2.2) and a cardinality constraint matcher that performs a look up in cardinality compatibility table in order to compare cardinalities of schema elements (§4.3.1). Structural similarities exploit previously computed basic similarities and are based on (i) similarity of paths, (ii) similarity of immediate descendants and (iii) similarity of leaves (§4.3.2). For example, similarity of paths is computed as a normalised sum of basic similarities between the sets of elements these paths are composed of, namely elements from the root to the node under consideration (§4.2.1). Structural similarities are aggregated as a weighted sum and then these aggregated similarities are used to choose the best match pairs by applying a threshold. These constitute the alignment for a pair of DTDs. Finally, for two DTDs, best match pairs are summed up and normalised, thereby resulting in a final similarity between these DTDs. The result of the first phase is the similarity matrix of a set of DTDs. During the second phase, based on the DTD similarity matrix, a hierarchical clustering [Everitt, 1993] is applied to group DTDs into clusters.

6.1.15 ToMAS (University of Toronto and IBM Almaden)

ToMAS (Toronto Mapping Adaptation System) is a system that automatically detects and adapts mappings that have become invalid or inconsistent when schemas evolve [Velegrakis et al., 2003, Velegrakis et al., 2004b, Velegrakis et al., 2004a]. It is assumed that (i) the matching step has already been performed, and (ii) correspondences have already been made operational, e.g., by using the Clio system (§6.3.2). Since in open environments, such as the web, schemas can evolve without prior notice, some correspondences may become invalid. This system aims at handling such cases, thereby preserving mapping consistency. In this sense, ToMAS complements the systems dealing with the problems mentioned in items (i) and (ii) above. In particular, it detects mappings affected by structural or constraint changes and it generates automatically the necessary rewritings that are consistent with the updates that have occurred. ToMAS handles relational and XML schemas. It takes two schemas and a set of mappings between them as input. The system works in two phases. First, as a preprocessing step, mappings are analysed and turned into logically valid mappings (if they are not already). During the second step, the result of the previous step is maintained through schema changes. In particular, mappings are modified one by one independently, as appropriate for each kind of change that may occur to the schemas. Three classes of (primitive) schema changes are addressed: (i) operations that change the schema semantics by adding or removing constraints, (ii) modifications to the schema structure by adding or removing elements, and (iii) modifications that reshape schema structure by moving, copying, or renaming elements. The final result of ToMAS is a set of *adapted* mappings which are consistent with the structure and semantics of the evolved schemas.

6.1.16 MapOnto (University of Toronto and Rutgers University)

MapOnto is a system for constructing complex mappings between ontologies and relational or XML schemas [An et al., 2005a, An et al., 2005b, An et al., 2006]. This system operates in a similar settings as the Clio tool (§6.3.2). In a sense, this work can be viewed as an extension of Clio when the target schema is an ontology which is treated as a relational schema consisting of unary and binary tables. MapOnto takes as input three arguments: (i) an ontology specified in an ontology representation language, e.g., OWL, (ii) relational or XML schema, and (iii) simple correspondences, e.g., between XML attributes and ontology datatype properties. Input schema and ontology are internally encoded as labelled graphs. Then, the approach looks for 'reasonable' connections among the graphs. The system produces in a semi-automatic way a set of complex mapping formulas expressed in a subset of first-order logic (Horn clauses). The list of logical formulas is also ordered by the tool, thereby suggesting the most reasonable mappings. Finally, users can inspect that list and choose the best ones.

6.1.17 OntoMerge (Yale University and University of Oregon)

OntoMerge [Dou et al., 2005] is a system for ontology translation on the semantic web. Ontology translation refers here to such tasks as (i) dataset translation, that is translating a set of facts expressed in one ontology to another; (ii) generating ontology extensions, that is given two ontologies o and o' and an extension (subontology) o_s of the first one, build the corresponding extension o'_s, and (iii) query answering from multiple ontologies. The main idea of the approach is to perform ontology translation by ontology merging and automated reasoning. Input ontologies are translated from a source knowledge representation formalism, e.g., OWL, to an internal representation, which is Web-PDDL [McDermott and Dou, 2002]. Merging two ontologies is performed by taking the union of the axioms defining them. Bridge axioms or bridge rules are then added to relate the terms in one ontology to the terms in the other. Once the merged ontology is constructed, the ontology translation tasks can be performed fully automatically by mechanised reasoning. In particular, inferences, depending on the task, are conducted either in a demand-driven (backward-chaining) or data-driven (forward chaining) way with the help of a first-order theorem prover, called *OntoEngine*. It is assumed that bridge rules are to be provided by domain experts, or by other matching algorithms, which are able to discover and interpret them with clear semantics. Finally, it is worth noting that OntoMerge supports bridge rules which can be expressed using the full power of predicate calculus.

6.1.18 CtxMatch and CtxMatch2 (University of Trento and ITC-IRST)

CtxMatch [Bouquet et al., 2003c, Bouquet et al., 2003b] uses a semantic matching approach (§4.5.2). It translates the ontology matching problem into the logical validity problem and computes logical relations, such as equivalence, subsumption between concepts and properties. CtxMatch is a sequential system. At the element level it uses only WordNet to find initial matches for classes (§4.2.2). CtxMatch2 [Bouquet et al., 2006] improves on CtxMatch by handling properties. At the structure level, it exploits description logic reasoners, such as Pellet [Sirin et al., 2007] or FaCT [Tsarkov and Horrocks, 2006] to compute the final alignment in a way similar to what is presented in Sect. 4.5.2.

6.1.19 S-Match (University of Trento)

S-Match implements the idea of semantic matching as initially described in [Giunchiglia and Shvaiko, 2003a]. The first version of the S-Match system was a rationalised re-implementation of CtxMatch with a few added functionalities [Giunchiglia et al., 2004]. Later the system has undergone several evolutions, including extensions of libraries of element- and structure-level matchers, adding alignment explanations as well as iterative semantic matching [Giunchiglia and Yatskevich, 2004, Shvaiko et al., 2005, Giunchiglia et al., 2005a, Giunchiglia et al., 2006c, Giunchiglia et al., 2007]. S-Match is limited to tree-like structures and does not consider properties or roles.

S-Match takes as input two graph-like structures, e.g., classifications, XML schemas, ontologies, and returns as output logic relations, e.g., equivalence, subsumption, which are supposed to hold between the nodes of the graphs. The relations are determined by (i) expressing the entities of the ontologies as logical formulas, and (ii) reducing the matching problem to a propositional validity problem. In particular, the entities are translated into propositional formulas which explicitly express the concept descriptions as encoded in the ontology structure and in external resources, such as WordNet. This allows for a translation of the matching problem into a propositional validity problem, which can then be efficiently resolved using (sound and complete) state of the art propositional satisfiability solvers.

S-Match was designed and developed as a platform for semantic matching, namely a modular system with the core of computing semantic relations where single components can be plugged, unplugged or suitably customised. It is a sequential system with a parallel composition at the element level, see Fig. 6.3. The input ontologies (tree-like structures) are codified in a standard internal XML format. The module taking input ontologies performs some preprocessing with the help of oracles which provide the necessary a priori lexical and domain knowledge. Examples of oracles include WordNet (§4.2.2) and UMLS[2]. The output of the module is an enriched tree. These enriched trees are stored in an internal database (PTrees) where they can be browsed, edited and manipulated. The Match manager coordinates the matching

---

[2] http://www.nlm.nih.gov/research/umls/

process. S-Match libraries contain around 20 basic element-level matchers representing three categories, namely string-based, such as n-gram, edit distance (§4.2.1), WordNet sense-based and WordNet gloss-based matchers (§4.2.2). Structure-level matchers include SAT solvers, e.g., those of SAT4J[3], and ad hoc reasoning methods [Giunchiglia et al., 2005b].

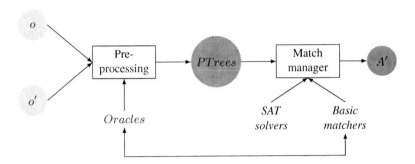

Fig. 6.3. S-Match architecture: ontology entities are converted to logic formulas by using the preprocessor and oracles. The Match manager then uses various basic element-level matchers and logic provers for finding relations between these formulas which, in turn, correspond to relations between the entities.

6.1.20 HCONE (University of the Aegean)

HCONE is an approach to domain ontology matching and merging by exploiting different levels of interaction with users [Kotis et al., 2006, Vouros and Kotis, 2005, Kotis and Vouros, 2004]. First, an alignment between two input ontologies is computed with the help of WordNet (§4.2.2). Then, the alignment is processed straightforwardly by using some merging rules, e.g., renaming, into a new merged ontology. The HCONE basic matching algorithm works in six steps:

1. Chose a concept from one ontology, denoted as c.
2. Obtain all the WordNet senses of c, denoted as $s_1, s_2, \ldots, s_m$. For example, the concept Facility has five senses in WordNet.
3. Obtain hypernyms and hyponyms of all the senses of c (§4.2.2). For example, Police is a hyponym of Facility.
4. Build the $n \times m$ *association matrix*. This relates the n most frequently occurring terms in the *vicinity* of the m senses determined in step 2. The vicinity terms include those from the same synsets of m senses, hypernyms and hyponyms from step 3. In the case of the Facility example this is a 93×5 matrix. For example, the number of occurrences of such a vicinity term as Police is 3.

---

[3] http://www.sat4j.org/

5. Build a query q by using terms which are subconcepts of c, e.g., TransportationSystem, or which are related to c via domain specific relations in the input ontology. If the terms considered for q also exist among the n terms from step 4, then q memorises that position with the help of flags. Thus, for the Facility concept, q is a 93 position vector, and, since the position of TransportationSystem is at the 35th place the value of $q[35]$ is 1.
6. Taking as input the association matrix computed at step 4 and the query computed at step 5, Latent Semantic Indexing (§4.2.2) is used to compute the grades for what is the correct WordNet sense to be used for the given context (query).

The highest graded sense expresses the most plausible meaning for the concept under consideration. Finally, the relationship between concepts is computed. For instance, equivalence between two concepts holds if the same WordNet sense has been chosen for those concepts based on the procedure described above. The subsumption relation is computed between two concepts if a hypernym relation holds between the WordNet senses corresponding to these concepts. Based on the level at which users are involved in the matching process, HCONE provides three algorithms to ontology matching. These are: fully automated, semi-automated and user-based. Users are involved in order to provide feedback on what is to be the correct WordNet sense on a one by one basis (user-based), or only in some limited number of cases, by exploiting some heuristics (semi-automated).

6.1.21 MoA (Electronics and Telecomunication Research Institute, ETRI)

MoA is an ontology merging and alignment tool [Kim et al., 2005]. It consists of: (i) a library of methods for importing, matching, modifying, merging ontologies, and (ii) a shell for using those methods. MoA handles ontologies specified in OWL-DL. It is able to compute equivalence and subsumption relations between entities (classes, properties) of the input ontologies. The matching approach is based on concept (dis)similarity derived from linguistic clues. The MoA tool is a sequential solution. The preprocessing step includes three phases: (i) names of classes and properties are tokenised (§4.2.1); (ii) tokens of entities are associated with their meaning by using WordNet senses; (iii) meanings of tokens of ancestors of the entity under consideration are also taken into account, thereby extending the local meanings. This step is essentially the same as some part of the preprocessing done within the S-Match system (§6.1.19). Matching itself is based on rules. It is performed in a double loop over all the pairs of entities from two input ontologies. For example, equivalence between two classes or properties holds when there is equivalence between these entities in either step (ii) or (iii). The equivalence, in turn, is decided via relations between the WordNet senses for one of the possible solutions (see Sect. 4.2.2). Thus, for example, author can be found to be equivalent to writer because they belong to the same synset in WordNet.

6.1.22 ASCO (INRIA Sophia-Antipolis)

ASCO is a system that automatically discovers pairs of corresponding elements in two input ontologies [Bach et al., 2004]. ASCO handles ontologies in RDF Schema and computes alignments between classes, relations, and classes and relations. A new version, ASCO2, deals with OWL ontologies [Bach and Dieng-Kuntz, 2005].

The matching is organised sequentially in three phases. During the first phase (linguistic matching) the system normalises terms and expressions, e.g., by punctuation, upper cases, special symbols. Depending on their use in the ontology or if they are bags of words, ASCO uses different string comparison metrics for comparing the terms. Single terms are compared by using Jaro–Winkler, Levenshtein or Monge–Elkan (§4.2.1) and external resources, such as WordNet. Based on token similarities, the similarity between sets of tokens is computed using TFIDF. The obtained values are aggregated through a weighted sum.

The second phase (structure matching), computes similarities between classes and relations by propagating the input of linguistic similarities. The algorithms is an iterative fixed point computation algorithm that propagates similarity to the neighbours (subclasses, superclasses and siblings). Similarities between sets of objects are computed through single linkage. The propagation terminates when the class similarities and the relation similarities do not change after an iteration or a certain number of iterations is reached.

In the third phase, the linguistic and structural similarity are aggregated through a weighted sum and, if the similarities between matching candidates exceed a threshold (§5.7.1), they are selected for the resulting alignment.

6.1.23 BayesOWL and BN mapping (University of Maryland)

BayesOWL is a probabilistic framework for modelling uncertainty in the semantic web. It includes the Bayesian Network mapping module (§5.5.1), which is in charge of automatic ontology matching [Pan et al., 2005]. The approach works in three steps. First, two input ontologies are translated into two Bayesian networks. Specifically, classes are translated into nodes in Bayesian network, while edges are created if the corresponding two classes are related by a *predicate* in the input ontologies. During the second step, matching candidates are generated between two Bayesian networks by learning joint probabilities from the web data. In particular, for each concept in an ontology, a group of sample text documents (called *exemplars*) is created by querying a search engine. The query contains all the terms, e.g., {product book science} (opposed to a single term, e.g., {science}), in the path from the root to the concept (term) under consideration in the given ontology, thereby enabling some word sense disambiguation (§4.2.2). A text classifier, e.g., Rainbow[4], is trained on the statistical information about exemplars from the first ontology. Then, concepts of the second ontology are classified with respect to the concepts of the first ontology by feeding their exemplars to the trained classifier. A similarity between two concepts is determined with the help of the Jaccard coefficient

---

[4] http://www.cs.umass.edu/~mccallum/bow/rainbow/

(§4.4) computed from the joint probabilities. These are used to construct the conditional probability tables. During the third step, the mappings are refined as an update (combination of the Jeffrey rule and Iterative Proportional Fitting Procedure [Jeffrey, 1983, Cramer, 2000]) on probability distributions of concepts in the second Bayesian network, by distributions of concepts in the first Bayesian network, in accordance with the given similarities. By performing Bayesian inference with the updated distribution of the second Bayesian network, the final alignment is produced.

6.1.24 OMEN (The Pennsylvania State University and Stanford University)

OMEN (Ontology Mapping ENhancer [Mitra *et al.*, 2005]) is a semi-automatic probabilistic ontology matching system based on a Bayesian network (§5.5.1). It takes as input two ontologies and an initial probability distribution derived, for instance, from basic (element level) linguistic matchers. In turn, OMEN provides a structure level matching algorithm, thereby deriving the new mappings or discarding the existing false mappings. The approach can be summarised in four logical steps. First, it creates a Bayesian network, where a node stands for a mapping between pairs of classes or properties of the input ontologies. Edges represent the influences between the nodes of the network. This encoding is different from the one described in Sect. 6.1.23. During the second step, OMEN uses a set of *meta-rules* that capture the influence of the structure of input ontologies in the neighbourhood of the input mappings in order to generate conditional probability tables for the given network. An example of a basic meta-rule is as follows. There are two conditions: (i) if the i-th concept from the first ontology, $c_{1,i} \in o_1$, matches the j-th concept from the second ontology, $c_{2,j} \in o_2$; (ii) if there is a relation q between concepts $c_{1,i}$ and $c_{1,k}$ in the first ontology, which matches a relation q' between concepts $c_{2,j}$ and $c_{2,m}$ in the second ontology. Then we can increase the probability of match between concepts $c_{1,k}$ and $c_{2,m}$. Other rules rely more heavily on the semantics of the language in which the input ontologies are encoded. During the third step, inferences are made (OMEN uses Bayesian Network tools in Java (BNJ)[5]) to generate *a posteriori* probabilities for each node. Finally, *a posteriori* probabilities, which are higher than a threshold (§5.7.1), are selected to generate the resulting alignment.

6.1.25 DCM framework (University of Illinois at Urbana-Champaign)

MetaQuerier [Chang *et al.*, 2005] is a middleware system that assists users in finding and quering multiple databases on the web. It exploits the Dual Correlation Mining (DCM) matching framework to facilitate source selection according to user search keywords [He and Chang, 2006]. Unlike other works, the given approach takes as input multiple schemas and returns alignments between all of them. This setting is called *holistic* schema matching. DCM automatically discovers complex correspondences, e.g., {author} corresponds to {first name, last name}, between attributes of the web query interfaces in the same domain of interest, e.g., books. As the name

---

[5] http://bnj.sourceforge.net

DCM indicates, schema matching is viewed as *correlation mining*. The idea is that co-occurrence patterns often suggest complex matches. That is, *grouping attributes*, such as first name and last name, tend to co-occur in query interfaces. Technically, this means that those attributes are positively correlated. Contrary, attribute names which are synonyms, e.g., quantity and amount, rarely co-occur, thus representing an example of negative correlation between them. Matching is performed in two phases. During the first phase (matching discovery), a set of matching candidates is generated by mining first positive and then negative correlations among attributes and attribute groups. Some thresholds and a specific correlation measure such as the H-measure are also used. During the second phase (matching construction), by applying ranking strategies, e.g., scoring function, rules, and selection, such as iterative greedy selection (§5.7.3), the final alignment is produced.

6.2 Instance-based systems

Instance-based systems are those taking advantage mostly of instances, i.e., of data expressed with regard to the ontology or data indexed by the ontology.

6.2.1 T-tree (INRIA Rhône-Alpes)

T-tree [Euzenat, 1994] is an environment for generating taxonomies and classes from objects (instances). It can, in particular, infer dependencies between classes, called bridges, of different ontologies sharing the same set of instances based only on the extension of classes (§4.4.1). The system, given a set of source taxonomies called viewpoints, and a destination viewpoint, returns all the bridges in a minimal fashion which are satisfied by the available data. That is the set of bridges for which the objects in every source class are indeed in the destination class. The algorithm compares the extension (set of instances) of the presumed destination to the intersection of those of the presumed source classes. If there is no inclusion of the latter in the former, the algorithm is re-iterated on all the sets of source classes which contain at least one class which is a subclass of the tested source classes. If the intersection of the extension of the presumed source classes is included in that of the presumed destination class, a bridge can be established from the latter (and also from any set of subclasses of the source classes) to the former (and also any superclass of the destination class). However, other bridges can also exist on the subclasses of the destination. The algorithm is thus re-iterated on them. It stops when the bridge is trivial, i.e., when the source is empty. Users validate the inferred bridges.

Bridge inference is the search for correlation between two sets of variables. This correlation is particular to a data analysis point of view since it does not need to be valid on the whole set of individuals (the algorithm looks for subsets under which the correlation is valid) and it is based on strict set equality (not similarity). However, even if the bridge inference algorithm has been described with set inclusion, it can be helped by other measurements which will narrow or broaden the search. More generally, the inclusion and emptiness tests can be replaced by tests based on the similarity of two sets of objects (as is usual in data analysis).

The bridge inference algorithm is not dependent on the instance-based interpretation: it depends on the meaning of the operators $\subseteq$, $\cap$ and $= \emptyset$-test (which are interpreted as their set-theoretic counterpart in the case of the instance-based algorithms). A second version of the system (with the same properties) uses structural comparison: $\subseteq$ is subtyping, $\cap$ is type intersection and $= \emptyset$-test is a subtyping test.

6.2.2 CAIMAN (Technische Universität München and Universität Kaiserslautern)

CAIMAN [Lacher and Groh, 2001] is a system for document exchange, which facilitates retrieval and publishing services among the communities of interest. These services are enabled by using semi-automatic ontology matching. The approach focuses on light-weight ontologies, such as web classifications. The main idea behind matching is to calculate a probability measure between the concepts of two ontologies, by applying machine learning techniques for text classification, e.g., the Rocchio classifier. In particular, based on the documents, a representative feature vector (a word-count, weighted by TFIDF feature vector, §4.2.1) is created for each concept in an ontology. Then, the cosine measure (§4.2.1) is computed for two of those class vectors. Finally, with the help of a threshold, the resulting alignment is produced.

6.2.3 FCA-merge (University of Karlsruhe)

FCA-merge uses formal concept analysis techniques to merge two ontologies sharing the same set of instances [Stumme and Mädche, 2001]. The overall process of merging two ontologies consists of three steps, namely (i) instance extraction, (ii) concept lattice computation, (iii) interactive generation of the final merged ontology. The approach provides, as a first step, methods for extracting instances of classes from documents. The extraction of instances from text documents circumvents the problem that in most applications there are no individuals which are simultaneously instances of the source ontologies and which could be used as a basis for identifying similar concepts. During the second step, the system uses formal concept analysis techniques (§4.4.1) in order to compute the concept lattice involving both ontologies. The last step consists of deriving the merged ontology from the concept lattice. The produced lattice is explored and transformed by users who further simplify it and generate the taxonomy of an ontology.

The result is a merge rather than an alignment. However, the concepts that are merged can be considered as exactly matched and those which are not can be considered in subsumption relation with their ancestors or siblings.

6.2.4 LSD (University of Washington)

Learning Source Descriptions (LSD) is a system for the semi-automatic discovery of one-to-one alignments between the (leaf) elements of source schemas and a mediated (global) schema in data integration [Doan et al., 2001]. The main idea behind the approach is to learn from the mappings created manually between the mediated schema and some of the source schemas, in order to propose in an automatic manner the mappings for the subsequent source schemas. LSD handles XML schemas. A schema is modelled as a tree, where the nodes are XML tag names. The approach works in two phases. During the first (training) phase, useful objects, such as element names and data values, are extracted from the input schemas. Then, from these objects and manually created alignments, the system trains multiple basic matchers (addressing different features of objects, such as formats, word frequencies, characteristics of value distributions) and a meta-matcher. Some examples of basic matchers are the WHIRL learner (§5.4.2), the naive Bayesian learner (§5.4.1). The meta-matcher combines the predictions of basic matchers. It is trained by using a stacked generalisation (learning) technique (§5.4.5). During the second (matching) phase LSD extracts the necessary objects from the remaining (new) source schemas. Then, by applying the trained basic matchers and the meta-matcher on the new objects (the *classification* operation), LSD obtains a prediction list of matching candidates. Finally, by taking into account integrity constraints and applying some thresholds, the final alignment is extracted.

6.2.5 GLUE (University of Washington)

GLUE [Doan et al., 2004], a successor of LSD, is a system that employs multiple machine learning techniques to semi-automatically discover one-to-one semantic mappings (which are sometimes called 'glue' for interoperability) between two taxonomies. The idea of the approach is to calculate the *joint distributions* of the classes, instead of committing to a particular definition of similarity. Thus, any particular similarity measure can be computed as a function over the joint distributions. As does its predecessor, LSD, GLUE follows a multistrategy learning approach, involving several basic matchers and a meta-matcher. The system works in three steps. First, it learns the joint probability distributions of classes of two taxonomies. In particular, it exploits two basic matchers: the *content learner* (naive Bayes technique, §5.4.1) and the *name learner* (a variation of the previous one). The meta-matcher, in turn, performs a linear combination of the basic matchers. Weights for these matchers are assigned manually. During the second step, the system estimates the similarity between two classes in a user-supplied function of their joint probability distributions. This results in a similarity matrix between terms of two taxonomies. Finally, some domain-dependent, e.g., subsumption, and domain-independent, e.g., if all children of node x match node y, then x also matches y, constraints (heuristics) are applied by using a *relaxation labelling* technique. These are used in order to filter some of the matches out of the similarity matrix and keep only the best ones.

6.2.6 iMAP (University of Illinois and University of Washington)

iMAP [Dhamankar et al., 2004] is a system that semi-automatically discovers one-to-one (e.g., amount ≡ quantity) and, most importantly, complex (e.g., address ≡

concat(city, street)) mappings between relational database schemas. The schema matching problem is reformulated as a search in a match space, which is often, very large or even infinite. To perform the search effectively, iMAP uses multiple basic matchers, called searches, e.g., text, numeric, category, unit conversion, each of which addresses a particular subset of the match space. For example, the text searcher considers the concatenation of text attributes, while the numeric searcher considers combining attributes with arithmetic expressions. The system works in three steps (see Fig. 6.4). First, matching candidates are generated by applying basic matchers (the match generator module). Even if a basic matcher, such as the text searcher, addresses only the space of concatenations, this space can still be very large. To this end, the search strategy is controlled by using the *beam search* technique [Russell and Norvig, 1995]. During the second step, for each target attribute, matching candidates of the source schema are evaluated by exploiting additional types of information, e.g., using the naive Bayes evaluator (§5.4.1), which would be computationally expensive to use during the first step. These yield additional scores. Then, all the scores are combined into a final one (the similarity estimator module). The result of this step is a similarity matrix between ⟨*target attribute, match candidate*⟩ pairs. Finally, by using a set of domain constraints and mappings from the previous match operations (if applicable and available), the similarity matrix is cleaned up, such that only the best matches for target attributes are returned as the result (the match selector module). The system is also able to explain the results it produces with the help of the explanation module, see for details Chap. 9.

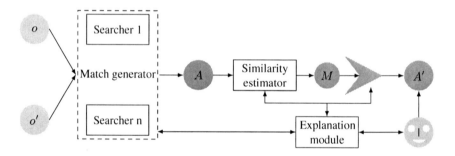

Fig. 6.4. iMAP architecture: several matchers, called searchers, are run in parallel. They provide candidate matches that can be complex. These candidates are further selected by applying the similarity estimator, and then, the final alignment is extracted. Additionally, the explanation module allows users to understand the results and control the process.

6.2.7 Automatch (George Mason University)

Automatch [Berlin and Motro, 2002] is a system for automatic discovery of mappings between the attributes of database schemas. The approach assumes that several schemas from the domain under consideration have already been manually matched by domain experts. This assumption is a realistic one for a data integration scenario. Then, by using Bayesian learning (§5.4.1), Automatch acquires probabilistic knowledge from the manually matched schemas, and creates the *attribute dictionary* which accumulates the knowledge about each attribute by means of its *possible values* and the *probability estimates* of these values. In order to avoid a rapid growth of the dictionary, the system also uses *statistical feature selection* techniques, such as mutual information, information gain, and likelihood ratio, to learn efficiently, i.e., only from the most informative values, such as 10% of the actually available input training data. A new pair of schemas is matched automatically via the precompiled attribute dictionary. The system first matches each attribute of the input schemas against the attribute dictionary, thereby producing individual match scores (a real number). Then, these individual scores are further combined by taking their sum to produce the scores between the attributes of the input schemas. Finally, the scores between the input schemas, in turn, are combined again, by using a *minimum cost maximum flow* graph algorithm and some thresholds in order to find the overall optimal matching between the input schemas with respect to the sum of the individual match scores.

6.2.8 SBI&NB (The Graduate University for Advanced Studies)

SBI (Similarity-Based Integration) is a system for automatic statistical matching among classifications [Ichise *et al.*, 2003, Ichise *et al.*, 2004]. SBI&NB is the extension of SBI by plugging into the system a naive Bayes classifier (§5.4.1). The idea of SBI is to determine correspondences between classes of two classifications by statistically comparing the membership of the documents to these classes. The pairs of similar classes are determined in a top-down fashion by using the κ-statistic method [Fleiss, 1973]. These pairs are considered to be the final alignment. SBI&NB combines sequentially the SBI and the naive Bayes classifier. The naive Bayes enables hierarchical classification of documents. Thus, the system takes also into account structural information of the input classifications. The exploited classifier is Pachinko Machine naive Bayes from the Rainbow system[4].

6.2.9 Kang and Naughton (University of Wisconsin-Madison)

Kang and Naughton proposed a structural instance-based approach for discovering correspondences among attributes of relational schemas with *opaque* column names [Kang and Naughton, 2003]. By opaque column names are meant names which are hard to interpret, such as A and B instead of Model and Color. The approach works in two phases. During the first phase, two table instances are taken as input and the corresponding (weighted) dependency graphs are constructed based on *mutual information* and *entropy*. The conditional entropy used here describes (with a non negative real number) the uncertainty of values in an attribute given knowledge (probability distribution) of another attribute. Mutual information, in turn, measures (with a non negative real number) the reduction in uncertainty of one attribute due to the knowledge of the other attribute, i.e., the amount of information captured in one attribute about the other. It is zero when two attributes are independent, and increases as the dependency between the two attributes grows. Mutual information is computed over all pairs of attributes in a table. Thus, in dependency graphs, a weight on an edge stands for mutual information between two adjacent attributes. A weight on a node stands for entropy of the attribute. During the second phase, matching node pairs are discovered between the dependency graphs by running a graph matching algorithm. The quality of matching is assessed by using metrics, such as the Euclidean distance (§4.2.1). The distance is assigned to each potential correspondence between attributes of two schemas and a one-to-one alignment which is a minimum weighted graph matching (§5.7.3) is returned.

6.2.10 Dumas (Technische Universität Berlin and Humboldt-Universität zu Berlin)

Dumas (DUplicate-based MAtching of Schemas) is an approach which identifies one-to-one alignments between attributes by analysing the duplicates in data instances of the relational schemas [Bilke and Naumann, 2005]. Unlike other instance-based approaches which look for similar properties of instances, such as distribution of characters, in columns of schemas under consideration, this approach looks for similar rows or tuples. The system works in two phases: (i) identify objects within databases with opaque schemas, and (ii) derive correspondences from a set of similar duplicates.

For object identification (§4.4.2), in Dumas, tuples are viewed as strings and a string comparison metric, such as cosine measure (§4.2.1), is used to compare two tuples. Specifically, tuples are tokenised and each token is assigned a weight based on TFIDF scheme (§4.2.1). In order to avoid complete pairwise comparison of tuples from two databases, the WHIRL algorithm (§5.4.2) is used. It performs a focused search based on those common values that have high TFIDF score. The algorithm ranks tuple pairs according to their similarity and identifies the k most similar tuple pairs.

During the second phase, based on the k duplicate pairs with highest confidence, correspondences between attributes are derived. The intuition is that if two field values are similar, then their respective attributes match. A field-wise similarity comparison is made for each of the k duplicates, thereby resulting in a similarity matrix. For comparing tuple fields, a variation of a TFIDF-based measure, called soft TFIDF [Cohen et al., 2003a], is used. It allows the consideration of similar terms as opposed to equal terms. The resulting alignment is extracted from the similarity matrix by finding the maximum weight matching. Finally, if based on the maximum matching, multiple alternative matches are possible, therefore the algorithm iterates back to the first phase in order to try to improve the result by discovering more duplicates.

6.2.11 Wang and colleagues (Hong Kong University of Science and Technology and Microsoft Research Asia)

Wang and colleagues propose an instance-based solution for discovering one-to-one alignments among the web databases [Wang et al., 2004] (see also Sect. 6.1.25).

These are query interfaces (HTML forms) and backend databases which dynamically provide information in response to user queries. Authors distinguish between (i) the query interface, which exposes attributes that can be queried in the web database and (ii) the result schema presenting the query results, which exposes attributes that are shown to users. Matching between different query interfaces (inter-site matching) is critical for data integration between web databases. Matching between the interface and result schema of a single web database (intra-site matching), in turn, is useful for automatic data annotation and database content crawling. The approach is based on the following observations (among others):

– The keywords of queries (whose semantics match the semantics of the input element of a query interface) that return results are likely to reappear in attributes of the returned result. For example, such keywords as Logic submitted to the input element title matches its intended use (while it is not the case with the field author which will unlikely produce expected results), and therefore, some results with books about logics will be returned. Moreover, part (Logic) of the value Introduction to logic of the title attribute should reappear in the result schema.
– Based on the work in [He and Chang, 2003], the authors assume the existence and availability of a populated global schema, that is a view capturing common attributes of data, for the web databases of the same domain of interest.

The approach presents a combined schema model that involves five kinds of schema matching for web databases in the same domain of interest: global-interface, global-result, interface-result, interface-interface, and result-result. The approach works in two phases: (i) query probing and (ii) instance-based matching.

The first phase deals with acquiring data instances from web databases by query probing. It exhaustively sends the attribute values of pre-known instances from a global schema and collects results from the web databases under consideration in a *query occurrence cube*. The cube height stands for the number of attributes in the given global schema. The cube width stands for the number of attributes in the interface schema. The cube depth is the number of attributes in the result schema. Finally, each cell in this cube stores an occurrence count associated with the three dimensions. This cube is further projected onto three *query occurrence matrices*, which represent relationships between pairs of three schemas, namely global-interface, global-result, and interface-result.

During the second phase, the re-occurrences of submitted query keywords in the returned results data are analysed. In order to perform intra-site matching, the *mutual information* between pairs of attributes from two schemas is computed (see also Sect. 6.2.9). In order to perform inter-site matching a vector-based similarity is used (§4.2.1). In particular, each attribute of an individual interface or result schema is viewed as a *document* and each attribute of the global schema is view as a *concept*. Each row in the occurrence matrix represents a corresponding document vector. The similarity between attributes from different schemas is computed by using the cosine measure (§4.2.1) between two vectors. Finally, for both intra-site matching and inter-site matching, the matrix element whose value is the largest both in its row and column represents a final correspondence (this is the greedy alignment extraction presented in Sect. 5.7.3).

6.2.12 sPLMap (University of Duisburg-Essen, and ISTI-CNR)

sPLMap (probabilistic, logic-based mapping between schemas) is a framework which combines logics with probability theory in order to support uncertain schema mappings [Nottelmann and Straccia, 2005, Nottelmann and Straccia, 2006]. In particular, it is a GLAV-like framework [Lenzerini, 2002] where the alignment is defined as uncertain rules in probabilistic Datalog. This allows the support for imprecise matches, e.g., between author and editor attributes and a more general attribute, such as creator, which is often the case in schemas with different levels of granularity. sPLMap matches only attributes of the same concept (typically documents). The system operates in three main steps. First, it evaluates the quality of all possible individual correspondences on the basis of probability distributions (called interpretation). It selects the set of correspondences that maximises probability on the basis of instance data.

Then, for each correspondence, matchers are used as quality estimators: they provide a measure of the plausibility of the correspondence. sPLMap has been tested with the following matchers: (i) same attribute names (§4.2.1), (ii) exact tuples (§4.4), (iii) the k-nearest neighbour classifier, and (iv) the naive Bayesian classifier (§5.4.1). The result of these matchers are aggregated by using linear or logistic functions, or their combinations (§5.2). Coefficients of the normalisation functions are learnt by regression in a system-training phase. Finally, the computed probabilities are transformed in correspondence weights (used as the probability of the corresponding Datalog clause) by using the Bayes theorem.

6.3 Mixed, schema-based and instance-based systems

The following systems take advantage of both schema-level and instance-level input information if they are both available.

6.3.1 SEMINT (Northwestern University, NEC and The MITRE Corporation)

SEMantic INTegrator (SEMINT) is a tool based on neural networks to assist in identifying attribute correspondences in heterogeneous databases [Li and Clifton, 1994, Li and Clifton, 2000]. It supports access to a variety of database systems and utilises both schema- and instance-level information to produce rules for matching corresponding attributes automatically. The approach works as follows. First, it extracts from two databases all the necessary information (features or discriminators) which is potentially available and useful for matching. This includes normalised schema information, e.g., field specifications, such as datatypes, length, constraints, and statistics about data values, e.g., character patterns, such as ratio of numerical characters, ratio of white spaces, and numerical patterns, such as mean, variance, standard deviation. Second, by using a neural network as a classifier with the self-organising map algorithm (§5.4.3), it groups the attributes based on similarity of the features for a single (the first) database. Then, it uses a back-propagation neural network for learning and recognition. Based on the previously obtained clusters, the learning is performed. Finally, using a trained neural network on the first database features and clusters, the system recognises and computes similarities between the categories of attributes from the first database and the features of attributes from the second database, thus, generating a list of match candidates, which are to be inspected and confirmed or discarded by users.

6.3.2 Clio (IBM Almaden and University of Toronto)

Clio is a system for managing and facilitating data transformation and integration tasks within heterogeneous environments [Miller et al., 2000, Miller et al., 2001, Naumann et al., 2002, Haas et al., 2005], see Fig. 6.5. Clio handles relational and XML schemas. As a first step, the system transforms the input schemas into an internal representation, which is a nested relational model. The Clio approach is focused on making the alignment operational. It is assumed that the matching step, namely, identification of the *value correspondences*, is performed with the help of a schema matching component or manually. The built-in schema matching algorithm of Clio combines in a sequential manner instance-based attribute classification via a variation of a naive Bayes classifier (§5.4.1) and string matching between elements names, e.g., by using an edit distance (§4.2.1). Then, taking the n-m value correspondences (the alignment) together with constraints coming from the input schemas, Clio compiles these into an internal query graph representation. In particular, an interpretation of the input correspondences is given. Thus, a set of logical mappings with formal semantics is produced. To this end, Clio also supports mappings composition [Fagin et al., 2004]. Finally, the query graph can be serialised into different query languages, e.g., SQL, XSLT, XQuery, thus enabling actual data to be moved from a source to a target, or to answer queries. The system, besides trivial transformations, aims at discovering complex ones, such as the generation of keys, references and join conditions.

6.3.3 IF-Map (University of Southampton and University of Edinburgh)

IF-Map (Information-Flow-based Map) [Kalfoglou and Schorlemmer, 2003a] is an ontology matching system based on the Barwise–Seligman theory of information flow [Barwise and Seligman, 1997]. The basic principle of IF-Map is to match two local ontologies by looking at how these are related to a common reference ontology. It is assumed that such a reference ontology represents an agreed understanding that facilitates the sharing of knowledge. This means that two local ontologies have significant fragments of them that conform to the reference ontology. It is also assumed that the local ontologies are populated with instances, while the reference ontology does not need to.

178   6 Overview of matching systems

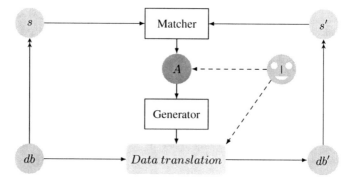

Fig. 6.5. Clio architecture: Clio goes all the way from matching schemas to translating data from one database to another one. It is made up of a classical matcher but also involves users at each step: input, matching control and translation execution.

Matching works as follows. If the reference ontology can be expressed in each of the local ontologies and instances of the local ontologies can be assigned concepts in the reference ontology (or be mapped to equivalent instances in the reference ontology), then IF-Map uses formal concept analysis (§4.4.1) between the three ontologies in order to find the Galois lattice from which it is possible to extract an alignment.

When the mappings are not available, IF-Map generates candidate pairs of mappings (called *infomorphism* in information flow theory) and artificial instances. They are generated through the enforcement of constraints that are induced by the definition of the reference ontology and by heuristics based on string-based (§4.2.1) and structure-based (§4.3) methods.

IF-Map deals with ontologies expressed in KIF or RDF. The IF-MAP method is declaratively specified in Horn logic and is executed with a Prolog interpreter, so the ontologies are translated into Prolog clauses beforehand. IF-Map produces concept-to-concept and relation-to-relation alignments.

6.3.4 NOM and QOM (University of Karlsruhe)

NOM (Naive Ontology Mapping) [Ehrig and Sure, 2004] and QOM (Quick Ontology Mapping) [Ehrig and Staab, 2004] are components of the FOAM framework (§8.2.5).

NOM adopts the idea of parallel composition of matchers from COMA (§6.1.12). Some innovations with respect to COMA are in the set of elementary matchers based on rules, exploiting explicitly codified knowledge in ontologies, such as information about super and subconcepts, super and subproperties, etc. As from [Ehrig and Sure, 2004], the system supports 17 rules related to those of Table 4.6 (p. 100). For example, a rule states that if superconcepts are the same, the actual concepts are similar to each other. These rules are based on various terminological and structural techniques.

6.3 Mixed, schema-based and instance-based systems 179

QOM (Quick Ontology Mapping) [Ehrig and Staab, 2004] is a variation of the NOM system dedicated to improve the efficiency of the system. The approach is based on the idea that the loss of quality in matching algorithms is marginal (to a standard baseline); however improvement in efficiency can be significant. This fact allows QOM to produce correspondences fast, even for large-size ontologies. QOM is grounded on matching rules of NOM. However, for the purpose of efficiency the use of some rules, e.g., the rules that traverse the taxonomy, has been restricted. QOM avoids the complete pairwise comparison of trees in favour of an incomplete top-down strategy, thereby focusing only on promising matching candidates.

The similarity measures produced by basic matchers (matching rules) are refined by using a sigmoïd function (§5.7.2), thereby emphasising high individual similarities and de-emphasising low individual similarities. They are then aggregated through weighted average (§5.2). Finally, with the help of thresholds, the final alignment is produced.

6.3.5 oMap (CNR Pisa)

oMap [Straccia and Troncy, 2005] is a system for matching OWL ontologies. It is built on top of the Alignment API (§8.2.4) and has been used for distributed information retrieval in [Straccia and Troncy, 2006]. oMap uses several matchers (called classifiers) that are used for giving a plausibility of a correspondence as a function of an input alignment between two ontologies. The matchers include (i) a classifier based on classic string similarity measure over normalised entity names (§4.2.1); (ii) a naive Bayes classifier (§5.4.1) used on instance data, and (iii) a 'semantic' matcher which propagates initial weights through the ontology constructors used in the definitions of ontology entities. This last one starts with an input alignment associating plausibility to correspondences between primitive entities and computes the plausibility of a new alignment by propagating these measures through the definitions of the considered entities. The propagation rules depend on the ontology constructions, e.g., when passing through a conjunction, the plausibility will be minimised. Each matcher has its own threshold and they are ordered among themselves.

There are two ways in which matchers can work: (i) in parallel, in which case their results are aggregated through a weighted average, such that the weights correspond to the credit accorded to each of the classifiers, (ii) in sequence, in which case each matcher only adds new correspondences to the input ontologies. A typical order starts with string similarity, before naive Bayes, and then the 'semantic' matcher is used.

6.3.6 Xu and Embley (Brigham Young University)

Xu and Embley proposed a parallel composition approach to discover, in addition to one-to-one alignments, also one-to-many and many-to-many correspondences between graph-like structures, e.g., XML schemas, classifications [Xu and Embley, 2003, Embley et al., 2004]. Schema matching is performed by a combination (an average function) of multiple matchers and with the help of external knowledge resources, such as domain ontologies. The basic element level matchers used in the approach include *name matcher* and *value-characteristic matcher*. The name matcher, besides string comparisons (§4.2.1), also performs some linguistic normalisation, such as stemming and removing stop words (§4.2.2). It also detects synonyms among node names with the help of WordNet (§4.2.2). In particular, matching rules are obtained via a C4.5 decision tree generator (§5.4.4) that has been trained over WordNet by using several hundreds synonym names found in the available databases from a domain of interest. The value-characteristic matcher determines where two values of schema elements share similar value characteristics, such as means or variances for numerical data. Similar to the name matcher, matching rules are obtained by training the C4.5 decision tree generator over value characteristics of the available databases from a domain of interest. Structure level matchers are used to suggest new correspondences as well as to confirm correspondences identified by element level matchers, for example, by considering similarities between the neighbour elements computed by element level matchers. Another example of a structural matcher makes use of a domain ontology. In particular, it tries to match both schemas A and B to the structure C, which is an external domain ontology, in order to decide if A corresponds to B.

6.3.7 Wise-Integrator (SUNY at Binghamton, University of Illinois at Chicago and University of Louisiana at Lafayette)

Wise-Integrator is a tool that performs automatic integration of Web Interfaces of Search Engines [He *et al.*, 2004, He *et al.*, 2005]. It provides a unified interface to e-commerce search engines of the same domain of interest, such as books and music. Therefore, users can pose queries by using this interface and the search mediator sends the translated subqueries to each site involved in handling this query and then the results of these sites are reconciled and presented to users. Wise-Integrator consists of two main subsystems: (i) an interface schema extractor, and (ii) an interface schema integrator. The first component, given a set of HTML pages with query interfaces, identifies logical attributes and derives some meta-information about them, e.g., datatype, thereby building an interface schema out of them. For example, the system can derive (guess) that the field Publication Date, is likely to be of date datatype. The second component discovers matching attributes among multiple query interfaces and then merges them, thereby resulting in global attributes. These are used, in turn, to produce a unified search interface.

Attribute matching in Wise-Integrator is based on two types of matches: *positive* and *predictive*. Positive matches are based on the following matching methods: exact name match, look up for synonymy, hypernymy and meronymy in WordNet (§4.2.2), and value-based matchers. When one of the positive matches occurs, the corresponding attributes are considered as matched. Predictive matches are based on the following matching methods: approximate name match, e.g., edit distance (§4.2.1), datatype compatibility (§4.3.1), value pattern matcher (§4.4.3). Predictive matches have to be strong enough (which is decided based on a threshold) in order to indicate that the attributes under consideration match.

Positive and predictive matches are carried out in two clustering steps: *positive match based clustering* and *predictive match based clustering*. In the first step all the interfaces are taken as input and attributes are grouped into clusters based on the positive matches between attributes. Clustering is done by following some pre-defined rules which govern the order of execution of underlying matchers and how to make groupings based on results of those matchers. For example, the first results of exact name matches are considered and then results of value-based and WordNet-based matchers. Finally, for each cluster a representative attribute name (RAN) is determined. For example, for the cluster with attribute names {Format, Binding type, Format} the RAN is Format. During the second step all local interfaces are reconsidered again. Clustering is done following some pre-defined rules which employ previously determined RANs and a simple weighting scheme over the results of predictive matching methods, if all else fail. When all potentially matching attributes are clustered together, the global attribute for each group of such attributes is generated.

6.3.8 OLA (INRIA Rhône-Alpes and Université de Montréal)

OLA (OWL Lite Aligner) [Euzenat and Valtchev, 2004] is an ontology matching system which is designed with the idea of balancing the contribution of each of the components that compose an ontology, e.g., classes, constraints, data instances. OLA handles ontologies in OWL. It first compiles the input ontologies into graph structures, unveiling all relationships between entities. These graph structures produce the constraints for expressing a similarity between the elements of the ontologies. The similarity between nodes of the graphs follows two principles: (i) it depends on the category of node considered, e.g., class, property, and (ii) it takes into account all the features of this category, e.g., superclasses, properties, as presented in Table 4.6.

The distance between nodes in the graph are expressed as a system of equations based on string-based (§4.2.1), language-based (§4.2.2) and structure-based (§4.3) similarities (as well as taking instances into account whenever necessary). These distances are almost linearly aggregated (they are linearly aggregated modulo local matches of entities). For computing these distances, the algorithm starts with base distance measures computed from labels and concrete datatypes. Then, it iterates a fixed point algorithm until no improvement is produced. From that solution, an alignment is generated which satisfies some additional criterion on the obtained alignment and the distance between matched entities. The algorithm is described in more detail in Sect. 5.3.2. The OLA architecture is typically the one displayed in Fig. 5.8 (p. 127).

6.3.9 Falcon-AO (China Southwest University)

Falcon-AO is a system for matching OWL ontologies. It is made of two components, namely those for performing linguistic and structure matching, see also Fig. 6.6.

LMO is a linguistic matcher. It associates with each ontology entity a bag of words which is built from the entity label, the entity annotations as well as the labels of connected entities. The similarity between entities is based on TFIDF (§4.2.1) [Qu et al., 2006].

GMO is a bipartite graph matcher [Hu et al., 2005]. It starts by considering the RDF representation of the ontologies as a bipartite graph which is represented by its adjacency matrix (A and A'). The distance between the ontologies is represented by a distance matrix (X) and the distance (or update) equations between two entities are simply a linear combination of all entities they are adjacent to, i.e., $X^{t+1} = AX^t A'^T + A^T X^t A'$. This process can be bootstrapped with an initial distance matrix. However, the real process is more complex than described here because it distinguishes between external and internal entities as well as between classes, relations and instances.

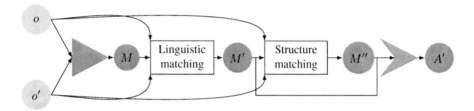

Fig. 6.6. Falcon-AO architecture: it is a sequential composition of two components, but if the output of the linguistic matcher is considered of sufficient quality, then no structure matching is performed.

First LMO is used for assessing the similarity between ontology entities on the basis of their name and text annotations. If the result has a high confidence, then it is directly returned for extracting an alignment. Otherwise, the result is used as input for the GMO matcher which tries to find an alignment on the basis of the relationships between entities [Jian et al., 2005].

6.3.10 RiMOM (Tsinghua University)

The RiMOM (Risk Minimisation based Ontology Mapping) approach, being inspired by Bayesian decision theory, formalises ontology matching as a decision making problem [Tang et al., 2006]. Given two ontologies, it aims at an optimal and automatic discovery of alignments which can be complex (such as including concatenation operators). The approach first searches for concept-to-concept correspondences and then for property-to-property correspondences. The RiMOM matching process is organised into the following phases [Li et al., 2006]:

1. Select matchers to use. This task can be performed either automatically or manually. The basic idea of automatic strategy selection is if two ontologies have high

label similarity factor, then RiMOM will rely more on linguistic based strategies; while if the two ontologies have a high structure similarity factor, RiMOM will exploit similarity-propagation based strategies on them.

2. Execute multiple independent matchers, given the input ontologies and, optionally, user input. Examples of matchers include linguistic normalisation of labels, such as tokenisation, expansion of abbreviations and acronyms (§4.2.2) based on GATE tools[6], edit-distance, matchers that look for label similarity based on WordNet (§4.2.2), k-nearest neighbours statistical learning, naive Bayes matcher (§5.4.1), as well as some other heuristics for data type similarity and taxonomic structure similarity. This results in a cube of similarity values in [0 1] for each pair of entities from the two ontologies (see also Sect. 6.1.12).

3. Combine the results by aggregating the values produced during the previous step into a single value. This is performed by using a linear-interpolation.

4. Similarity propagation. If the two ontologies have high structure similarity factor, RiMOM employs an algorithm called similarity propagation to refine the found alignments and to find new alignments that cannot be discovered using the other strategies. Similarity propagation makes use of structure information.

5. Extract alignment for a pair of ontologies based on thresholds (§5.7.1) and some refinement heuristics to eliminate unreasonable correspondences, e.g., use concept-to-concept correspondences to refine property-to-property correspondences.

6. Iterate the above described process by taking the output of one iteration as input into the next iteration until no new correspondences are produced. At each iteration, users can select matchers, and approve and discard correspondences from the returned alignment

RiMOM offers three possible structural propagation strategies: concept-to-concept propagation strategy (CCP), property-to-property propagation strategy (PPP), and concept-to-property propagation strategy (CPP). For choosing between them, RiMOM uses heuristic rules. For example, if the structure similarity factor is lower than some threshold then RiMOM does not use the CCP and PPP strategies, only CPP is used.

6.3.11 Corpus-based matching (University of Washington, Microsoft Research and University of Illinois)

Madhavan and colleagues [Madhavan *et al.*, 2005] proposed an approach to schema matching which, besides input information available from schemas under consideration, also exploits some domain specific knowledge via an external corpus of schemas and mappings. The approach is inspired from the use of corpus in information retrieval, where similarity between queries and concepts is determined based on analysing large corpora of text. In schema matching, such a corpus can be initialised with a small number of schemas obtained, for example, by using available

---

[6] http://gate.ac.uk/

standard schemas in the domain of interest, and should eventually evolve in time with new matching tasks.

Since the corpus is intended to have different representations of each concept in the domain, it should facilitate learning these variations in the elements and their properties. The corpus is exploited in two ways. First, to obtain an additional evidence about each element being matched by including evidence from similar elements in the corpus. Second, in the corpus, similar elements are clustered and some statistics for clusters are computed, such as neighbourhood and ordering of elements. These statistics are ultimately used to build constraints that facilitate selection of the correspondences in the resulting alignment.

The approach handles web forms and relational schemas and focuses on one-to-one alignments. It works in two phases. Firstly, schemas under consideration are matched against the corpus, thereby augmenting these with possible variations of their elements based on knowledge available from the corpus. Secondly, augmented schemas are matched against each other. In both cases the same set of matchers is applied. In particular, basic matchers, called learners, include (i) a *name learner*, (ii) a *text learner*, (iii) a *data instance learner*, and (iv) a *context learner*. These matchers mostly follow the ideas of techniques used in LSD (§6.2.4) and Cupid (§6.1.11). For example, the *name learner* exploits names of elements. It applies tokenisation and n-grams (§4.2.1) to the names in order to create training examples. The matcher itself is a text classifier, such as naive Bayes (§5.4.1). In addition, the name learner, uses edit distance (§4.2.1), in order to determine similarity between element names string. The *data instance learner* determines whether the values of instances share common patterns, same words, etc. A matcher, called *meta-learner*, combines the results produced by basic matchers. It uses *logistic regression* with the help of the stacking technique (§5.4.5) in order to learn its parameters. Finally, by using constraints based on the statistics obtained from the corpus, candidate correspondences are filtered in order to produce the final alignment.

6.4 Meta-matching systems

Meta-matching systems are systems whose originality is in the way they use and combine other matching systems rather than in the matchers themselves.

6.4.1 APFEL (University of Karlsruhe and University of Koblenz-Landau)

APFEL (Alignment Process Feature Estimation and Learning) is a machine learning approach that explores user validation of initial alignments for optimising automatically the configuration parameters of some of the matching strategies of the system, e.g., weights, thresholds, for the given matching task [Ehrig *et al.*, 2005]. It is a component of the FOAM framework (§8.2.5). The overall architecture of APFEL is given in Fig. 6.7.

APFEL parameterises the FOAM steps by using declarative representations of the (i) features engineered, Q_F; (ii) similarities assessed, Q_S; (iii) weight schemas,

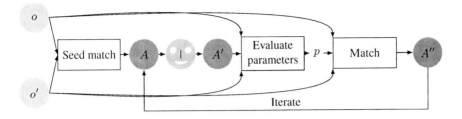

Fig. 6.7. APFEL architecture (adapted from [Ehrig, 2007]): it generates alignments and asks users for feedback. Then it adjusts methods and aggregation parameters in order to minimise the error and iterate, if necessary.

e.g., for similarity aggregation, Q_W; and (iv) thresholds, Q_T. For that purpose, the interfaces of matching systems are unified as Parameterisable Alignment Methods (PAM), which accept these parameters. First, given a matching system, for example QOM (§6.3.4) or Prompt (§6.1.9), a PAM is initialised with it, e.g., PAM(QOM). Then, once an initial alignment is obtained, this alignment is validated by users. Finally, by analysing the validated alignment and the above parameters, with the help of machine learning techniques (§5.4), e.g., decision tree learner, neural networks, support vector machines of the WEKA machine learning environment[7], a tuned weighting scheme and thresholds are produced for the given matching task. This process can be iterated.

6.4.2 eTuner (University of Illinois and The MITRE Corporation)

eTuner [Sayyadian *et al.*, 2005] is a system which, given a particular matching task, automatically tunes a schema matching system (computing one-to-one alignments). For that purpose, it chooses the most effective basic matchers, and the best parameters to be used, e.g., thresholds. eTuner models a matching system as a triple: $\langle L, G, K \rangle$, such that:

- L is a library of matching components, including basic matchers, e.g., edit distance, n-gram, combiners, e.g., modules taking average, minimum and maximum of the results produced by basic matchers, constraint enforcers, e.g., pre-defined domain constraints or heuristics which are computationally expensive to be used as basic matchers, and match selectors, e.g., modules applying thresholds for determining the final alignment.
- G is a directed graph which encodes the execution flow among the components of the given matching system.
- K is a set of *knobs* to be set (and named *knob configuration*). Matching components are viewed as black boxes which expose a set of adjustable knobs, such as thresholds, weights, or coefficients.

---

[7] http://www.cs.waikato.ac.nz/ml/weka/

186   6 Overview of matching systems

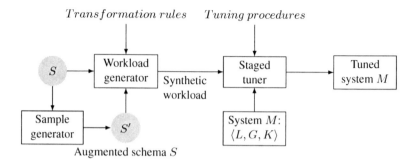

Fig. 6.8. eTuner architecture: eTuner generates a set of schemas to match with an initial schema. Then, it generates a plan for learning parameters. Finally, it tunes the method parameters and aggregation parameters.

The system works in two phases (see Fig. 6.8). During the first phase, in which the workload is synthesised with a known ground truth, given a single schema S, the system synthetises several schemas $(S_1, S_2,\ldots,S_n)$ out of S by altering it (for instance by modifying names of attributes, e.g., authors becomes aut). Thus, by taking a set of pairs $\{\langle S, S_1\rangle, \langle S, S_2\rangle, \ldots \langle S, S_n\rangle\}$ together with the reference correspondences available for free by construction of the synthetic schemas, the F-measure (§7.3) can be computed over any knob configuration. The second phase consists of searching the best parameters. Since the space of knob configurations can be large, the system uses a sequential, greedy approach, called *staged tuning*. In particular, by using the synthetic workload, it first tunes each of the basic matchers in isolation. Then, it tunes the combination of the basic matchers and the combiner, having the knobs of the basic matchers fixed, and so on and so forth. Once the entire system is tuned, it can be applied to match schema S with any subsequent schemas.

6.5 Summary

The panorama of systems considered in this chapter has multiplied the diversity of basic techniques of Chap. 4 by the variety of strategies for combining them introduced in Chap. 5. Moreover, usually each individual system innovates on a particular aspect. However, there are several constant features that are shared by the majority of systems. In summary, the following can be observed concerning the presented systems:

- Based on the number of systems considered in the various sections of this chapter, we can conclude that schema-based matching solutions have been so far investigated more intensively than the instance-based solutions. We believe that this is an objective trend, since we have striven to cover state of the art systems without bias towards any particular kind of solutions.

6.5 Summary

- Most of the systems under consideration focus on specific application domains, such as books and music, as well as on dealing with particular ontology types, such as DTDs, relational schemas and OWL ontologies. Only few systems aim at being general, i.e., suit various application domains, and generic, i.e., handle multiple types of ontologies. Some examples of the latter include Cupid (§6.1.11), COMA and COMA++ (§6.1.12), Similarity flooding (§6.1.13), and S-Match (§6.1.19).
- Most of the approaches take as input a pair of ontologies, while only few systems take as input multiple ontologies. Some examples of the latter include DCM (§6.1.25) and Wise-Integrator (§6.3.7).
- Most of the approaches handle only tree-like structures, while only few systems handle graphs. Some examples of the latter include Cupid (§6.1.11), COMA and COMA++ (§6.1.12), and OLA (§6.3.8).
- Most of the systems focus on discovery of one-to-one alignments, while only few systems have tried to address the problem of discovering more complex correspondences, such as one-to-many and many-to-many, e.g., iMAP (§6.2.6) and DCM (§6.1.25).
- Most of the systems focus on computing confidence measures in the [0 1] range, most often standing for the fact that the equivalence relation holds between ontology entities. Only few systems compute logical relations between ontology entities, such as equivalence, subsumption. Some examples of the latter include CtxMatch (§6.1.18) and S-Match (§6.1.19).

Table 6.1 summarises how the matching systems cover the solution space in terms of the classifications of Chap. 3. For example, S-Match (§6.1.19) exploits string-based element-level matchers, external matchers based on WordNet, propositional satisfiability techniques, etc. OLA (§6.3.8), in turn, exploits, besides string-based element-level matchers, also a matcher based on WordNet, iterative fixed point computation, etc. Table 6.1 also testifies that ontology matching research so far was mainly focused on syntactic and external techniques. In fact, many systems rely on the same string-based techniques. Similar observation can be also made concerning the use of WordNet as an external resource of common knowledge. In turn, semantic techniques have rarely been exploited, this is only done by CtxMatch (§6.1.18), S-Match (§6.1.19) and OntoMerge (§6.1.17). Concerning instance-based system, techniques based on naive Bayes classifier and common value patterns are the most prominent.

Table 6.1. Basic matchers used by the different systems.

	Element-level Syntactic	External	Structure-level Syntactic	Semantic
DELTA §6.1.1	String-based	-	-	-
Hovy §6.1.2	String-based, Language-based	-	Taxonomic structure	-
TranScm	String-based	Built-in thesaurus	Taxonomic structure,	-

Table 6.1. Basic matchers used by the different systems (continued).

	Element-level Syntactic	External	Structure-level Syntactic	Semantic
§6.1.3			Matching of neighbourhood	
DIKE §6.1.4	String-based, Domain compatibility	WordNet	Matching of neighbourhood	-
SKAT §6.1.5	String-based	Auxiliary thesaurus, Corpus-based	Taxonomic structure, Matching of neighbourhood	-
Artemis §6.1.6	Domain compatibility, Language-based	Common thesaurus	Matching of neighbours via thesaurus, Clustering	-
H-Match §6.1.7	Domain compatibility, Language-based, Domains and ranges	Common thesaurus	Matching of neighbours via thesaurus, Relations	-
Tess §6.1.8	String-based, domain compatibility	-	Matching of neighbours	-
Anchor-Prompt §6.1.9	String-based, Domains and ranges	-	Bounded paths matching: (arbitrary links), Taxonomic structure	-
OntoBuilder §6.1.10	String-based, Language-based	Thesaurus look up	-	-
Cupid §6.1.11	String-based, Language-based, Datatypes, Key properties	Auxiliary thesauri	Tree matching weighted by leaves	-
COMA & COMA++ §6.1.12	String-based, Language-based, Datatypes	Auxiliary thesauri, Alignment reuse, Repository of structures	DAG (tree) matching with a bias towards various structures, e.g., leaves	-
Similarity flooding §6.1.13	String-based, Datatypes, Key properties	-	Iterative fixed point computation	-
XClust §6.1.14	Cardinality constraints	WordNet	Paths, Children, Leaves, Clustering	-
ToMAS §6.1.15	-	External alignments	Preserving consistency, Structure comparison	-
MapOnto §6.1.16	-	External alignments	Structure comparison	-
OntoMerge §6.1.17	-	External alignments	-	-
CtxMatch §6.1.18	String-based, Language-based	WordNet	-	Based on description logics
S-Match §6.1.19	String-based, Language-based	WordNet	-	Propositional SAT
HCONE §6.1.20	Language-based (LSI)	WordNet	-	-
MoA §6.1.21	Language-based	WordNet	-	-
ASCO §6.1.22	String-based, Language-based	WordNet	Iterative similarity propagation	-
BayesOWL §6.1.23	Text classifier	Google	Bayesian inference	-
OMEN §6.1.24	-	External alignment	Bayesian inference, Meta-rules	-
DCM §6.1.25	-	-	Correlation mining, Statistics	-

6.5 Summary

Table 6.1. Basic matchers used by the different systems (continued).

	Element-level Syntactic	External	Structure-level Syntactic	Semantic
T-tree §6.2.1	-	-	Correlation mining	-
CAIMAN §6.2.2	String-based (Rocchio classifier)	-	-	-
FCA-merge §6.2.3	-	-	Formal concept analysis	-
LSD/GLUE/ iMAP §6.2.4-6.2.6	WHIRL, Naive Bayes	Domain constraints	Hierarchical structure	-
Automatch §6.2.7	Naive Bayes	-	Internal structure, Statistics	-
SBI&NB §6.2.8	Statistics, Naive Bayes	-	Pachinko Machine naive Bayes	-
Kang & Naughton §6.2.9	Information entropy	-	Mutual information, Dependency graph matching	-
Dumas §6.2.10	String-based WHIRL	-	Instance identification	-
Wang & al. §6.2.11	Language-based	-	Mutual information	-
sPLMap §6.2.12	Naive Bayes, kNN classifier, String-based	-	-	-
SEMINT §6.3.1	Neural network, Datatypes, Value patterns	-	-	-
Clio §6.3.2	String-based, Language-based, Naive Bayes	-	Structure comparison	-
IF-Map §6.3.3	String-based	-	Formal concept analysis	-
NOM & QOM §6.3.4	String-based, Domains and ranges	Application-specific vocabulary	Matching of neighbours, Taxonomic structure	-
oMap §6.3.5	Naive Bayes, String-based	-	Similarity propagation	-
Xu & al. §6.3.6	String-based, Language-based	WordNet, Domain ontology	Decision trees	-
Wise-Integrator §6.3.7	String-based, Language-based, Datatypes, Value patterns	WordNet	Clustering	-
OLA §6.3.8	String-based, Language-based, Datatypes	WordNet	Iterative fixed point computation, Matching of neighbours, Taxonomic structure	-
Falcon-AO §6.3.9	String-based	WordNet	Structural affinity	-
RiMOM §6.3.10	String-based, Naive Bayes	WordNet	Taxonomic structure, Similarity propagation	-
Corpus-based matching §6.3.11	String-based, Language-based, Naive Bayes, Value patterns	Corpus schemas, Domain constraints	-	-

Table 6.2 summarises the position of these systems with regard to some of the requirements of Sect. 1.7 (namely those requirements that can be given in the specification of the system rather than being measured). In Table 6.2, the *Input* column stands for the input taken by the systems. In particular, it mentions the languages that the systems are able to handle (if this information was not available form the articles describing the corresponding systems we used general terms, such as database schema and ontology instead). This is, of course, very important for someone who has a certain type of ontology to match and is looking for a system. The *Needs* column stands for the resources that must be available for the system to work. This covers the automatic aspect of Sect. 1.7, which is here denoted as *user* when user feedback is required, *semi* when the system can take advantage of user feedback but can operate without it and *auto* when the system works without user intervention (of course, users can influence the system by providing the initial input or evaluating the results afterwards, but this is not taken into account here). Similarly, the *instances* value specifies that the system requires data instances to work. In addition, some systems may require *training* before the actual matching as well as *alignment* to be improved. The *Output* column denotes the form of the results given by the system: *Alignment* means that the system returns a set of correspondences, *merge* that it merges the input ontologies or schemas, *axioms* or *rules* that it provides rules for querying or completing the ontologies, etc.

Table 6.2. Position of the presented systems with regard to the requirements of Chap. 1.

System	Input	Needs	Output	Operation
DELTA §6.1.1	Relational schema, EER	User	Alignment	-
Hovy §6.1.2	Ontology	Semi	Alignment	-
TranScm §6.1.3	SGML, OO	Semi	Translator	Data translation
DIKE §6.1.4	ER	Semi	Merge	Query mediation
SKAT §6.1.5	RDF	Semi	Bridge rules	Data translation
Artemis §6.1.6	Relational schema, OO, ER	Auto	Views	Query mediation
H-Match §6.1.7	OWL	Auto	Alignment	P2P query mediation
Tess §6.1.8	Database schema	Auto	Rules	Version matching
Anchor-Prompt §6.1.9	OWL, RDF	User	Axioms (OWL/RDF)	Ontology merging
OntoBuilder §6.1.10	Web form, XML schema	User	Mediator	Query mediation
Cupid §6.1.11	XML schema, Relational schema	Auto	Alignment	-
COMA & COMA++ §6.1.12	Relational schema, XML schema, OWL	User	Alignment	Data translation

Table 6.2. Position of these systems with regard to the requirements of Chap. 1 (continued).

System	Input	Needs	Output	Operation
Similarity flooding §6.1.13	XML schema, Relational schema	User	Alignment	-
XClust §6.1.14	DTD	Auto	Alignment	-
ToMAS §6.1.15	Relational schema, XML schema	Query, Alignment	Query, Alignment	Data transformation
MapOnto §6.1.16	Relational schema, XML schema, OWL	Alignment	Rules	Data translation
OntoMerge §6.1.17	OWL	Alignment	Ontology	Ontology merging
CtxMatch/CtxMatch2 §6.1.18	Classification, OWL	User	Alignment	-
S-Match §6.1.19	Classification, XML schema, OWL	Auto	Alignment	-
HCONE §6.1.20	OWL	Auto, Semi, User	Ontology	Ontology merging
MoA §6.1.21	OWL	Auto	Axioms, OWL	-
ASCO §6.1.22	RDFS, OWL	Auto	Alignment	-
BayesOWL §6.1.23	Classification, OWL	Auto	Alignment	-
OMEN §6.1.24	OWL	Auto, Alignment	Alignment	-
DCM §6.1.25	Web form	Auto	Alignment	Data integration
T-tree §6.2.1	Ontology	Auto, Instances	Alignment	-
CAIMAN §6.2.2	Classification	Semi, Instances, Training	Alignment	-
FCA-merge §6.2.3	Ontology	User, Instances	Ontology	Ontology merging
LSD/GLUE §6.2.4, §6.2.5	Relational schema, XML schema, Taxonomy	Auto, Instances, Training	Alignment	-
iMAP §6.2.6	Relational schema	Auto, Instances, Training	Alignment	-
Automatch §6.2.7	Relational schema	Auto, Instances, Training	Alignment	-
SBI&NB §6.2.8	Classification	Auto, Instances, Training	Alignment	-
Kang & Naughton §6.2.9	Relational schema	Instances	Alignment	-
Dumas §6.2.10	Relational schema	Instances	Alignment	-
Wang & al. §6.2.11	Web form	Instances	Alignment	Data integration
sPLMap §6.2.12	Database schema	Auto, Instances, Training	Rules	Data translation
SEMINT §6.3.1	Relational schema	Auto, Instances (opt.), Training	Alignment	-
Clio	Relational schema,	Semi,	Query	Data

192   6 Overview of matching systems

Table 6.2. Position of these systems with regard to the requirements of Chap. 1 (continued).

System	Input	Needs	Output	Operation
§6.3.2	XML schema	Instances (opt.)	transformation	translation
IF-Map §6.3.3	KIF, RDF	Auto, Instances, Common reference	Alignment	-
NOM & QOM §6.3.4	RDF, OWL	Auto, Instances (opt.)	Alignment	-
oMap §6.3.5	OWL	Auto, Instances (opt.), Training	Alignment	Query answering
Xu & al. §6.3.6	XML schema, Taxonomy	Auto, Instances (opt.), Training	Alignment	-
Wise-Integrator §6.3.7	Web form	Auto	Mediator	Data integration
OLA §6.3.8	RDF, OWL	Auto, Instances (opt.)	Alignment	-
Falcon-AO §6.3.9	RDF, OWL	Auto Instances (opt.)	Alignment	-
RiMOM §6.3.10	OWL	Auto Instances (opt.)	Alignment	-
Corpus-based matching §6.3.11	Relational schema, Web form	Text corpora, Instances, Training	Alignment	-
APFEL §6.4.1	RDF, OWL	User	Alignment	-
eTuner §6.4.2	Relational schema, Taxonomy	Auto	Alignment	-

The *Output* delivered by a system is very important because it shows the capability of the system to be used for some applications, e.g., a system delivering views and data translators cannot be used for merging ontologies as is. It is remarkable that many systems deliver alignments. As such, they are not fully committed to any kind of operation to be performed and can be used in a variety of applications. This could be viewed as a sign of possible interoperability between systems. However, due to lack of a common alignment format, each system uses its own way to deliver alignments (as lists of URIs, tables, etc.). Finally, the *Operation* column describes the ways in which a system can process alignments.

Not all the requirements are addressed in Table 6.2. Indeed, completeness, correctness, run time should not be judged from the claims of system developers. No meaningful system can be proved to be complete, correct or as fast as possible in a task like ontology matching. Therefore, the degree of fulfillment of these requirements must be evaluated and compared across systems. Moreover, different applications have different priorities regarding these requirements, hence, they may need different systems. Thus, this evaluation depends on an application in which the system is to be used.

It is difficult to evaluate and compare systems without commonly agreed testbenches, principles and available implementations. The next chapter presents methods for empirical evaluation and comparison of matching systems.

7
Evaluation of matching systems

The increasing number of methods available for ontology matching suggests the need for evaluating these methods. However, very few extensive experimental comparisons of algorithms are available. Matching systems are difficult to compare, but we believe that the ontology matching field can only evolve if evaluation criteria are provided. These should help system designers to assess the strengths and weaknesses of their systems as well as help application developers to choose the most appropriate algorithm.

In this chapter we first consider the main motivations for evaluating matching systems and the principles that could guide such an evaluation (§7.1). We also discuss existing evaluation resources, different available data sets and the structure of some of these data sets (§7.2). Then, we overview the measures used for the evaluation of matching systems (§7.3). Finally, we consider in more detail the settings of an evaluation protocol for a particular application, as opposed to evaluation for comparing matching systems in general (§7.4).

7.1 Evaluation principles

All evaluation activities must be carried out with a clear procedure. So we first recall here the goal of evaluating ontology matching systems (§7.1.1), the principles on which evaluation should be based (§7.1.2) and some examples of evaluation initiatives (§7.1.3).

7.1.1 Goals

A major and long term purpose of the evaluation of ontology matching methods is to help designers and developers of such methods to improve them and to help users to evaluate the suitability of the proposed methods to their needs. The evaluation should thus be run over several years in order to allow for adequate measurement of the evolution of the field.

Evaluation should also help assess absolute results produced by the matching systems, i.e., what are the properties achieved by a system, and relative results, i.e., how these compare to the results of other systems.

One particular kind of evaluation is benchmarking. A benchmark is a well-defined set of tests on which the results of a system or subsystem can be measured [Castro et al., 2004]. It should enable the measure of the degree of achievement of proposed tasks on a well-defined scale (that can be achieved or not). It should be reproducible and stable, so that it can be used repeatedly for (i) testing the improvement or degradation of a system with certainty, (ii) situating a system among others.

A medium term goal for evaluation efforts is to set up a collection of reference sets of tests, or benchmark suites of the available tools and to compare their evolution with regard to this reference. Building benchmark suites is highly valuable not just for groups of people that participate in planned evaluations but for all the research community, since system designers can make use of these at any time and compare themselves to others.

7.1.2 Principles

We describe below several principles that must guide the evaluation process:

Systematic procedure. Evaluation results have to be non ambiguous and their procedure should be reproducible. Thus, the application of the procedure to different systems or to the same system at different moments of time should be comparable.

Continuity. Evaluation, and most particularly benchmarking, must not be a one-shot exercise but a continuous effort in order to identify the progress made by the field and eventually stop when no more progress is made anymore.

Quality and equity. The evaluation rules must be precise and defined beforehand. In order to be worthwhile, the evaluation material must be of the best possible quality. This also means that the evaluation material must not be biased towards a particular kind of algorithm but driven only by the tasks to be solved.

Dissemination. In order to have a high impact, the evaluation activity must be disseminated without excessive barriers. To that extent the data sets and results produced must be published and made as freely available as possible. The evaluation campaigns must be open to participants worldwide.

Intelligibility. It is of high importance that the evaluation results could be analysed by the stakeholders and understood by everyone. For that purpose, it is useful not only that the final results are published but also the alignments themselves. Finally, of high importance is the archival explanation of the results to the stakeholders.

Each evaluation must be carried out according to some methodology. It is usually based on three successive steps [Castro et al., 2004]:

Planning involves defining the task to be performed as well as its constraints, e.g., resources allowed, computer environment, required output, finding data sets on which to perform the tasks and setting the measures to be used.

Processing consists of executing the plan.
Analysing evaluates the results achieved according to planned measurements.

These three steps can be complemented by a recalibration loop that helps in redefining the plan for the next evaluation from lesson learnt in the current one.

7.1.3 Examples of evaluations

In order to illustrate what can be done as evaluation, we briefly discuss a model evaluation initiative, called TREC, and the Ontology Alignment Evaluation Initiative.

Text REtrieval Conference

TREC[1] is the Text REtrieval Conference organised by the National Institute of Standards and Technology (NIST) in the USA. It has been run yearly since 1992. It is a very good model for evaluation in a focussed computer science research field, especially because it has been very successful in helping the field to progress.

The goals of TREC are to:

– increase research in information retrieval based on large-scale collections;
– provide a forum for stakeholders;
– facilitate technology transfer;
– improve evaluation methodology;
– create a series of test collections on various aspects of information retrieval.

It is now organised in several tracks, each of which corresponding to one kind of evaluation, which, in turn, is organised over several years. Five years is now the accepted standard in order to be able to compare the results. Tracks organised so far include:

– static text retrieval;
– interactive retrieval;
– information retrieval in a narrow domain, e.g., genomics, using ad hoc resources;
– media retrieval (other than text);
– answer finding.

Typically each track has between 8 and 20 participants. While each track is precisely defined, TREC has now a track record on investigating the evaluation of many different features of the retrieval tasks.

Ontology Alignment Evaluation Initiative

Since 2004, a group of researchers on ontology matching have run several evaluation campaigns which are identified as Ontology Alignment Evaluation Initiative (OAEI)[2]. OAEI campaigns which have been organised so far include:

---

[1] http://trec.nist.gov
[2] http://oaei.ontologymatching.org

196   7 Evaluation of matching systems

- The Information Interpretation and Integration Conference (I3CON), held at the NIST Performance Metrics for Intelligent Systems (PerMIS) workshop, is an ontology matching demonstration competition on the model of TREC. This evaluation focused on comparison of various matching algorithms on real-life test cases.
- The Ontology Alignment Contest [Euzenat et al., 2004b] at the third Evaluation of Ontology-based Tools (EON) workshop [Sure et al., 2004], held at the annual International Semantic Web Conference (ISWC), is targeted at the characterisation of matching methods with regard to particular ontology features. This contest defined a proper set of benchmark tests for assessing feature-related behaviour.
- The Ontology Alignment Evaluation Initiative 2005 [Euzenat et al., 2005b] held at the Integrating Ontologies workshop of the International Conference on Knowledge Capture (K-Cap) [Ashpole et al., 2005]. It has improved on the previous ones by multiplying the number of tests considered as a benchmark and by introducing two real-world test cases.
- The Ontology Alignment Evaluation Initiative 2006 held during the Ontology Matching workshop of the annual International Semantic Web Conference [Shvaiko et al., 2006a]. It introduced several tracks distinguishing between particular data sets from various domains, e.g., medicine and food, and a matching consensus workshop aimed at studying the process by which people can find a consensus on what is a reference alignment. New tracks were devoted to large expressive ontologies and matching thesauri and directories.

In each of these campaigns, the participants are required to provide their resulting alignments in the Alignment format (§8.1.5). They are equipped with the Alignment API (§8.2.4) for helping them to produce and to assess the results before the meeting. Results to all tests are compulsory as well as a fixed-format paper describing experiences with tests processing. Participants are also expected to present their results at the meeting. The results of these tests are evaluated by clearly announced measures, typically precision and recall (§7.3). Finally, the results of these evaluation campaigns as well as the full data sets are available for download from the OAEI web site[2].

The OAEI campaigns tend to set a solid basis for evaluating the progress in matching algorithms by providing a stable benchmark suite, thus allowing progress to be monitored year after year and facilitating the calibration of the participating matching algorithms.

7.1.4 Types of evaluations

In Chap. 2, we characterised an alignment as a set of pairs of entities e and e', coming from each ontology o and o', related by a particular relation r. Also many algorithms add some confidence measure n for the fact that a particular relation holds. From this characterisation it is possible to require any matching method to output an alignment [Noy and Musen, 2002a, Euzenat, 2003], given

- two ontologies to be matched;

- an input partial alignment, which can be possibly empty;
- a characterisation of the desired alignment, e.g., 1:+, ?:?.

From this output, the quality of the matching process could be assessed with the help of measurements of the difference between the output and the reference alignment.

From this basic setting there are several ways of planning the evaluation depending, in part, of its purpose. There can be several classifications depending on the criteria used. Let us consider a classification of evaluations with regard to what they are supposed to evaluate:

Competence benchmarks allow the characterisation of the level of competence and performance of a particular system with regard to a set of well defined tasks. Usually, tasks are designed to isolate particular characteristics. This kind of benchmarking is reminiscent to kernel benchmarks or unit tests, such as the Standard Performance Evaluation Corporation (SPEC) benchmarks[3].

Competence benchmarks aim at characterising the kind of task each method is good for or the kind of input it can handle well. There are many different areas in which methods can be evaluated. One approach is to look at the kind of features they use for finding matching entities, for example, following one of the classifications mentioned in Chap. 3.

Benchmark suites must be stable so that they enable the monitoring of the evolution of the field over time. Moreover, they do not need to be run blindly since they are run several times. Thus, they can be freely distributed and designers of new systems can take advantage of them at any time.

Comparative evaluation allows the comparison of the results of various systems on a common task. A comparative evaluation constitutes a competition targeted at finding the best system, and thus, the best practices, among several ones. Since the goal is to compare the systems, it is of utmost importance that the rules and the evaluation criteria are clearly specified.

Because it is difficult to guarantee that the systems are not tuned for the evaluation, it is preferable to run blind tests or nearly blind tests. This means that the participants become aware of the data set very shortly before the evaluation and that the data set must be changed at each evaluation.

Finally, it is worth noting that such an evaluation, because it is run in a limited time span with relatively similar resources, requires a well defined processing mode. In counterpart, this allows acquiring more accurate non functional measurements, such as run time and memory.

Application-specific evaluation allows the comparison of the results of various systems on the output of a particular application instead of considering the alignments in general. They are useful for a company that has a real application and wants to find the best system to use in this application. It can also be useful for a competitive evaluation.

---

[3] http://www.spec.org

The goals of these three kinds of evaluations are different. Competence benchmarks aim at helping system designers to evaluate their systems and to situate them with regard to a common stable framework. It is also helpful for improving individual systems. The comparative evaluation enables the comparison of systems on general purpose tasks. Its goal is mainly to help the improvement of the field as a whole, rather than individual systems. It can also help users in selecting an appropriate system. Application-specific evaluations aim at identifying an adequate system for one particular application at one particular moment (see also Sect. 7.4).

7.2 Data sets for evaluation

One very important aspect of evaluation is the data set used for performing it. Finding a suitable data set is a critical issue because of the differences in form and quality of the possible data sets. We present first different factors that can influence evaluation (§7.2.1) and then discuss various data set categories (§7.2.2).

7.2.1 Dimensions and variability of alignment evaluation

Each of the elements featured in the matching process definition (Chap. 2) can have specific characteristics that influence the difficulty of the matching task. It is thus necessary to identify and control these characteristics. We called them dimensions because they define a space of possible tests.

Characterising the variability of matching tasks helps in assessing the limitations of benchmark suites and designing benchmarks spanning the whole spectrum of matching. Indeed, for each point in this variability space a specific test could be designed. However, there could be too many of them and it is thus necessary, for each data set, to choose among the most representative values for most of these possible parameters.

These dimensions and the questions they raise are a refinement of the requirements that have been studied in Sect. 1.7. These requirements only considered general categories called *input* and *process*. Such categories need to be refined for precisely defining what a particular application can expect from a matching system, while the initial requirements concern application classes. Knowing the relations between a data set and the dimensions can be used by the application designer for finding a suitable data set with which to evaluate the systems.

We review below the dimensions and justify some choices in designing benchmarks. This extends the typology introduced in [Noy and Musen, 2002a, Do et al., 2002] with regard to our definition of the matching process in Sect. 2.5.1.

Input ontologies

Input ontologies o and o' can be characterised by at least three different dimensions:

Heterogeneity: of the input languages: are they described in the same knowledge representation languages?

Languages: what are the languages of the ontologies? Some examples of languages include KIF, OWL, RDFS, F-Logic, UML, SQL DDL, and XML Schema.
Number: is this an alignment between two ontologies or should it match more ontologies?

As mentioned in Chap. 2, we consider here matching between homogeneous languages. The language used should be adapted to the kinds of features to be assessed by the evaluation. Thus, for example, a data set about directory matching should not be expressed in UML. However, the choice of language will also determine the systems that can be evaluated. It is perfectly admissible that not all the evaluation campaigns use the same languages.

Tasks involving multiple matching are very specific at the moment and only a small number of algorithms are considering them [He and Chang, 2006, Su et al., 2006]. Therefore, we consider here only evaluation of the matching results between two ontologies.

Input alignment

The input alignment A can have the following characteristics:

Complete or update: Is the matching process required to complete an existing input alignment or is it authorised to change it?
Multiplicity: How many entities of one ontology can correspond to one entity of the other ontology?

It is reasonable to start with tests without input alignment, especially since this helps focus on the intrinsic capabilities of matchers instead of capabilities of matchers helped by input.

Parameters and resources

Parameters p and resources r of the matching process are identified as:

Oracles and resources: Are oracles permitted? If so, which ones (the answer can be 'any resource')? Is user input permitted?
Training: Can training be performed on a sample?
Proper parameters: Are some parameters necessary? If so, what are they? This point is quite important when a method is very sensitive to the variation of parameters. A good tuning of these must be available.

Many systems take advantage of some external resources, such as WordNet, sets of morphological rules or a previous alignment of general purpose catalogues, e.g., Yahoo and Google. It is perfectly possible to use these resources as long as they have not been specifically tuned for the evaluation, for instance, using a sublexicon which is dedicated to the domain considered by the tests. In addition, it is perfectly acceptable that the algorithms prune or adapt these resources to the actual ontologies, as long as it is the normal process of the algorithm. However, this processing time must be considered within the running time.

Some algorithms can take advantage of the web for selecting resources that are adapted to the considered ontology. This is acceptable. However, this may compromise the replicability of the evaluation results.

In the current state, there is no consensus or valuable methods for handling and evaluating the contribution of user input to the matching process, so allowing it is hard to account for.

Training on some sample is very often used by methods for matching ontologies. Thus, providing a training set can be useful for comparing algorithms based on machine learning. The training set can also be considered as some partial input alignment.

Of course, some parameters can be provided to the methods participating in the evaluation. However, these parameters must be the same for all tests. Only automatic tuning, as part of the matching process, is acceptable.

Output alignment

We identify the following possible constraints on the output alignment A' of the algorithm:

Multiplicity: How many entities of one ontology can correspond to one entity of the others, e.g., injective, total, one-to-one?
Justification: Should a justification of the results be provided?
Relations: Should the relations involved in the correspondences be only the equivalence relations or could they be of other types, such as subsumption ($\leq$) or incompatibility ($\perp$).
Strictness: Can the result be expressed with trust-degrees different from $\top$ and $\perp$ or should they be hardened before?

In real life, there is no reason why two independently developed ontologies should have a particular alignment multiplicity other than *:*. This should be the default (non) constraint on the output alignment. However, if all tests provide some particular type of alignment, e.g., ?:? in the EON ontology tests, this introduces a bias. This bias can be suppressed by having each type of alignment equally represented. This is neither easy to find nor realistic. What would be realistic is to have a statistical evaluation of the proportion of each type of alignment.

Another worthwhile feature for users is the availability of meaningful explanations or justifications of the correspondences. However, in the absence of a standard for explanations (see Chap. 9) it is not possible to evaluate them at the moment.

As mentioned in Chap. 6, some systems associate a relation between the entities that is different from equivalence, e.g., specificity, and some of them associate a degree of confidence to the correspondence. Concerning the relation, not all algorithms deliver the same structure, however, they can deliver equivalence. Thus, this is a common ground for the evaluation. When the set of relations becomes a standard, it will be useful to introduce new relations. As far as the degree of confidence is concerned, reference alignments should express correspondences that hold or not. It is natural that reference alignments contain only $\top$ confidence measure. For the resulting alignment, it is appropriate that an algorithm delivers a weighted alignment. In particular, some useful measures take these weights into account (§7.3.1). However, the lack of consensus on the interpretation of these weights renders such alignments difficult to evaluate.

Matching process

The matching process f itself can be constrained by:

Resource constraints: Is there a maximal amount of time or space available for computing the alignment?
Language restrictions: Is the matching scope limited to some kind of entities, e.g., only classes?
Property: Must some property be true of the alignment? For instance, one might want that the alignment is satisfiable (as defined in Chap. 2) or that it preserves consequences or that the initial alignment is preserved, i.e., $o, o', A' \models A$.

Resource constraints can be considered either as constraining the amount of a resource or a measure of the amount consumed (§7.3.3). It is an important factor, at least for comparative evaluation, and must be measured. It can also be measured for competence benchmarks, even if it is difficult due to the heterogeneity of the environments in which benchmarking is performed.

Constraints on the kind of language construct to be found in alignments can be designed. However, currently very few matching algorithms can align complex expressions. Most of them align the identified (named) entities and some of them are only restricted to concepts. With regard to its importance and its coverage by current matching systems, it makes sense to ask for matching of named entities and consider complex expressions later.

The properties of the alignments provided by the matching algorithms are not very often mentioned and they are very heterogeneous depending of the implemented techniques. It is thus difficult to ask for particular properties.

7.2.2 Examples of data sets

Datasets for matching ontologies are not easy to find. The first problem is that they require pairs of public and well-designed ontologies with a meaningful overlap. There are not that many such ontologies around. Moreover, for evaluating the matching algorithms they should also provide reference alignments, making this number even lower.

In addition, it is necessary to take into account the quality of the ontologies and alignments: the ontologies are more interesting if their matching reflects realistic matching problems and the alignments must be correct or, at least, be the expected ones. To these conditions, [Avesani *et al.*, 2005] added that the data sets have to discriminate among various approaches and have to help in identifying weaknesses of matching systems.

Most of the reference alignments used for evaluation involve human judgements. However, humans are not usually very good at matching ontologies manually. Thus, it would be useful to evaluate the task in which alignments are embedded, e.g., information retrieval, web service invocation, instead of matching itself.

Let us consider some examples of data sets that have been used so far.

OAEI systematic benchmark suite

The data set made for the first OAEI campaign is an artificial data set built from one OWL ontology on the bibliography topic. It contains 33 named classes, 24 object properties, 40 data properties, 56 named individuals and 20 anonymous individuals.

This initial ontology is systematically and automatically altered by distorting the features of ontology languages, e.g., names, properties, subclass relations. The alteration results in a set of more than 50 pairs of ontologies. The kind of proposed alignments is still limited: they only match named classes and properties and they mostly use the equivalence relation with confidence of 1.

This data set can be considered as correct by construction. It is not realistic nor very hard: it is based on small tests and offers some easy ways to reach the correct result. It is especially made for evaluating the strengths and weaknesses of the matching systems. This data set is used for every OAEI campaign in order to measure the evolution of the field.

Large scale ontology sets

There is a need for large scale ontologies to be matched. One early attempt to do this has been reported in [Zhang *et al.*, 2004]. It consists in matching two large ontologies from the domain of anatomy. The two considered ontologies are the Foundational Model of Anatomy (FMA) developed at the University of Washington and Galen developed at the University of Manchester.

These are huge real world ontologies which contain classes, relations, text documentation and labelling. These ontologies contain thousands of classes and barely no instance data. They have been made available in OWL [Euzenat *et al.*, 2005b]. Unfortunately, FMA is not freely available and the permission to use it must be obtained from the University of Washington.

This test case is obviously realistic. The experiments reported in [Zhang *et al.*, 2004] show that it is difficult, and help in finding the weaknesses. The main problem with the task is that it does not have a widely acknowledged reference alignment.

Another candidate test set has been used in [Lambrix and Edberg, 2003]. It consists of matching the Gene Ontology and the Signal ontology. One important point about these two ontologies is that they do not concern exactly the same domain: they only overlap to a certain extent.

These tests have been used for evaluating interactive tools for merging ontologies but they could also be used for evaluating ontology matching systems independently. However, so far, the reference alignments for these tests have not been made available.

Directory sets

[Avesani *et al.*, 2005] proposed a practical way to build test sets for matching web directories or classifications. This has the advantage of generating automatically a test set that changes with time. The idea is to consider two web directories, i.e., a hierarchy of topical web pages that indexes web pages, and to take advantage of the corresponding indexed pages for deciding if two topics are equivalent. This can already be considered as a matching technique in itself (§6.2.8).

This technique has been used in the OAEI-2005 and 2006 campaigns. The collection reported in [Avesani *et al.*, 2005] contained from 300 000 to 800 000 topics. It has been characterised by the authors as a difficult test.

Thesauri

There is currently a wealth of resources available from the fields of libraries, museums and more generally cultural heritage. These are thesauri covering hierarchies of concepts considered as terms and large amounts of textual knowledge, usually about pieces of art. A huge interest in using different thesauri together has created the need for matching them.

For the OAEI-2006, a pair of thesauri on the topic of food has been provided: AGROVOC is a thesaurus for the Food and Agriculture Organisation and NAL is a thesaurus from the US Agricultural department. They respectively contain 16 000 and 41 000 terms. The two test sets have been made available in SKOS (§8.1.7).

This task can be considered as a representative of real world matching problems. These thesauri are certainly challenging with respect to their size. However, at the moment, there is no reference alignment for these tests.

Other test collections

The Illinois Semantic Integration Archivehad offered until recently some of the data sets that have been used for evaluating various systems presented in Chap. 6. In addition, the ontology matching web page refers to test cases that have been built and published by matching tools designers[4].

7.3 Evaluation measures

In order to evaluate the results of matching algorithms, it is necessary to confront them with ontologies to be matched and to compare the alignments produced with a reference alignment based on some criteria.

---
[4] http://www.ontologymatching.org/evaluation.html

This section is concerned with the question of how to measure the results returned by ontology matchers. It considers different possible measures for evaluating matching algorithms and systems. They include both qualitative and quantitative measures. We divide them into compliance measures and performance measures. Compliance measures evaluate the degree of conformance of returned alignments to what is expected. We will present some classical evaluation measures (§7.3.1) and measures especially designed for ontology matching evaluation (§7.3.2). Performance measures account for such features of algorithms as speed and memory consumption (§7.3.3). User-related measures focus on evaluation of user interaction with a matching system (§7.3.4).

7.3.1 Compliance measures

Compliance measures evaluate the degree of compliance of a system with regard to some standard. They can be used for computing the quality of the output provided by a system compared to a reference output. As noted before (§7.2), such a reference output is not always available, not always useful and not always consensual. However, for the purpose of benchmarking, we can assume that it is desirable to provide such a reference.

There are many ways to qualitatively evaluate returned results. One common possibility consists of proposing a reference alignment R to which the result from the evaluated matching algorithm A is compared. In what follows, the alignments A and R are considered to be sets of correspondences, being pairs of entities.

Let us consider the case of the evaluation of two alignments on classes with equivalence relations that must be compared to the alignment of Fig. 2.9 (p. 48). This alignment is considered to be the reference alignment (R) and is made up of three correspondences:

Book = Volume	Person = Human	Science = Essay	(R)

The alignments to be compared are nearmiss (A_1) and farone (A_2). These are presented in Fig. 7.1. They are made up of the following correspondences:

Product = Volume	Person = Writer	Science = Essay	(A_1)
Book = Volume	Children = Literature	Pocket = Essay	(A_2)

A first simple distance between two sets is the Hamming distance. It measures the dissimilarity between two alignments by counting the joint correspondences with regard to the correspondences of both sets.

Definition 7.1 (Hamming distance). *Given a reference alignment R, the Hamming distance between R and some alignment A is a dissimilarity $H : \Lambda \times \Lambda \to [0\ 1]$ defined as follows:*

$$H(A, R) = 1 - \frac{|A \cap R|}{|A \cup R|}.$$

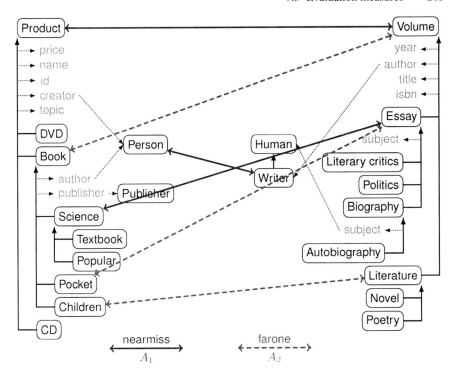

Fig. 7.1. Two class alignments between the ontologies of Fig. 2.7. These are to be compared with the alignment of Fig. 2.9 restricted to class correspondences.

Example 7.2 (Hamming distance between alignments). Taking the class part of the alignment of Fig. 2.9, as the reference alignment, we can compare it with the result given by the alignments of Fig. 7.1. In both cases, the Hamming distance between these alignments is very high: .8. The shorter the distance, the better. Indeed, both alignments only found one correct correspondence out of three. Thus, this results in two inaccurate correspondences and two missed correspondences.

The most prominent criteria are *precision* and *recall* originating from information retrieval [van Rijsbergen, 1975] and adapted to ontology matching [Do et al., 2002]. Precision and recall are based on the comparison of the resulting alignment A with a reference alignment R, effectively comparing which correspondences are discovered and which are not. These criteria are well understood and widely accepted.

Precision measures the ratio of correctly found correspondences (true positives) over the total number of returned correspondences (true positives and false positives), see Fig. 7.2. In logical terms, precision is meant to measure the degree of correctness of the method.

Definition 7.3 (Precision). *Given a reference alignment R, the precision of some alignment A is a function $P : \Lambda \times \Lambda \to [0\ 1]$ such that:*

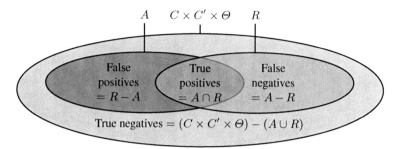

Fig. 7.2. Two alignments as sets of correspondences and relations between them.

$$P(A, R) = \frac{|R \cap A|}{|A|}.$$

Precision can also be determined without explicitly having a complete reference alignment. In this case only the correct alignments among the retrieved alignments have to be determined, namely $R \cap A$.

Recall measures the ratio of correctly found correspondences (true positives) over the total number of expected correspondences (true positives and false negatives). In logical terms, recall is meant to measure the degree of completeness of the alignment.

Definition 7.4 (Recall). *Given a reference alignment R, the recall of some alignment A is a function $R: \Lambda \times \Lambda \to [0\ 1]$ such that:*

$$R(A, R) = \frac{|R \cap A|}{|R|}.$$

Notice that in the definition above, the letter R stands for both the recall function and the reference alignment. Since one is a function and the other is a set, these are easy to be distinguished by their use even if referred by the same letter.

Table 7.1. The results of the evaluation measures for the two alignments of Fig. 7.1 as well as for the reference alignment itself.

Alignment	Precision	Recall	F-measure	Fallout	Overall	SBS
refalign	1.00	1.00	1.00	0.00	1.00	1.00
nearmiss (A_1)	0.33	0.33	0.33	0.67	-0.33	0.27
farone (A_2)	0.33	0.33	0.33	0.67	-0.33	0.27

The fallout measures the percentage of retrieved pairs which are false positives.

7.3 Evaluation measures

Definition 7.5 (Fallout). *Given a reference alignment R, the fallout of some alignment A is a function $F : \Lambda \times \Lambda \to [0\ 1]$ such that:*

$$F(A, R) = \frac{|A| - |A \cap R|}{|A|} = \frac{|A \setminus R|}{|A|}.$$

Noise and silence are the complement measures of precision and recall. These are defined as follows: $Noise(A, R) = 1 - P(A, R)$ and $Silence(A, R) = 1 - R(A, R)$.

Although precision and recall are the most widely and commonly used measures, when comparing systems one may prefer to have only a unique measure. Moreover, systems are often not comparable based solely on precision and recall. The one which has higher recall may have a lower precision and vice versa. So, it is not a good idea to compare systems on precision and recall alone. For this purpose, two measures are introduced which aggregate precision and recall: F-measure and overall.

Definition 7.6 (F-measure). *Given a reference alignment R and a number α between 0 and 1, the F-measure of some alignment A is a function $M_\alpha : \Lambda \times \Lambda \to [0\ 1]$ such that:*

$$M_\alpha(A, R) = \frac{P(A, R) \times R(A, R)}{(1 - \alpha) \times P(A, R) + \alpha \times R(A, R)}.$$

If $\alpha = 1$, then the F-measure is equal to precision and if $\alpha = 0$, the F-measure is equal to recall. In between, the higher the value of α, the more importance is given to precision with regard to recall. Very often, the value $\alpha = 0.5$ is used, i.e.,

$$M_{0.5}(A, R) = \frac{2 \times P(A, R) \times R(A, R)}{P(A, R) + R(A, R)}.$$

This is the harmonic mean of precision and recall. Such a measure can be used for selecting the parameters, in particular, a threshold to put on the results, such that the F-measure is optimal. Moreover, it allows comparing systems by their precision and recall at the point where their F-measure is maximal.

The overall measure, also defined in [Melnik et al., 2002] as matching accuracy, is the ratio of the number of errors on the size of the expected alignment. It is considered as an edit distance between an alignment and a reference alignment in which the only operation is 'error correction'. In this respect, it is considered as a measure of the effort required to fix the alignment. The overall is always lower than the F-measure and it ranges between $[-1\ 1]$. In fact, if precision is under .5 overall has a negative value denoting that repairing the alignment is not worth the effort.

Definition 7.7 (Overall). *Given a reference alignment R, the overall measure of some alignment A is a function $O : \Lambda \times \Lambda \to [-1\ 1]$ such that:*

$$O(A, R) = R(A, R) \times \left(2 - \frac{1}{P(A, R)}\right).$$

208   7 Evaluation of matching systems

Alternatively, it can also be defined as follows:

$$O(A, R) = 1 - \frac{|(A \cup R) - (A \cap R)|}{|R|} = 1 - \frac{|R - A| + |A - R|}{|R|}.$$

When comparing systems in which precision and recall can be continuously determined, it is convenient to draw the precision/recall curve and compare these curves (see Fig. 7.3). There are two advantages of these curves: (i) they allow the comparison of alignments with confidence measures, (ii) they are independent of the interpretation of the confidence, only their induced order is relevant. This kind of measure is widespread when presenting results (in the TREC competitions for instance).

Non equal correspondences

Currently, the proposed compliance measures are purely related to the identity of the correspondences, including or ignoring the degree of confidence (strength) and relation.

This is not satisfactory because this does not account for the semantics of the relations and strengths. In order to provide more accurate comparisons, it is necessary to be able to measure some distance between strengths and relations. The distance between the strengths of two correspondences can be considered to be the absolute value between the two strength values. This can be used for comparing two sets of correspondences on the basis of the strengths attributed to each correspondence.

Definition 7.8 (Strength-based similarity). *Given a reference alignment R, the strength-based similarity between R and some alignment A is defined as follows:*

$$SBS(A, R) = \frac{2 \times \sum_{c \in A \dot{\cap} R} |strength_A(c) - strength_R(c)|}{|A| + |B|},$$

where $A \dot{\cap} R = \{\langle e, e', r, n_A, n_R \rangle; \langle e, e', r, n_A \rangle \in A \wedge \langle e, e', r, n_R \rangle \in R\}$, *such that* $\forall e, e', r, \langle e, e', r, 0 \rangle \in A \cup R$.

The denominator of strength-based similarity can be used instead of the intersection in each of the definitions of Sect. 7.3.1. [Ehrig, 2007] also pointed out that these measures must be handled carefully since their results are to be judged with regard to statistical significance.

Some examples of measures introduced so far are given in Table 7.1.

7.3.2 Generalising precision and recall

Although precision and recall are well understood and widely accepted, they have a drawback: whatever correspondences have not been discovered are definitely not considered (all-or-nothing). As a result, they do not discriminate between an alignment that may be very close to the expected result and another quite remote from it and they do not measure the effort required from users to correct alignments. In fact,

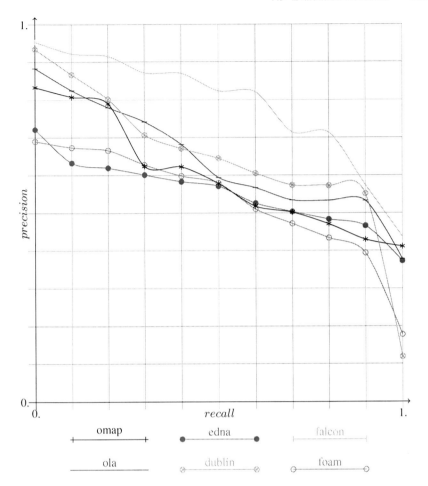

Fig. 7.3. Six precision/recall curves of the OAEI-2005 participants who scored above the simple Edit distance measure (edna) on class names. The curves are obtained from the provided alignments in which correspondences are ordered by decreasing confidence. For each decimal value v between 0 and 100, the algorithm selects the first correspondences in order to have $v\%$ recall, i.e., the corresponding percentage of correspondences from the reference alignment, and it reports the precision at that point, i.e., with only these correspondences. More details can be found in [Euzenat *et al.*, 2005b].

the alignment A_1 of Fig. 7.1 is arguably better than the alignment A_2. However, as testified in Table 7.1, they score exactly the same for all presented measures.

Often, it makes sense to not only have a decision whether a particular correspondence has been found or not, but also a measure of the proximity of the found alignments. This implies that *near misses* are also taken into consideration instead of only the exact matches.

Moreover, the alignments have to go through users scrutiny and correction before being used. Therefore, it is worth measuring the effort required from users for correcting the provided alignment instead of only if some correction is needed. This also calls for a relaxation of precision and recall.

Similar remarks have been made in computational linguistics and some solutions have been proposed in [Langlais et al., 1998, Sun and Lin, 2001]. In the context of alignment evaluation, [Ehrig and Euzenat, 2005] investigated generalising precision and recall in order to overcome these problems. As precision and recall are easily explained measures, it is good to extend them. This also ensures that measures derived from precision and recall, e.g., F-measure, still can be computed easily. Relaxing precision and recall amounts to measuring the proximity of alignments rather than the strict size of their overlap. Instead of taking the cardinality of the intersection of the two sets $|R \cap A|$, natural relaxations of precision and recall measure their proximity $\omega(A, R)$. Moreover, the relaxed measure is required to preserve positiveness, maximality and boundedness satisfied by $|R \cap A|$.

Definition 7.9 (Relaxed precision and recall). *Given a reference alignment R and an overlap function ω between alignments, the precision of an alignment A is defined as follows:*

$$P_\omega(A, R) = \frac{\omega(A, R)}{|A|},$$

and recall is defined as follows:

$$R_\omega(A, R) = \frac{\omega(A, R)}{|R|},$$

with ω satisfying the following conditions:

$$\forall A, B, \omega(A, B) \geq 0 \qquad \text{(positiveness)}$$
$$\forall A, B, \omega(A, B) \leq \min(|A|, |B|) \qquad \text{(maximality)}$$
$$\forall A, B, \omega(A, B) \geq |A \cap B| \qquad \text{(boundedness)}$$

We do not require symmetry, especially since R and A are not in symmetrical positions: the former is the reference and the latter is judged against it. There are many different ways to design a proximity between two sets satisfying these properties. The most obvious one, which we have retained, consists of finding correspondences that match each other and computing the sum of their proximity. This can be defined as an overlap proximity:

Definition 7.10 (Overlap proximity). *The overlap proximity between two alignments A and R is defined as follows:*

$$\omega(A, R) = \sum_{\langle a, r \rangle \in M(A, R)} \sigma(a, r),$$

such that

- $M(A, R)$ is the best match between the correspondences of A and R with regard to σ;
- $\sigma(a, r)$ is a similarity measure between two correspondences.

The standard measure $|A \cap R|$ used in precision and recall is such an overlap proximity with σ which provides the value 1 if the two correspondences are equal and 0 otherwise.

From this simple set of constraints, there have been designed several concrete measures detailed in [Ehrig and Euzenat, 2005]:

Symmetric measure calculates the distance in the ontologies between the found entities and the reference ones. The closer they are, the higher the similarity.

Effort-based computes the effort necessary to modify the errors found in the alignments: it is based on a model of what is involved in modifying an alignment through an alignment editor for retrieving the reference alignment. This measure is arguably better than the overall [Melnik et al., 2002] presented before because it can weight differently different errors depending on the difficulty to correct it which itself depends on the editing environment used.

Oriented is a specific measure which uses a different ω for precision and recall depending on the impact an error has on these measures. For example, when one wants to retrieve instances of some class, a subclass of the expected one is correct but not complete and thus affects recall but not precision. This measure is targeted at application-specific evaluation.

If these proposed precision and recall measures are applied to the alignments of Fig. 7.1, they yield the results of Table 7.2. They mainly illustrate entity pair similarities, as relations and confidences are always identical. For the oriented measure we assume that the query is given in ontology o and the answer has to be retrieved from ontology o'. Since the oriented measure is non symmetric, one has to define the direction beforehand. The results are the same between the three proposed generalisations due to the simple nature of the example. They show a better discrimination between the nearmiss and the farone alignments.

Table 7.2. Precision and recall results on the alignments of Fig. 7.1.

ω	(R, R)		(R, A_1)		(R, A_2)	
	P	R	P	R	P	R
standard	1.00	1.00	0.33	0.33	0.33	0.33
symmetric	1.00	1.00	0.50	0.50	0.33	0.33
effort-based	1.00	1.00	0.50	0.50	0.33	0.33
oriented	1.00	1.00	0.50	0.50	0.33	0.33

The measures which have been introduced address the problems raised in the introduction and fulfil the requirements:

- They keep precision and recall untouched for the best alignment R;

- They help discriminate between irrelevant alignments, such as A_2, and those, which are not far from target ones, like A_1;

Another development currently under investigation consists of developing similar measures to account for the semantics of the language used for ontologies [Euzenat, 2007].

7.3.3 Performance measures

Performance measures assess the resource consumption for matching two ontologies. They can be used when the algorithms are 100% compliant or balanced against compliance [Ehrig and Staab, 2004]. We mention some of these criteria below.

Unlike the compliance measures, performance measures depend on the processing environment and the underlying ontology management system. Thus, it is difficult to obtain objective evaluations, because they are based on the usual measures, namely processing time in seconds and memory in bytes. The important point here is that algorithms that are being compared should be run under the same conditions.

Speed

Speed is measured by the amount of time taken by the algorithms for performing their matching tasks. It should be measured in the same conditions, i.e., same processor, same memory consumption, for all the systems. If user interaction is required, one has to ensure that only the processing time of the matching algorithm is measured.

Memory

The amount of memory used for performing the matching task marks another performance measure. Due to the dependency with underlying systems, it could also make sense to measure only the extra memory required in addition to that of the ontology management system, but it still remains highly dependent.

Scalability

There are two possibilities for measuring scalability, at least in terms of speed and memory requirements. Firstly, it can be assessed by a theoretical study. Secondly, it can be assessed by evaluation campaigns with quantified tests of increasing complexity. From the results, the relationship between the complexity of the test and the required amount of resources can be represented graphically and the mathematical relationship can be approximated.

7.3.4 User-related measures

So far the measures have been machine focused. In some cases, algorithms or applications require some kind of user interaction. This can range from users using the alignment results to concrete user input during the matching process. In this case, it is even more difficult to obtain an objective evaluation. Below we discuss measures which involve users in the evaluation loop.

Level of user input effort

In cases where algorithms require user intervention, this intervention could be measured in terms of some elementary information the users provide to the system, e.g., the number of correspondences. When comparing systems which require different input or no input from users, it is necessary to consider a standard for elementary information to be measured. This is not an easy task.

General subjective satisfaction

From a use case point of view it makes sense to directly measure user satisfaction. As this is a subjective measure it cannot be assessed easily. Extensive preparations have to be made to ensure a valid evaluation. Almost all of the objective measures mentioned so far have a subjective counterpart. Possible measurements include:

– input effort,
– speed,
– resource consumption, e.g., memory,
– output exactness, related to precision,
– output completeness, related to recall, and
– understandability of results, e.g., explanations.

Due to its subjective nature numerical ranges as evaluation results are less appropriate than qualitative values, such as very good, good, satisfactory.

7.4 Application-specific evaluation

So far evaluation has been considered in general. However, the evaluation could also be considered in the context of a particular application or a particular kind of applications. Application-specific evaluation is dedicated to find a suitable system for a particular task. This is especially useful for application designers who need to integrate a matching system and this complements the requirement satisfaction analysis as presented in Sect. 6.5.

There are two main complementary ways to design application-specific evaluations: (i) using a specific test set and experiment design; (ii) interpreting the results with an application-oriented bias. As a matter of fact, there are tasks which require high recall (for instance, matching as a first step of an interactive merge process) and others which require high precision, e.g., automatic matching for autonomously connecting two web services.

[Ehrig, 2007] provided an analysis of the different needs for evaluation depending of the considered applications. We have applied this technique to the requirements of Table 1.1 (see Chap. 1). As a matter of fact, it can be rewritten with respect to the measurements developed in this chapter. We used this technique to design Table 7.3. This table is slightly more detailed than Table 1.1 because it uses three values instead of two. Here 'low' corresponds to not relevant, 'high' corresponds to relevant and 'medium' corresponds to an in-between position. Therefore, different *application profiles* could be established to explicitly compare matching algorithms with respect to certain tasks.

Table 7.3. Application requirements of Table 1.1 reinterpreted as measurement weights.

Application	speed	automatic	precision	recall
Ontology evolution (§1.1)	medium	low	high	high
Schema integration (§1.2)	low	low	high	high
Catalogue integration (§1.2)	low	low	high	high
Data integration (§1.2)	low	low	high	high
P2P information sharing (§1.3)	high	low	medium	medium
Web service composition (§1.4)	high	high	high	low
Multi agent communication (§1.5)	high	high	high	medium
Context matching in ambient computing (§1.5)	high	high	high	medium
Semantic web browsing (§1.6)	high	medium	medium	low
Query answering (§1.6)	high	medium	high	medium

Such a table can be useful for aggregating the measures corresponding to each of these aspects with different weights or to have an ordered way to interpret evaluation results.

7.4.1 Aggregating measures

Different measures suit different evaluation goals. If we want to improve systems, it is best to have as many indicators as possible. However, in order to single out the best system, it is generally better to focus on the very relevant factors. This can be achieved by only selecting the few very relevant factors, e.g., speed and precision for query answering, or by aggregating measures in relation with their relevance.

For aggregating measures depending on a particular application, its is possible to use weights corresponding to the values of Table 7.3, and thus respecting the importance of each factor. Weighted aggregation measures, such as those presented in Sect. 5.2 (weighted sum, product or average), can be used.

7.4 Application-specific evaluation

F-measure is already an aggregation of precision and recall. It can be generalised as a harmonic mean, for any number of measures. This requires us to assign every measurement a weight, such that these weights sum to 1. Obviously the weights have to be chosen carefully, again depending on the goal.

Definition 7.11 (Weighted harmonic mean). *Given a reference alignment R, a set of measures $(M_i)_{i \in I}$ provided with a set of weights $(w_i)_{i \in I}$ between 0 and 1 such that their sum is 1, the weighted harmonic mean of some alignment A is as follows:*

$$W(A, R) = \frac{\prod_{i \in I} M_i(A, R)}{\sum_{i \in I} w_i \times M_i(A, R)}.$$

Example 7.12 (Weighted aggregation of evaluation measures). Consider that we need to choose among two available systems S_1 and S_2, for an application, such as schema integration, peer-to-peer system or query answering. We apply weights corresponding to the criteria of Table 7.3. The weights are $high = 5$, $medium = 3$ and $low = 1$. They are normalised (so as to sum to 1.) for each kind of application. The performance of the systems with regard to the criteria are given in the following table as well as the resulting harmonic means for each pair system×application:

					S_1	S_2
				Speed	.8	.5
				Automatic	1.	1.
				Precision	.7	.9
	Speed	Automatic	Precision	Recall	.8	.8
Schema integration	.08	.08	.42	.42	.77	.81
Peer-to-peer system	.42	.08	.25	.25	.79	.66
Query answering	.31	.19	.31	.19	.80	.72

Those who need a matching system for a peer-to-peer or query answering application should choose system S_1 (.79 and .80 are better than .66 and .72) and those who want to use it for schema integration should use system S_2 (.81 is better than .77). The importance of speed in the two first systems outweights the relatively lower precision.

7.4.2 Evaluation setting

Application-specific evaluation can also be carried out by having a specific evaluation setting. This has the advantage of being more realistic than artificial test beds and of providing very specific information, but the drawback is that it has to be changed for each different application.

An application-specific evaluation has to start with a selection of the task as described in Sect. 7.2.1 corresponding to the application. It is moreover useful to set up experiments which do not stop at the delivery of alignments but carry on with the particular task. This is especially true when there is a clear measure of success of the overall task. Such a setting assists in focusing on the most useful issues for the task. For instance, it may be the case that the gain in accuracy in one algorithm over another is not useful for the task while the gain in speed of the latter really matters. If no clear measure is available, then using a weighted aggregation measure as suggested above would help.

Nevertheless, it is extremely difficult to determine the evaluation value of the matching process independently. The effects of other components of the overall application have to be carefully filtered out.

There are several problems associated with this approach:

- It will be difficult to account for the performances of matching algorithms if the systems which carry the task are different. If these systems provide an end-to-end integration, then the evaluation would be simpler in comparing the alignments and applying a specific metric.
- Very often the matching systems are considered as semi-automatic, i.e., users must control the result. The task cannot be accomplished in isolation (and this brings back the issue of involving users in the evaluation loop).
- This would require a specific setting for each task.

7.5 Summary

As noted before, it is very difficult to know a priori the quality to expect from a matching system. For that purpose, evaluation of matching systems must be performed. The evaluation must apply to all aspects of these systems: their availability, their capacity to provide accurate alignments in a desirable time, etc.

At least two difficulties arise when evaluating matching systems. Matching tasks are so different that a system can perform very well on some data and not that well on some other. It is thus necessary that evaluation data sets are as different as possible and that results are kept separate so that someone with a particular task can choose a system that performs adequately on this task. The second difficulty is the choice of evaluation criteria. As mentioned in Sect. 7.3.2, precision and recall are not always appropriate for ontology matching and must be improved to account for the semantics of ontologies.

Over several years we have implemented an ongoing effort for evaluating ontology matching algorithms under the Ontology Alignment Evaluation Initiative[5]. The main purpose of OAEI is to improve the quality of ontology matching algorithms by continuous comparison with other new methods. It aims at providing quality benchmark suites that system designers can use for training their systems as well as for organising evaluation campaigns for comparing systems.

---

[5] http://oaei.ontologymatching.org

Part IV

Representing, explaining, and processing alignments

8
Frameworks and formats: representing alignments

Once matching is performed, the resulting alignments are usually used in a wider context than a matching system itself. To that extent, several proposals have been made for representing the alignments and exchanging them among tools. This chapter is concerned with these topics.

Alignments can be stored, exchanged and manipulated in a variety of ways. We present some frameworks and formats that help do so. In particular, we address the following aspects:

- Formats that enable the syntactic expression and the manipulation of alignments and that can be used for exchanging the alignments across applications (§8.1).
- Frameworks that provide a wider set of operations for manipulating alignments. These frameworks and languages are usually not concerned with the way alignments are found. They, at best, define a match operation which generates alignments, however, they allow plugging in matching methods whose alignments can be manipulated (§8.2).
- Finally, ontology editors provide environments for either manually or automatically creating alignments and then using them, for instance, for importing data and merging ontologies (§8.3).

The first category follows an open philosophy in which alignments can be used in any context, while the latter categories impose a closed interpretation of alignments, however, they can be open to the integration of new tools.

8.1 Alignment formats

As can be noted from Chap. 6, many matching systems deliver alignments. Applications could use this property for replacing a matcher by another one or combining several of them. Unfortunately, the alignment is very often output as a mere list of pairs, HTML table or similarity matrix. We consider here what could be a proper format in order for these systems to interoperate.

8 Frameworks and formats: representing alignments

We briefly present various formats that have been proposed so far for expressing relations between ontologies. We mostly compare these formats on the basis of their syntax. A deeper analysis of some of these in terms of semantics and expressiveness is provided in [Serafini *et al.*, 2005].

For the purpose of a uniform comparison of these formats, we use an example that extends the one of Fig. 2.9. It consists of expressing that:

- a Science book in the left-hand side ontology corresponds to an Essay whose subject is an instance of Science in the right-hand side ontology,
- a Pocket book in the left-hand side ontology corresponds to a Volume whose size is less than 14 in the right-hand side ontology,
- a Book which has politics as a topic in the left-hand side ontology corresponds to a Politics essay in the right-hand side ontology, and
- a Writer in the left-hand side ontology is someone who have authored a volume in the right-hand side ontology.

We will consider that the first ontology is identified by the XML &onto1; entity or the http://book.ontologymatching.org/example/culture-shop.owl URL and the second ontology is identified by the XML &onto2; entity standing for the http://book.ontologymatching.org/example/library.owl URL.

Three of these extra correspondences are displayed in Fig. 8.1. When translated into first-order logic, these can be represented as follows:

$$\forall x, Pocket(x) \equiv Volume(x) \land size(x,y) \land y \leq 14$$
$$\forall x, Science(x) \equiv Essay(x) \land (\forall y, subject(x,y) \Rightarrow Science(y))$$
$$\forall x, Book(x) \land topic(x, politics) \equiv Politics(x)$$
$$\forall x, Writer(x) \Leftarrow \exists y, author(y,x)$$

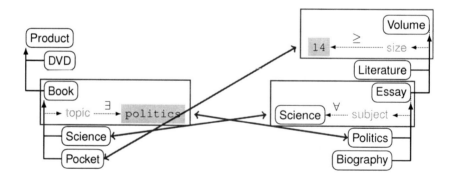

Fig. 8.1. Three correspondences that can be set in an elaborate alignment format.

Let us discuss the main formats available for expressing such alignments.

8.1.1 MAFRA Semantic bridge ontology (SBO)

MAFRA [da Silva, 2004, Mädche et al., 2002] stands for MApping FRAmework[1] (see also Sect. 8.2.3). It is a system for extracting mappings from ontologies and executing them as data transformation from one ontology to another one. The system was first designed to work with the DAML+OIL language, an ancestor of OWL.

MAFRA does not define a real exchange format for ontology alignment. Rather, it provides an ontology, called the Semantic Bridge Ontology. The instantiation of this ontology constitutes an ontology mapping document. The serialisation of this format has not been described in detail in documents so we freely use our own transcription[2].

The main concepts in this ontology are SemanticBridges and Services. A SemanticBridge is tied to the Services that are able to implement the bridge as a data transformation. A Service can be thought of as a function: $f : Arg^n \longrightarrow Arg^m$ that maps tuples of arguments into tuples of arguments. It can be identified by a URI, for instance. The arguments are typed and can be ontology concepts, property paths, literals or arrays of these.

SemanticBridges, which in turn can be ConceptBridges or PropertyBridges, express a relation between two sets of entities by composing elementary services that are applied to them. For instance, a SemanticBridge between two ontologies can map those Volumes with size larger than 14 to Pocket books in the following way:

```
ConceptBridge: Volume2Pocket
    x: <o2#Volume>; o2:size >= 14 -> <o1#Pocket>
ConceptBridge: Book2Politics
    x: <o1#Book>; o1:topic == 'politics' -> <o2#Politics>
```

Entities to be mapped are identified within the ontology (instances) through a path. Paths serve the dual purposes (i) of navigating within the ontology structure and (ii) of providing the context for further characterising the concerned entities. In this context paths play exactly the same role as in Xpath [Clark and DeRose (ed.), 2001]. They are further enriched with conditions. In the example above, `<o2#Volume>; o2:size >= 14` is a path with condition that the final step size has a value inferior to 14.

An ontology mapping document satisfying the semantic bridge ontology is a collection of such bridges plus information on the concerned ontologies, as well as constraints: for example, the *exclusivity* conditions that can be used for expressing that an entity cannot be mapped by more than one bridge rule.

The semantic bridge ontology provides a framework and a format for expressing alignments. This format is used as output from ontology matchers and input for data transformations.

The format provided by the Semantic Bridge Ontology is not very clear since the language is described in UML. This is a minor problem that could be solved by

---

[1] This is also the name of a city in Portugal featuring a rich palace.
[2] [Mädche et al., 2002] presents the Semantic Bridge Ontology as a DAML+OIL ontology, but it turns out to have evolved a lot since then and we have not been able to find a serialisation that could stand as a proper format.

222   8 Frameworks and formats: representing alignments

exposing some RDF/XML format (a previous version of the framework had been described as a DAML ontology [Mädche *et al.*, 2002]). Moreover, this format is a relatively complex language that is tied to the MAFRA architecture (§8.2.3). It does not separate the declarative aspect of relations from the more operational aspects of services: the relations are described with regard to the services able to implement them. The services can be arbitrary small, such as string concatenation, or large, such as implementing a complete alignment by a program. On the one hand, this guarantees that these alignments can be used: SBO-documents can readily be used as data translation. On the other hand, this does not favour the use of these alignments in other ways, for instance, for merging ontologies or mediating queries.

8.1.2 OWL

OWL can be considered as a language for expressing correspondences between ontologies. As a matter of fact, the equivalentClass and equivalentProperty primitives have been introduced for relating elements in ontologies describing the same domain. This use has been documented by the W3C best practices working group [Uschold, 2005]. Moreover, these primitives are only shorthands for other primitives, e.g., subClassOf, subPropertyOf, that already allow the relation of entities. For example, the following OWL ontology fragment

```
<owl:Property rdf:about="&onto1;#author">
  <owl:equivalentProperty rdf:resource="&onto2;#author"/>
</owl:Property>

<owl:Class rdf:about="&onto1;#Book">
  <owl:equivalentClass rdf:resource="&onto2;#Volume"/>
</owl:Class>

<owl:Class rdf:about="&onto2;#title">
  <owl:subClass ="&onto1;#name"/>
</owl:Class>
```

can be seen as an alignment expressing the equivalence of properties author and author, the equivalence of classes Book and Volume and the coverage of property title in the second ontology by name in the first one. Moreover, any ontology, as soon as it involves entities from different ontologies, expresses alignments. For instance, the following OWL ontology fragment

```
<owl:Class rdf:ID="&onto1;#Science">
  <owl:equivalentClass>
    <owl:Class>
      <owl:subClassOf rdf:resource="&onto2;#Essay"/>
      <owl:subClassOf>
        <owl:Restriction>
          <owl:onProperty rdf:resource="&onto2;#subject"/>
          <owl:allValuesFrom rdf:resource="&onto2;#Science"/>
        </owl:Restriction>
      </owl:subClassOf>
    </owl:Class>
  </owl:equivalentClass>
</owl:Class>
<owl:Class rdf:ID="&onto2;#Writer">
  <owl:subClassOf>
```

```
    <owl:Restriction>
      <owl:onProperty rdf:resource="&onto1;#hasWritten"/>
      <owl:minCardinality
        rdf:datatype="&xsd;nonNegativeInteger">1</owl:minCardinality>
    </owl:Restriction>
  </owl:subClassOf>
</owl:Class>
```

expresses that a Science book in the first ontology is equivalent to an Essay whose subject is an instance of the Science class in the second one and that those who have written at least one thing in the first ontology are Writer in the second one.

Not surprisingly, the OWL language can be used as an alignment expression language. However, using it this way has some drawbacks:

1. It forces the use of a particular ontology language: OWL. It is still possible to relate in this way ontologies that are expressed in other languages without benefiting from the construction of complex terms. However, the alignment will not benefit from the content of the ontologies themselves.
2. It mixes correspondences and definitions. This is a problem for the clarity of alignments as well as for lightweight applications which do not want to interpret the OWL language.
3. It is interpreted only in the framework of a global interpretation of one OWL theory. It is difficult to use this expression for only importing data expressed under one ontology into another one because this application requires sorting out definitions from correspondences.

Other languages have been designed for overcoming these problems. For example, SKOS solves the first problem, but introduces its own language, SWRL solves the second problem and C-OWL attempts to solve the third problem. These languages are presented hereafter.

8.1.3 Contextualized OWL (C-OWL)

C-OWL is an extension of OWL to express mappings between heterogeneous ontologies [Bouquet *et al.*, 2003a, Bouquet *et al.*, 2004b]. The new constructs in C-OWL, with respect to OWL, are called *bridge rules*, and they allow the expression of relations between classes, relations and individuals interpreted in heterogeneous domains.

Bridge rules are oriented correspondences, from a source ontology o to a target ontology o'. They use a set of five relations: more general ($\sqsupseteq$), more specific ($\sqsubseteq$), equivalent ($\equiv$), disjoint ($\perp$) and overlap ($*$). These relations are always applied to named entities. Bridge rules are always interpreted from the standpoint of the target ontology. They express how the target ontology translates the source ontology in itself.

Bridge rules are expressed separately from the ontologies they refer to. The examples considered here are given below in C-OWL XML syntax:

```xml
<?xml version="1.0"?>
<!DOCTYPE rdf:RDF [
    <!ENTITY rdf   "http://www.w3.org/1999/02/22-rdf-syntax-ns#" >
    <!ENTITY owl   "http://www.w3.org/2002/07/owl#" >
    <!ENTITY cowl  "http://www.itc.it/cowl#" >
    <!ENTITY onto1 "http://book.ontologymatching.org/example/culture-shop.owl" >
    <!ENTITY onto2 "http://book.ontologymatching.org/example/library.owl" >
  ]>
<rdf:RDF
  xmlns      ="&cowl;"
  xmlns:cowl ="&cowl;"
  xmlns:owl  ="&owl;"
  xmlns:rdf  ="&rdf;"
  >
<cowl:Mapping>
    <cowl:sourceOntology>
      <owl:Ontology rdf:about="&onto1;"/>
    </cowl:sourceOntology>
    <cowl:targetOntology>
      <owl:Ontology rdf:about="&onto2;"/>
    </cowl:targetOntology>
    <cowl:bridgeRule>
      <cowl:Into>
        <cowl:source>
          <owl:Class rdf:about="&onto1;#Book"/>
        </cowl:source>
        <cowl:target>
          <owl:Class rdf:about="&onto2;#Volume"/>
        </cowl:target>
      </cowl:Into>
    </cowl:bridgeRule>
    <cowl:bridgeRule>
      <cowl:Onto>
        <cowl:source>
          <owl:Class rdf:about="&onto1;#name"/>
        </cowl:source>
        <cowl:target>
          <owl:Class rdf:about="&onto2;#title"/>
        </cowl:target>
      </cowl:Onto>
    </cowl:bridgeRule>
    <cowl:bridgeRule>
      <cowl:Equivalent>
        <cowl:source>
          <owl:Class rdf:about="&onto1;#author"/>
        </cowl:source>
        <cowl:target>
          <owl:Class rdf:about="&onto2;#author"/>
        </cowl:target>
      </cowl:Equivalent>
    </cowl:bridgeRule>
    <cowl:bridgeRule>
</cowl:Mapping>
</rdf:RDF>
```

The C-OWL proposal can express relatively simple alignments: no constructed classes are expressed, only named classes are used. The more expressive part resides in the relations used by the mapping. These alignments have a clear semantics, however it is given from a particular standpoint: that of the target ontology. C-OWL is based on the OWL language but relatively independent from this language which is confined at expressing the entities (the alignment part being specific).

8.1.4 SWRL

It is sometimes not enough to be able to express entity definitions; sometimes expressing rules is a more convenient expression mean. Moreover, rules can bring more expressiveness. SWRL (Semantic Web Rule Language) [Horrocks *et al.*, 2004] is a rule language for the semantic web. It extends OWL with an explicit notion of rule (from RuleML) that is interpreted as first order Horn clauses. These rules can be understood as correspondences between ontologies, especially when elements from the head and the body are from different ontologies, oriented as in C-OWL.

SWRL mixes the vocabulary from RuleML for exchanging rules with the OWL vocabulary for expressing knowledge. It defines a rule (ruleml:imp) with a body (ruleml:_body) and head (ruleml:_head) parts.

```
<ruleml:imp>
  <ruleml:_body>
    <swrlx:classAtom>
      <owlx:Class owlx:name="&onto1;#Book" />
      <ruleml:var>p</ruleml:var>
    </swrlx:classAtom>
    <swrlx:datavaluedPropertyAtom  swrlx:property="&onto1;#topic">
      <ruleml:var>p</ruleml:var>
      <owlx:DataValue owlx:datatype="&xsd;#string">politics</owlx:DataValue>
    </swrlx:datavaluedPropertyAtom>
  </ruleml:_body>
  <ruleml:_head>
    <swrlx:classAtom  swrlx:property="&onto2;#Politics">
      <ruleml:var>p</ruleml:var>
    </swrlx:classAtom>
  </ruleml:_head>
</ruleml:imp>

<ruleml:imp>
  <ruleml:_body>
    <swrlx:classAtom>
      <owlx:Class owlx:name="&onto2;#Volume" />
      <ruleml:var>p</ruleml:var>
    </swrlx:classAtom>
    <swrlx:datavaluedPropertyAtom  swrlx:property="&onto2;#size">
      <ruleml:var>p</ruleml:var>
      <ruleml:var>q</ruleml:var>
    </swrlx:datavaluedPropertyAtom>
    <swrlx:builtinAtom  swrlx:builtin="&swrlb;#greaterThanOrEqual">
      <owlx:DataValue owlx:datatype="&xsd;#int">14</owlx:DataValue>
      <ruleml:var>q</ruleml:var>
    </swrlx:builtinAtom>
  </ruleml:_body>
  <ruleml:_head>
    <swrlx:classAtom  swrlx:property="&onto1;#Pocket">
      <ruleml:var>p</ruleml:var>
    </swrlx:classAtom>
  </ruleml:_head>
</ruleml:imp>
```

The first rule expresses that a Book in the first ontology with politics as a value of its topic attribute is a Politics book in the second ontology. The second rule expresses that Volumes in the second ontology whose size is less than 14 are Pocket books in the first ontology.

The introduction of variables within constructs of the OWL language provides more expressiveness to the language. In particular, it allows the expression of what was called role-value maps in description logics or feature path equations in feature algebras [Smolka, 1992]. Of course, all the constructions available in OWL are usable in SWRL as well. SWRL also provides a set of built-in predicates on various datatypes provided by XML Schema as well as operators on collections, such as count.

SWRL rules can be used for expressing the correspondences between ontologies. These correspondences are expressed between formulas and interpreted as Horn clauses. They have the advantage over genuine OWL of being well identified as rules and are easier to manipulate as an alignment format than OWL, which is also used to express ontologies.

As in the OWL case, these rules have the drawback of forcing the use of OWL and are interpreted as merging ontologies. Again, the expression of a rule, such as the one above, freezes the use that can be made: the rule will help in considering some Books of the first ontology as Politics books in the second ontology. However, the rules work as a set of logical rules, not rewrite rules, so they can be used for merging, but not transforming ontologies.

8.1.5 Alignment format

Following [Euzenat, 2003], [Euzenat, 2004] provides an Alignment format on several levels, which can handle complex alignment definitions. This format is simpler than most of the alignment representations presented here. It also aims to be producible by most matching tools.

The alignment description is an envelope in which the correspondences are grouped. It expresses metadata about the alignment and features. These are:

References to matched ontologies;
A set of correspondences which expresses the relation holding between entities of the first ontology and entities of the second ontology;
Level used for characterising the type of correspondence (see next);
Arity (default 1:1) (denoted as 1 for injective and total, ? for injective, + for total and * for none, with each sign concerning one mapping and its converse): ?:?, ?:1, 1:?, 1:1, ?:+, +:?, 1:+, +:1, +:+, ?:*, *:?, 1:*, *:1, +:*, *:+, *:* (see also Sect. 2.5.2). These assertions could be provided as input (or constraint) for the alignment algorithm or as a result by the same algorithm.

More metadata can be added, in particular when the format is expressed in RDF, such as:

- the generating algorithm;
- the date of creation;
- whether the alignment is homogeneous (in language or entity).

Support for correspondences follows the Definition 2.11. They are expressed by:

entity1: the first matched entity;
entity2: the second matched entity;

relation: the relation holding between the two entities. It is not restricted to the equivalence relation, but can be more sophisticated, e.g., subsumption, incompatibility [Giunchiglia and Shvaiko, 2003b], or even some fuzzy relation. The default relation is equivalence.
strength: the confidence that the correspondence under consideration holds. The measure should belong to an ordered set Ξ including a maximum element $\top$ and a minimum element $\bot$; for instance, a float value between 0 and 1. The default strength is $\top$.
id: an identifier for the correspondence.

A full example of the level 0 Alignment format in RDF is as follows:

```
<?xml version='1.0' encoding='utf-8' standalone='no'?>
<!DOCTYPE rdf:RDF [
    <!ENTITY rdf   "http://www.w3.org/1999/02/22-rdf-syntax-ns#" >
    <!ENTITY xsd   "http://www.w3.org/2001/XMLSchema#" >
    <!ENTITY onto1 "http://book.ontologymatching.org/example/culture-shop.owl" >
    <!ENTITY onto2 "http://book.ontologymatching.org/example/library.owl" >
  ]>
<rdf:RDF xmlns='http://knowledgeweb.semanticweb.org/heterogeneity/alignment'
         xmlns:rdf='&rdf;'>
<Alignment>
  <xml>yes</xml>
  <level>0</level>
  <type>**</type>
  <onto1>&onto1;</onto1>
  <onto2>&onto2;</onto2>
  <map>
    <Cell>
      <entity1 rdf:resource='&onto1;#Book'/>
      <entity2 rdf:resource='&onto2;#Volume'/>
      <measure rdf:datatype='&xsd;float'>0.6363636363636364</measure>
      <relation>&lt;</relation>
    </Cell>
  </map>
  <map>
    <Cell>
      <entity1 rdf:resource='&onto1;#name'/>
      <entity2 rdf:resource='&onto2;#title'/>
      <measure rdf:datatype='&xsd;float'>1.0</measure>
      <relation>></relation>
    </Cell>
  </map>
</Alignment>
</rdf:RDF>
```

It describes a many-to-many level 0 alignment between two bibliographic ontologies. It contains two correspondences in which Book in the first ontology is less general than Volume in the second one and name in the first ontology is more general than title in the second one. These correspondences use the less and more general relations and a confidence measure .64 in the former case and 1. in the latter.

The Alignment format has been designed for offering a common format to different needs. Depending on the expressiveness of the matched entities, it offers several alignment levels which correspond to different options for expressing entities:

Level 0: These alignments relate entities identified by URIs. Any algorithm can deal with such alignments. This first level of alignment has the advantage to not depend on a particular language for expressing these entities. On this level, the matched entities may be classes, properties or individuals. However, they also can be any kind of a complex term that is used by the target language as soon as it is identified by a URI.

Level 1: These alignments replace pairs of entities by pairs of sets (or lists) of entities. A level 1 correspondence is thus a slight refinement of level 0. It can be easily parsed and is still language independent.

Level 2 (L): More general correspondence expressions can be useful. For instance, [Masolo et al., 2003] provides bridges from an ontology of services to the currently existing semantic web service description languages in first order logic. These kinds of correspondences can be expressed with level 2 alignments. These are no longer language independent and require the knowledge of the language used for parsing the format. In this case correspondences can be expressed between formulas, queries, etc.

The following two correspondences use OWL and a pseudo-query language in level 2 alignments:

```
<Cell>
  <entity1 rdf:resource='&onto2;#Writer'/>
  <entity2>
    <owl:Restriction>
      <owl:onProperty rdf:resource="&onto1;#hasWritten"/>
      <owl:minCardinality
          rdf:datatype="&xsd;nonNegativeInteger">1</owl:minCardinality>
    </owl:Restriction>
  </entity2>
  <measure rdf:datatype='&xsd;float'>0.6363636363636364</measure>
  <relation>&lt;</relation>
</Cell>
```

The example above describes the fact that a Writer in the second ontology is someone that hasWritten something in the first one.

```
<Cell>
  <entity1 rdf:resource='&onto1;#name'/>
  <entity2>
    <Apply rdf:resource='string-concatenate'>
      <args rdf:parseType="collection">
        <Path>
          <relation rdf:resource="&onto2;#firstname" />
        </Path>
        <Path>
          <relation rdf:resource="&onto2;#lastname" />
        </Path>
      </args>
    </Apply>
  </entity2>
  <measure rdf:datatype='&xsd;float'>1.0</measure>
  <relation>=</relation>
</Cell>
```

The example above describes the fact that the name in the first ontology is equivalent to the concatenation of firstname and lastname in the second one.

These kinds of correspondences are commonly used in logic-based languages or in the database world for defining the views in global-as-view or local-as-view approaches [Calvanese et al., 2002b]. It also resembles the SWRL rule language

[Horrocks *et al.*, 2004] when used with OWL (see also Sect. 8.1.4 for a simple example of such rules).

The Alignment format has been given an OWL ontology and a DTD for validating it in RDF/XML. It can be manipulated through the Alignment API which is presented next in Sect. 8.2.4. It has been used as the format for the OAEI evaluation campaigns (§7.1.3) and many different tools are able to output it, e.g., oMap (§6.3.5), FOAM (§8.2.5), OLA (§6.3.8), Falcon-AO (§6.3.9), and HCONE (§6.1.20).

Finally, the Alignment format allows the expression of alignments without commitment to a particular language. It is not targeted towards a particular use of the alignments and offers generators for a number of other formats. However, in contrast to the languages presented so far, this format does not offer much expressiveness. One of the good features of this format is its openness which allows the introduction of new relations and if necessary new types of expressions while keeping the compatibility with poorly expressive languages.

8.1.6 SEKT mapping language

The SEKT European project has developed an ontology mapping language [de Bruijn *et al.*, 2004] to specify mediators in semantic web services as defined in the Web Services Modeling Ontology, WSMO[3] [Roman *et al.*, 2004]. This language is both expressive and language independent. It needs to be grounded in another representation language in order to be interpreted. The alignments can be expressed in a human-readable language (used here) as well as with the help of an RDF vocabulary defined in [Scharffe, 2005].

The relationships between entities of this language are shown in Table 8.1. They allow the specific identification of which kind of entity is mapped into which other.

Table 8.1. SEKT-ML mapping types.

Language Construct	Description
ClassMapping	Mapping between two classes
PropertyMapping	Mapping between two properties
RelationMapping	Mapping between two relations
ClassPropertyMapping	Mapping between a class and a property
ClassRelationMapping	Mapping between a class and a relation
ClassInstanceMapping	Mapping between a class and an instance
IndividualMapping	Mapping between two instances

The entities to be mapped can be classes, properties, corresponding to data valued relations, relations, corresponding to object valued relations, instances of an ontology identified by their URIs or terms constructed from such entities through a set of operators, as described in Table 8.2.

---

[3] http://www.wsmo.org

Table 8.2. SEKT-ML logical constructors.

Entity	Operator
Class	and, or, not, join
Property	and, or, not, join
Relation	and, or, not, inverse, symetric, reflexive, transitive, join

The interpretation of class and property and, or and not operators follows the classical set-theoretic interpretation. The join operator allows the composition of classes or relations with conditions on which entities can be composed. The inverse operator provides the converse relation while the symetric, reflexive and transitive operations provide the respective closures of the relation.

It is, of course, possible to express the correspondences between named classes and relations. As presented in the following example, Book in the first ontology is less general than Volume in the second one and the authors relations are the same in both ontologies.

```
classMapping(
  annotation(<"rdfs:label"> 'Book to Volume')
  annotation(<"http://purl.org/dc/elements/1.1/description">
    'Map the Book concept to the Volume concept')
  unidirectional
  <"&onto1;#Book">
  <"&onto2;#Volume">)

relationMapping(
  annotation(<"rdfs:label"> 'authors to authors')
  bidirectional
  <"&onto1;#authors">
  <"&onto2;#authors">)
```

Conditions can be introduced in order to constrain the objects to which the mapping can be applied. These conditions are applied to classes or attributes (being applied to both properties and relations). In both cases, they can constrain the type, value or multiplicity of attribute values.

The language allows the expression of more complex correspondences comprising conditions. For instance, the Books whose topic is politics in the first ontology are Politics books in the second one or the Volumes in the second ontology whose size is inferior to 14 are Pocket books in the first ontology:

```
classMapping(
  annotation(<"rdfs:label"> 'conditional Book to Politics')
  unidirectional
  <"&onto1;#Book">
  <"&onto2;#Politics">
  attributeValuecondition(<"&onto1;#topic"> '= "politics"))

classMapping(
  annotation(<"rdfs:label"> 'conditional Volume to Pocket')
  unidirectional
  <"&onto2;#Volume">
  <"&onto1;#Pocket">
  attributeValuecondition(<"&onto2;#size"> '< 14))
```

The SEKT mapping language is an expressive alignment format offering many kinds of relations and entity constructors to users. One of its main advantages is its ontology language independence, giving a common format for expressing mappings. Thus, this proposal has a middle man position: it is independent from any particular language but expressive enough for covering a large part of the other languages.

8.1.7 SKOS

SKOS [Miles and Brickley, 2005b, Miles and Brickley, 2005a] stands for Simple Knowledge Organisation System. The SKOS core vocabulary is an RDF Schema aiming at expressing relationships between lightweight ontologies, e.g., folksonomies or thesauri. It is currently under development.

The goal of SKOS is to be a layer on top of other formalisms able to express the links between entities in these formalisms.

Concept and relation descriptions

SKOS allows the identification of the concepts that are present in other ontologies. The concept description part of SKOS is redundant with respect to other languages of that family; it has been designed for taking advantage of these concepts, for instance, in a graphical user interface, rather than only expressing the alignments.

Below are such concept descriptions dedicated to the description of the Book concept. SKOS defines various ways of presenting the concept with labels in several languages and alternate labels and symbols. It also provides the opportunity to add various notes and informal definitions to the concept:

```
<skos:Concept rdf:about="&onto1;#Book">
  <skos:prefLabel>Book</skos:prefLabel>
  <skos:altLabel xml:lang="fr">Livre</skos:altLabel>
  <skos:definition>A book is a set of sheets of papers bound together so that a the content printed on them can be consulted in sequence.</skos:definition>
  <skos:editorialNote>This is not an official definition</skos:editorialNote>
</skos:Concept>
```

In the above example Book is a concept with some information about it, such as the way it is named in different languages and its definition (similar to glosses in WordNet). SKOS also allows the description of collections of concepts given by their enumeration:

```
<skos:Collection rdf:about="&onto2;#Topics">
  <rdfs:label>Topics</rdfs:label>
  <skos:member rdf:resource="&onto2;#Critics"/>
  <skos:member rdf:resource="&onto2;#Literature"/>
  <skos:member rdf:resource="&onto2;#Science"/>
  <skos:member rdf:resource="&onto2;#Politics"/>
</skos:Collection>
```

This describes the collection of Topics featuring Critics, Literature, Science and Politics.

Concept relations

SKOS defines the so-called *semantic relationships* that express relations between the SKOS concepts. For instance, a term used in a thesaurus can be broader than another. There are three such relations as defined in Table 8.3.

Table 8.3. SKOS relation properties.

property	domain	range	inverse	property
broader	concept	concept	narrower	transitive
related	concept	concept	related	symmetric

The relations between concepts enable the assertion of the relative inclusion of concepts as broader or narrower terms as well as other informal relations. The following example displays the Book concept that is narrower than Volume but broader than Critics. It is also related to Work.

```
<skos:Concept rdf:about="&onto1;#Book">
  <skos:broader rdf:resource="&onto2;#Volume"/>
  <skos:narrower rdf:resource="&onto2;#Critics"/>
  <skos:related rdf:resource="&onto2;#Work"/>
<skos:Concept/>
```

Broader and narrower are transitive properties while related is symmetric.

Annotations

In addition, but of least interest here, SKOS defines annotation properties that enable users to use SKOS concepts for describing resources, i.e., to annotate them directly with SKOS. It defines the vocabulary displayed in Table 8.4.

Table 8.4. SKOS annotation properties.

property	domain	range	inverse
subject	resources	concept	isSubjectOf
primarySubject	resource	concept	isPrimarySubjectOf

This is used below to express that Justice is the primarySubjectOf Camus' La Chute and a subjectOf Russell's My Life.

```
<skos:Concept rdf:about="&onto2;#Justice">
  <skos:isPrimarySubjectOf rdf:resource="&onto2;#LaChute" />
  <skos:isSubjectOf rdf:resource="&onto2;#MyLife" />
</skod:Concept>
```

SKOS has the advantage of being a lightweight vocabulary defining from the ground a rich collection of relations between entities. Since it uses URIs for referring to objects it is fully integrated in the semantic web architecture and is not committed to a particular language. In fact, one of the main advantage of SKOS is that it lifts any kind of organised description into an easily usable set of classes. The relation part has the advantage of being very general but the drawback is that it lacks formal semantics. However, more semantics on these terms can be introduced by using the OWL vocabulary.

Like other formats which do not separate the ontologies from the correspondences, SKOS, in its most convenient form, mixes the highest power of RDF Schema and the expression of the alignments. This form of extensibility, through RDF Schema, prevents any non RDF Schema understanding application from fully grasping the SKOS content.

8.1.8 Comparison of existing formats

The formats we have considered so far can be globally compared (see Table 8.5). For that purpose, a set of criteria is applied to these different formats:

Web compatibility is the capacity of the format to be manipulated on the web. This involves its possible expression in XML, RDF and/or RDF/XML, as well as the possibility to identify entities by URIs. This should, in principle, enable the extensibility of the format by introducing new properties as well as the referencing of particular correspondences individually. This aspect is covered by the RDF/XML and URI criteria of Table 8.5.

Language independence is the ability to express alignments between entities described in different languages. This is often related to the use of URIs. In fact, language independence is mostly related to simplicity. This aspect is covered by the Language and Model criteria of Table 8.5.

Simplicity is the capacity to be dealt with in a simple form by simple tools. In particular, for example, requiring inference for correctly manipulating the alignment or requiring that the format covers an important part of some ontology representation language is not a sign of simplicity. A well structured format should help achieve this goal. This aspect is inverse to expressiveness.

Expressiveness is the capacity of the format to express complex alignments. This means that alignments are not restricted to matching entities identified by URIs but can create new ones. The constructions for expressing alignments can be arbitrary complex. In fact, it can be more complex than the knowledge expressed in the ontologies. This aspect is covered by the Relations, Terms, Type rest, Cardinality, Variables and Built-in criteria of Table 8.5.

Extendibility is the capacity to extend the format with specific purpose information in such a way that the tools which use this format are not disturbed by the extensions. Most of the systems presented here exhibit one kind of extendibility tied to the use of RDF which allows any new relation and object to be added. This aspect is covered by the '+' signs in Table 8.5.

8 Frameworks and formats: representing alignments

Purpose independence is a consequence of various factors expressed. We mention the intended use of each of the formats. It is clear, for instance, that a format designed for data integration, with very precise selection constraints, will rather be difficult to re-use in transforming ontologies. This aspect is covered by the Target application criterion of Table 8.5.

Executability is the capacity to be directly usable in mediators. This means that there are tools available for directly interpreting the format as a program processing knowledge. Executability is rather opposed to language independence. This aspect is covered by the Execution criterion of Table 8.5.

Table 8.5 provides the values of all the formats presented so far for each of the criteria.

Table 8.5. Summary of characteristics of the presented formats. + means that the system can be extended; Transf stands for transformation. The possible relations for the formats are subclass (sc), subproperty (sp), implication between formulas (imp). The terms concerned by the alignments can be classes (C), properties (P) or individuals (I).

Format	OWL	SBO	C-OWL	SWRL	Alignment	SEKT-ML	SKOS
Target app.	Merging	Data transf.	Data int.	Data int.	Generic	Data transf.	Merging
Language	OWL	UML	OWL	OWL	+		RDFS
Model	OWL		OWL+	OWL			
Execution	Logical	Transf	Logical	Logical		Logical	Alg.
RDF/XML	√		√	√	√	√	√
URI	√		√	√	√	√	√
Measures					√	√	
Relations	sc/sp		sc/sp	imp	sc/sp+	sc/sp/...	sc/sp
Terms	C/P/I	C/P/I	C/P/I	C/P/I	URI	C/P/I	C/P
Type rest	√	√	√	√		√	
Cardinality	√		√	√			
Variables				√			
Built-in		√+		√	+	+	

The difference between these formats lies in the continuum between: (i) very general languages that are easy to understand but which are unable to express complex alignments, e.g., SKOS, level 0 Alignment format, and (ii) very expressive languages whose semantics dictates their use and which requires deep understanding of the language, e.g., OWL, SWRL, C-OWL, MAFRA. The SEKT Mapping language stands in the middle of this continuum.

While most of the matching algorithms are only able to express the first kind of alignments, both kinds of languages are useful. Most of the expressive formats have a surface heterogeneity due to the languages on which they are based, e.g., UML, OWL, WSML. However, they have very similar features for: referring to ontology constructs, such as classes, properties; using logical formula constructors, such as conjunctions, implications, quantifiers; using datatypes and collections of built-in operators. Finally, it is surprising that there is no more heterogeneity in these expressive languages given that complexity is a factor of: the language used for expressing the ontologies, the language used for expressing the related entities, the semantics given to alignments and the language used for expressing relations.

As a summary, there is no universal format for expressing alignments. The choice of a format depends on the characteristics of the application. We think that there are two factors which can influence this choice: (i) the expressiveness required for the alignments, and (ii) the need to exchange with other applications, especially if they involve different ontology languages, or the ontology language used by the application.

We now consider frameworks that allow taking advantage of such alignment formats by offering openness to alignment manipulation, including matchers.

8.2 Alignment frameworks

The matching operation is typically only one of the steps towards the ultimate goal of ontology integration or web services composition, for instance. There exist infrastructures which use alignments as one of their components. The goal of such infrastructures is to enable users to perform high-level tasks which involve generating, manipulating, composing and applying alignments within the same environment. Similar to Chap. 6 in this section we did not enforce the terminology of Sect. 2.4 but kept that one as used by system designers.

8.2.1 Model management

Model management [Bernstein *et al.*, 2000, Madhavan *et al.*, 2002, Melnik, 2004] aims at providing a metadata manipulation infrastructure in order to reduce the amount of programming required to build metadata driven applications. It deals with *models* which can be related by *mappings*. A *model* is an information structure, such as XML schema, relational database schema, UML model. Similarly, *mappings* are oriented alignments from one model into another. Technically, a key idea of generic model management is to solve metadata intensive tasks at a high level of abstraction using a concise script. It is generic in the sense that a single implementation should be applicable to the majority of data models and scenarios, e.g., data translation, data integration. However, it is primarily targeted at databases. It provides an algebra to manipulate models and mappings. In [Melnik *et al.*, 2005], the following operators are defined:

- Match(m, m') which returns the mapping a between models m and m';
- Compose(a, a') which composes mappings a and a' into a new one a'', given that the range of a' is the domain of a;
- Confluence(a, a') which merges alignments by union of non conflicting correspondences, provided as a and a' that have the same domain and range;
- Merge(a, m, m') which merges two models m and m' according to mapping a;

- Extract(a, m) which extracts the portion of model m which is involved in mapping a;
- Diff(a, m) which extracts the portion of model m which is not involved in mapping a.

A mapping in this context is a function from m to m'. [Melnik et al., 2005] also provides axioms governing these operations. For instance, the merge operation between two models m' and m'' through a mapping a, returns a new model $m = \text{Domain}(a') \cup \text{Domain}(a'')$ and a pair of surjective mappings a' and a'' from m to m' and m'' respectively, such that: $a = \text{Compose}(\text{Invert}(a'), a'')$.

A typical example of the model management script is as follows:

```
A1 := Match( O1, O2 );
A2 := Match( O2, O3 );
O4 := Diff( O1, A1 );
A3 := Compose( A1, A2 )
O5 := Merge( Extract( O1, A1), O3, A3 );
O6 := Merge( O4, O5, ∅ );
```

The above example operates with three ontologies. It merges the first one and the last one on the basis of the composition of their alignment with the intermediate one. Finally, it adds the part of the first one that was not brought in the first merge.

There are some model management systems available. In particular, Rondo is a programming platform implementing generic model management [Melnik et al., 2003b, Melnik et al., 2003a]. It is based on conceptual structures which constitute the main Rondo abstractions:

Models such as relational schemas, XML schemas, are internally represented as directed labelled graphs, where nodes denote model elements, e.g., relations and attributes. Each such element is identified by an object identifier (OID).

Morphisms are binary relations over two, possibly overlapping, sets of OIDs. The morphism is typically used to represent a mapping between different kinds of models. Morphisms can always be inverted and composed.

Selectors are sets of node identifiers from a single or multiple models. These are denoted as S. A selector can be viewed as a relation with a single attribute, $S(V : OID)$, where V is a unique key.

The operators presented above, e.g., match, merge, are implemented upon these conceptual structures. Match is implemented in Rondo by using the Similarity flooding algorithm (§5.3.1).

Another system, called Moda, is described in [Melnik et al., 2005] in which correspondences are expressed as logical formulas. This system is more expressive than Rondo. Examples of some other model management systems include: GeRoMe [Kensche et al., 2005] and ModelGen [Atzeni et al., 2005, Atzeni et al., 2006].

8.2.2 COMA++ (University of Leipzig)

COMA++ [Do and Rahm, 2002, Do, 2005] is a schema matching infrastructure built on top of COMA (§6.1.12). It provides an extensible library of matching algorithms, a framework for combining obtained results, and a platform for the evaluation of the effectiveness of the different matchers. COMA++ enables importing, storing and editing schemas (or models). It also allows various operations on the alignments among which compose, merge and compare. Finally, alignments can be applied to schemas for transforming or merging them.

Contrary to Rondo, the matching operation is not described as atomic but rather described as workflow that can be graphically edited and processed. Users can control the execution of the workflow in a stepwise manner and dynamically change execution parameters. The possibility of performing iterations in the matching process assumes interaction with users who approve obtained matches and mismatches to gradually refine and improve the accuracy of match (see Fig. 8.2). The matching operation is performed by the Execution engine based on the settings provided by the Match customiser, including matchers to be used and match strategies.

The data structures are defined in a homogeneous proprietary format. The Schema pool provides various functions to import and export schemas and ontologies and save them to and from the internal Repository. Similarly, the Mapping pool provides functions to manipulate mappings. COMA++ can also export and import the matching workflows as executable scripts (similar to those manipulated in Rondo).

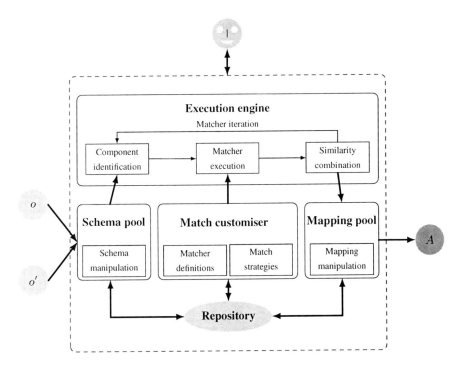

Fig. 8.2. COMA++ architecture (adapted from [Do, 2005]).

Finally, according to [Do, 2005], there are some other tools built on top of COMA++. For example, the CMC system provides a new weighting strategy to automatically combine multiple matchers [Tu and Yu, 2005], while the work of [Dragut and Lawrence, 2004] has adapted COMA to compute correspondences between schemas by performing a composition of the correspondences between individual schemas and a reference ontology.

8.2.3 MAFRA (Instituto Politecnico do Porto and University of Karlsruhe)

MAFRA (MApping FRAmework, already mentionned for its format in Sect. 8.1.1) is an interactive, incremental and dynamic framework for mapping distributed ontologies [da Silva, 2004, Mädche *et al.*, 2002]. The framework consists of horizontal and vertical dimensions. The horizontal dimension covers the mapping process. It is organised according to the following components:

- *Lift and Normalisation.* This module handles syntactic, structural, and language heterogeneity. In particular, the lifting process includes translation of input ontologies into an internal knowledge representation formalism, which is RDF Schema. Normalisation (§4.2), in turn, includes (i) tokenisation of entities, (ii) elimination of stop words, (iii) expansion of acronyms.
- *Similarity.* This module calculates similarities between ontology entities by exploiting a combination of multiple matchers. First, *lexical similarity* between each entity from the source ontology and all entities from the target ontology is determined based on WordNet and altered Resnik measure (§4.2.2). Second, the *property similarity* is computed. This measures similarity between concepts based on how similar the properties they are involved in are. Finally, *bottom-up* and *top-down similarities* are computed. For example, bottom-up matchers take as input the property (dis)similarity and propagate it from lower parts of the ontology to the upper concepts, thus yielding an overall view of similarity between ontologies.
- *Semantic Bridging.* Based on the similarities determined previously, the correspondences (bridges) between the entities of the source and target ontologies are established. Bridges, in turn, can be executed for the data translation task. The internals of bridges are discussed in detail in Sect. 8.1.1.
- *Execution.* The actual processing of bridges is performed in the execution module. This module translates instances from the source ontology to the target ontology. This translation can either be performed off-line, i.e., one time transformation, or on-line, i.e., dynamically, thus taking into account the 'fresh' data, if any.
- *Post-processing.* This module is in charge of the analysis and improvement of the transformation results, for instance, by recognising that two instances represent the same real-world object.

Components of the vertical dimension interact with horizontal modules during the whole mapping process. There are four vertical components. The *Evolution* module, in a user-assisted way, synchronises bridges obtained with the Semantic Bridging module according to the changes in the source and target ontologies. The *Cooperative Consensus Building* module helps users to select the correct mappings, when multiple mapping alternatives exist. The *Domain Constraints and Background Knowledge* module stores common and domain specific knowledge, e.g., WordNet, precompiled domain thesauri, which are used to facilitate the similarity computation. Finally, a graphical user interface assists users in accomplishing the mapping process with a desired quality.

8.2.4 Alignment API (INRIA Rhône-Alpes)

A Java API [Euzenat, 2004] is available for manipulating alignments in the Alignment format. It defines a set of interfaces and a set of functions that they can perform.

Classes

The OWL API is extended with the org.semanticweb.owl.align package which describes the Alignment API. This package name is used for historical reasons. In fact, the API itself is fully independent from OWL or the OWL API.

The Alignment API is essentially made of three interfaces:

Alignment describes a particular alignment. It contains a specification of the alignment and a list of cells.
Cell describes a particular correspondence between entities.
Relation does not mandate any particular feature.

To these interfaces implementing the Alignment format, are added a couple of other interfaces:

AlignmentProcess extends the Alignment interface by providing an align method. So this interface is used for implementing matching algorithms (Alignment can be used for representing and manipulating alignments independently of algorithms).
Evaluator describes the comparison of two alignments (the first one could serve as a reference). Each implementation measure must implement the eval method.

An additional AlignmentException class specifies the kind of exceptions that are raised by alignment algorithms and can be used by alignment implementations.

Functions

The Alignment API provides support for manipulating alignments. As in [Bechhofer et al., 2003], these functions are separated in their implementation. It offers the following functions:

Parsing and serialising an alignment from a file in RDF/XML (AlignmentParser.read(), Alignment.write());

Computing the alignment, with input alignment (Alignment.align(Alignment, Parameters));

Thresholding an alignment with threshold as argument (Alignment.cut(double));

Hardening an alignment by considering that all correspondences whose strength is strictly greater than the argument are converted to $\top$, while the others are converted to $\bot$ (Alignment.harden(double));

Comparing one alignment with another (Evaluator.eval(Parameters)) and serialising the result (Evaluator.write());

Outputting alignments in a particular format, e.g., SWRL, OWL, XSLT, RDF. (Alignment.render(visitor));

Matching and evaluation algorithms accept parameters. These are put in a structure that allows storing and retrieving them. The parameters can be various weights used by some algorithms, some intermediate thresholds or the tolerance of some iterative algorithms. There is no restriction on the kind of parameters to be used.

The Alignment API has been implemented in Java. This implementation has been used for various purposes: on-line alignment [Zhdanova and Shvaiko, 2006] and Evaluation tool in the Ontology Alignment Evaluation Initiative (§7.1.3). Also many extensions use it for implementing matching algorithms, such as oMap [Straccia and Troncy, 2005], FOAM [Ehrig et al., 2005], and OLA [Euzenat and Valtchev, 2004].

8.2.5 FOAM (University of Karlsruhe)

FOAM [Ehrig, 2007] is a general tool for processing similarity-based ontology matching (see also Sect. 6.3.4 and Sect. 6.4.1). It follows a general process which is presented in Fig. 8.3. It is made of the following steps:

Feature engineering selects the features of the ontologies that will be used for comparing the entities.

Search step selection selects the pairs of elements from both ontologies that will be compared.

Similarity computation computes the similarity between the selected pairs using the selected features.

Similarity aggregation combines the similarities obtained as the result of the previous step for each pair of entities.

Interpretation extracts an alignment from the computed similarity.

Iteration iterates this process, if necessary, taking advantage of the current computation.

The FOAM framework bundles several algorithms and strategies developed by its authors. Within this framework have been cast matching systems such as NOM, QOM (§6.3.4), and APFEL (§6.4.1). More systems can be integrated simply by changing any of the modules above. The global behaviour of the system can be parameterised through different scenarios, e.g., data integration, ontology merging,

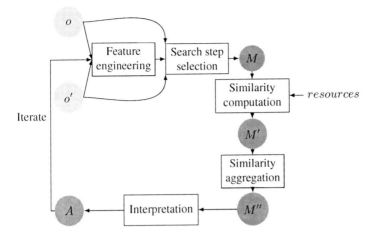

Fig. 8.3. FOAM architecture (adapted from [Ehrig, 2007]).

ontology evolution, query rewriting and reasoning, which offer default parameters adapted to these tasks.

FOAM itself is based on the KAON2 [Oberle et al., 2004] suite of tools and accepts ontologies in the OWL-DLP fragment. It offers a web-based interface. Finally, it also offers translation tools from and to the Alignment format (§8.1.5) and other formats.

Platforms for integrating matchers and alignment manipulation operations are relatively new, however, they constitute a promising perspective to knowledge engineers and application developers. Another, type of useful alignment manipulation systems are alignment editors which offer human users the opportunity to be involved in the matching process.

8.3 Ontology editors with alignment manipulation capabilities

Other tools for dealing with ontology matching are ontology edition environments provided with support for matching and importing ontologies. These tools are primarily made for creating ontologies, but they also provide tools for comparing ontologies and relating them.

8.3.1 Chimaera (Stanford University)

Chimaera is a browser-based environment for editing, merging and testing (diagnosing) large ontologies [McGuinness et al., 2000]. It aims to be a standard-based and generic tool. Users are provided with a graphical user interface (the Ontolingua ontology editor) for editing taxonomy and properties. They also can use various diagnosis commands, which provide a systematic support for pervasive tests and changes, e.g., tests for redundant super classes, slot value or type mismatch. Matching in the system is performed as one of the subtasks of a merge operation. Chimaera searches for merging candidates as pairs of matching terms, with terminological resources such as term names, term definitions, possible acronym and expanded forms, names that appear as suffixes of other names. It generates name resolution lists that help users in the merging task by suggesting terms which are candidates to be merged or to have taxonomic relationships not yet included in the merged ontology. The suggested candidates can be names of classes or slots. The result is output in OWL descriptions similar to those presented in the OWL format (§8.1). Chimaera also suggests taxonomy areas that are candidates for reorganisation. These edit points are identified by using heuristics, e.g., looking for classes that have direct subclasses from more than one ontology.

8.3.2 The Protégé Prompt Suite (Stanford University)

The Prompt Suite[4] is an interactive framework for comparing, matching, merging, maintaining versions, and translating between different knowledge representation formalisms [Noy and Musen, 2003, Noy, 2004b]. The main tools of the suite include: an interactive ontology merging tool, called iPrompt [Noy and Musen, 2000] (formerly known as Prompt), an ontology matching tool, called Anchor-Prompt (see Sect. 6.1.9 and [Noy and Musen, 2001]), an ontology-versioning tool, called PromptDiff [Noy and Musen, 2002b], and a tool for factoring out semantically complete subontologies, called PromptFactor.

Prompt is implemented as an extension to the Protégé[5] ontology editing environment. Thus, the Protégé browser provides overall capabilities for managing multiple ontologies. Prompt and Protégé are based on a frame-based knowledge model. Three types of frames are distinguished, namely classes, slots (properties) and instances. Below, we discuss the main Prompt Suite tools. We omit Anchor-Prompt, since it is presented in Sect. 6.1.9.

iPrompt takes as input two ontologies and leads users towards one merged ontology as output. First, iPrompt creates an initial set of matches based on lexical similarity between class names. Here can be used any name-based technique (§4.2). Then, it proceeds through the following cycle:

- Users choose an operation to perform, for instance, by selecting one from the iPrompt suggestion list. Some examples of operations are: merge classes, slots, instances, perform a shallow (deep) copy of a class.
- iPrompt performs the operation chosen at the previous step. It also identifies inconsistencies, e.g., name conflicts, redundancy in the class hierarchy, that the operation introduced, as well as possible strategies to resolve them. Finally, it generates a list of suggestions (concerning the next actions) for users.

---
[4] http://protege.stanford.edu/plugins/prompt/prompt.html
[5] http://protege.stanford.edu/

Some techniques, such as rearranging lists of suggestions, are used in order to keep users focused on the most important aspects of the given state of the process, thus converging on a desired merged ontology efficiently.

PromptDiff compares ontology versions and identifies the changes. This operation can be viewed as computing a Diff in model management (§8.2.1). PromptDiff produces a structural diff between two versions based on heuristics. It borrows many of them from iPrompt in order to identify what has changed from one version of an ontology to another. These heuristics include various techniques such as analysis and comparison of concept names, slots attached to concepts. The PromptDiff approach has two parts: (i) an extensible set of heuristic matchers; and (ii) a fixed point algorithm which combines the results of the matchers until they produce no more changes in the diff.

PromptFactor is a tool that uses the infrastructure of the framework thereby enabling users to factor out part of their ontology into a new subontology. Users specify concepts they are interested in the new ontology, then the tool performs the transitive closure of the superclass relation and all the relations defined by attributes. The tool does not employ any sophisticated algorithms as such, it just serves to simplify some of the tasks that ontology engineers usually perform.

Similarly to Protégé, Prompt can be extended through plug-ins. For example, there is a Prompt plug-in for FOAM.

Other examples of environments that allow the expression of alignments in ontology editing tools include KAON2 [Oberle et al., 2004] which is the basis for a tool called OntoStudio and the WSMX editor [Mocan et al., 2006].

8.4 Summary

When considering ontology matching, one must decide how the matching result will be used and delivered as well as how it will be produced. We have presented several contexts in which matching algorithms can be embedded. In particular, alignment formats allow the exchange of alignments among tools in well documented formats. They guarantee the openness of the result and everyone can take advantage of this result. Some frameworks offer more than just a format, such as the ability to manipulate the alignment and sometimes to apply it to data. The drawback of these environments, with the exception of the Alignment API, is that they cannot be easily embedded within other applications. Finally, some environments permit editing alignments as well as ontologies and can integrate matching algorithms.

The choice to be made is dependent on the matching purpose of the matching tool. If its goal is to establish alignments that will be used in unknown and multiple contexts, then choosing a format as neutral as possible is certainly the solution. If the matching result is used for processing some already known data, following a precise process, then a framework allowing this manipulation can be used. Finally, if the matching algorithm is to be integrated in a user driven environment, then an ontology and alignment editing tool is a good choice.

The two latter categories of tools have not been designed as open systems. In consequence, exporting alignments to other forms, e.g., one of the formats presented in Sect. 8.1 or an executable form, is required for integrating ontology matching to larger applications. The two next chapters develop these themes by considering first the explanation of matching results to users (Chap. 9) and then by discussing how alignments can be processed in applications (Chap. 10).

9
Explaining alignments

Matching systems may produce effective alignments that may not be intuitively obvious to human users. In order for users to trust the alignments, and thus use them, they need information about them, e.g., they need access to the sources that were used to determine semantic correspondences between ontology entities. Explanations are also useful when matching large applications with thousands of entities, e.g., business product classifications, such as UNSPSC and eCl@ss. In such cases, automatic matching solutions will find many plausible correspondences, and hence user input is required for performing cleaning-up of the alignment. Finally, explanations can also be viewed and applied as argumentation schemas for negotiating alignments between agents.

In this chapter we describe how a matching system can explain its answers, thus making the matching result intelligible. The material of this chapter is mainly based on the work in [Shvaiko *et al.*, 2005, McGuinness and Pinheiro da Silva, 2004, Dhamankar *et al.*, 2004, Laera *et al.*, 2006]. We first present the information required for providing explanations of matching and alignments (§9.1). Then, we discuss approaches to explanations of matching by examples of existing systems (§9.2). In turn, details of these approaches are provided in sequel, including default explanations (§9.3), explaining the basic matchers (§9.4), explaining the matching process (§9.5), and negotiating alignments by argumentation (§9.6).

9.1 Justifications

We have presented the matching process as the use of basic matchers combined by strategies. In order to provide explanations to users it is necessary to have information on both matters. In particular, this information involves justifications on the reason why a correspondence should hold or not.

Each correspondence can be assigned one or several justifications that support or infirm the correspondence. We call them *justified correspondences*. For instance, the justified correspondence:

$$\langle e, e', n, \leq, `I(e) \subseteq I(e')\text{'}\rangle$$

expresses that the correspondence $\langle e, e', n, \leq \rangle$ is thought to hold because '$I(e) \subseteq I(e')$' is verified. Similarly:

$$\langle e, e', n, =, `DPLL\ entailed\text{'}\ \rangle$$

expresses that the correspondence $\langle e, e', n, = \rangle$ is thought to hold because it has been proved by the Davis–Putnam–Longemann–Loveland (DPLL) procedure [Davis and Putnam, 1960, Davis et al., 1962].

In fact, justifications can be largely more complex than presented above. For instance, the second justification may involve a full proof of the correspondence and the axioms involved in that proof. This justification information can be found directly within the correspondences or provided on-demand by the matchers to the system requiring explanation.

We explore below what can be found in this justification part.

9.1.1 Information about basic matchers

When matching systems return alignments, users may not know which external sources of background knowledge were used, when these sources were updated, or whether the resulting correspondences was looked up or derived. However, ultimately, human users or agents have to make decisions about the alignments in a principled way. So, even when basic matchers simply rely on some external source of knowledge, users may need to understand where the information comes from, with different levels of detail.

Following [McGuinness and Pinheiro da Silva, 2004], we call information about the origins of asserted facts the provenance information. Some examples of this kind of information include:

- external knowledge source name, e.g., WordNet;
- date and authors of original information;
- authoritativeness of the source, that is whether it is certified as reliable by a third party;
- name of a basic matcher, version, authors, etc. If the basic matcher relies on a logical reasoner, such as a SAT solver, some more meta-information about the reasoner may be made available:
  - the reasoning method, e.g., the Davis–Putnam–Longemann–Loveland procedure;
  - properties, e.g., soundness and completeness characteristics of the result returned by the reasoner;
  - reasoner assumptions, e.g., closed world vs. open world.

Additional types of information may also be provided, such as a degree of belief for an external source of knowledge from a particular community, computed by using some social network analysis techniques.

9.1.2 Process traces

Matching systems typically combine multiple matchers (see Chap. 5). The final alignment is usually a result of synthesis, abstraction, deduction, and some other manipulations of their results. Thus, users may want to see a trace of the performed manipulations. We refer to them as process traces. Some examples of this kind of information include:

- a trace of rules or strategies applied;
- support for alternative paths leading to a single conclusion;
- support for accessing the implicit information that can be made explicit from any particular reasoning path.

Users may also want to understand why a particular correspondence was not discovered, or why a discovered correspondence was ranked in a particular place, thereby being included in or excluded from the final alignment.

9.2 Explanation approaches

The goal of explanations is to take advantage of the above mentioned types of information for rendering the matching process intelligible to the users. A key issue is to represent explanations in a simple and clear way [Léger *et al.*, 2005].

In fact, while knowledge provenance and process traces may be enough for experts when they attempt to understand why a correspondence was returned, usually they are inadequate for ordinary users. Thus, raw justifications have to be transformed into an understandable explanation for each of the correspondences. Techniques are required for transforming raw justifications and rewriting them into abstractions that produce the foundation for what is presented to users. Presentation support also needs to be provided for users to better understand explanations. Human users will need help in asking questions and obtaining answers of a manageable size. Additionally, agents may even need some control over requests, such as the ability to break large process traces into appropriate size portions, etc. Based on [McGuinness and Pinheiro da Silva, 2004], requirements for process presentation may include:

- methods for breaking up process traces into manageable pieces;
- methods for pruning process traces and explanations to help users find relevant information;
- methods for explanation navigation, including the ability to ask follow-up questions;
- methods for obtaining alternative justifications for answers;
- different presentation formats, e.g., natural language, graphs, and associated translation techniques;
- methods for obtaining justifications for conflicting answers;
- abstraction techniques.

There are several approaches to provide explanations of the answers from matching systems. We describe below three such approaches. There are, however, few works on the topic in the literature and even fewer implemented systems. So, this chapter more specifically describes the explanation approaches as implemented in two systems, namely S-Match (§6.1.19) and iMAP (§6.2.6).

9.2.1 The proof presentation approach

Semantic matchers usually produce formal proofs of their inferences as the basis for a correspondence. They can thus benefit from work developed for displaying and explaining proofs.

For instance, S-Match [Shvaiko et al., 2005] has been extended to use the Inference Web infrastructure as well as the Proof Markup Language (PML) [McGuinness and Pinheiro da Silva, 2003, Pinheiro da Silva et al., 2006]. Thus, meaningful fragments of S-Match proofs can be loaded on demand. Users can browse an entire proof or they can restrict their view and refer only to specific, relevant parts of proofs. The proof elements are also connected to information about basic matchers that generated the hypotheses.

9.2.2 The strategic flow approach

Many matchers are composed of other matchers and have to decide in favour of some particular results over others. This composition and decision flow can be recorded in a dependency graph and used for providing explanation to users.

For instance, iMAP [Dhamankar et al., 2004] records dependencies at a very precise level (correspondence per correspondence) and can provide users with justifications for (i) existing correspondences, (ii) absent correspondences, and (iii) correspondence ranking. It provides explanations by extracting in the dependency graph the part that has an influence on the choice of a correspondence and generates an explanation in English from this extracted subgraph.

9.2.3 The argumentation approach

The argumentation approach considers the justifications or arguments in favour or against specific correspondences. Argumentation theories can determine, from a set of arguments, the correspondences which will be considered to hold and those which will not.

Argumentation can be applied to justify the matching results to users on the basis of the arguments and counter-arguments or to negotiate the correspondences that should be in an alignment. So far, this approach has mainly been applied to agents negotiating the alignments [Laera et al., 2006] rather than for explaining them. The argumentative approach is different from the proof presentation approach because it does not follow the formal proof of the correspondences. It is also more suitable when no such a proof exists.

9.3 A default explanation

A default explanation of alignments should be a short, natural language, high-level explanation without any technical details. It is designed to be intuitive and understandable by ordinary users.

9.3.1 The S-Match example

We concentrate on class matching and motivate the problem by the simple catalogue matching example shown in Fig. 9.1. Let us suppose that an agent wants to exchange or to search for documents with another agent. The documents of both agents are stored in catalogues according to class hierarchies o and o', respectively. S-Match takes as input these hierarchies, decomposes the tree matching problem into a set of node matching problems, which are, in turn, translated into a propositional validity problem, which can then be efficiently resolved using sound and complete SAT solver (§4.5.2).

Fig. 9.1. Simple catalogue matching problem.

From the example in Fig. 9.1, trying to prove that the node with label Europe in o (denoted as Europe) is equivalent to the node with label Pictures in o' (denoted as Pictures'), requires constructing the following formula (see Sect. 4.5.2 for details of formula construction):

$$\underbrace{((\text{Images} \equiv \text{Pictures'}) \wedge (\text{Europe} \equiv \text{Europe'}))}_{Axioms} \rightarrow$$

$$\underbrace{((\text{Images} \wedge \text{Europe})}_{Context_c} \equiv \underbrace{(\text{Europe'} \wedge \text{Pictures'}))}_{Context_{c'}}$$

In this example, the negated formula is unsatisfiable, thus the equivalence relation holds between the nodes under consideration.

Let us suppose that agent o' is interested in knowing why S-Match suggested a set of documents stored under the node with label Europe in o as the result to the query – 'find European pictures'. A default explanation is presented in Fig. 9.2. To simplify

250   9 Explaining alignments

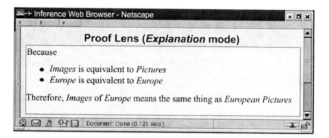

Fig. 9.2. S-Match explanation in English.

the presentation, whenever it is clear from the context to which classification a label under consideration belongs, we do not tag it with the prime symbol (').

From the explanation in Fig. 9.2, users may learn that Images in o and Pictures in o' can be interchanged, in the context of the query. Users may also learn that Europe in o denotes the same concept as Europe (European) in o'. Therefore, they can conclude that Images of Europe means the same thing as European Pictures.

9.3.2 The iMAP example

iMAP differs substantially from S-Match. It is based on a combination of constraint- and instance-based basic matchers. Once the matchers have produced the candidate correspondences, a *similarity estimator* computes, for each candidate, its similarity score. Finally, by applying the *match selector* the best matches are returned as the final alignment.

Let us consider how iMAP explains why pname = last-name is ranked higher than concat(first-name, last-name). Fig. 9.3 shows the explanation as produced by iMAP [Dhamankar et al., 2004].

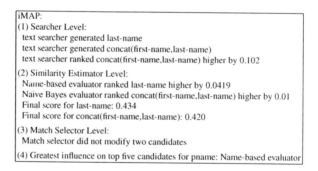

Fig. 9.3. iMAP explanation in English.

At the matcher level, concat(first-name, last-name) was ranked higher than the element with label last-name. It also clearly shows that things went wrong at the similarity estimator level. The naive Bayes evaluator still ranked matches correctly, but the name-based evaluator flipped the ranking, which was the cause of the ranking mistake.

The last line of the explanation also confirmed the above conclusion, since it states that the name-based evaluator has the greatest influence on the top five match candidates for pname. Thus, the main reason for the incorrect ranking for pname appears to be that the name-based evaluator has too much influence. This explanation would allow users to fine tune the system, possibly by reducing the weight of the name-based evaluator in the score combination step.

Users may not be satisfied with this level of explanations. Let us therefore discuss how they can investigate the details of the matching process by exploiting more verbose explanations, which are discussed in the forthcoming sections.

9.4 Explaining basic matchers

Explaining basic matchers requires only to formulate the justification information. It is illustrated through S-Match.

Let us suppose that an agent wants to see the sources of background knowledge used in order to determine the correspondence. For example, which applications, publications, other sources, have been used to determine that Images is equivalent to Pictures. Fig. 9.4 presents the source metadata for the default explanation of Fig. 9.2.

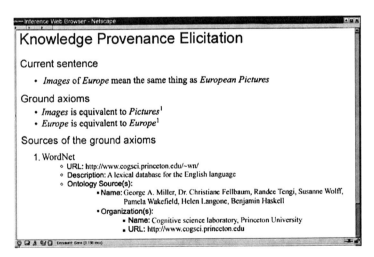

Fig. 9.4. S-Match source metadata information.

In this case, both (all) the ground sentences used in the S-Match proof came from WordNet. Using WorldNet, S-Match learnt that the first sense of the word Pictures is a synonym to the second sense of the word Images. Therefore, S-Match can conclude that these two words are equivalent words in the context of the answer (§4.2.2). The meta-information about WordNet is also presented in Fig. 9.4 as *sources of the ground axioms*. Further examples of explanations include providing meta information about the S-Match library of element-level matchers, i.e., those which are based not only on WordNet, or the order in which the matchers are used. This use of metadata is not restricted to S-Match and can be applied to any resource used in matching.

9.5 Explaining the matching process

S-Match and iMAP follow different matching strategies. iMAP follows a learning-based solution, while S-Match reduces the matching problem to a propositional validity problem. Let us discuss how they explain the matching process.

9.5.1 Dependency graphs

Explanations of alignments in the iMAP system are based on the idea of a *dependency graph*, which traces the matchers, memorising relevant slices of the graph used to determine a particular correspondence. Finally, exploiting the dependency graph, explanations are presented to users as shown in Fig. 9.3.

The dependency graph is constructed during the matching process. It records the flow of matches, data and assumptions into and out of system components. The nodes of the graph are schema attributes, assumptions made by system components, candidate correspondences, etc. Two nodes in the graph are connected by a directed edge if one of them is the successor of the other in the decision process. The label of the edge is the system component that was responsible for the decision.

Fig. 9.5 shows a dependency graph fragment that records the creation and flow for the correspondence month-posted = monthly-fee-rate. The *preprocessor* finds that both month-posted and monthly-fee-rate have values between 1 and 12 and hence makes the assumptions that they represent months. The date matcher takes these assumptions and generates month-posted = monthly-fee-rate as a candidate correspondence. This candidate is then scored by the name-based evaluator and the naive Bayes evaluator. The scores are merged by a *combining module* to produce a single score. The *match selector* acts upon the several alignment candidates generated to produce the final list of alignments. Here, for the target attribute list-price, the selector reduces the rank of the candidate correspondence price $*$ (1 + monthly-fee-rate) since it discovers that monthly-fee-rate maps to month-posted.

In each case, the system synthesises an explanation in English for the users. To provide explanations, iMAP selects the relevant slices of dependency graph that record the creation and processing of a particular correspondence. For example, the slice for month-posted = monthly-fee-rate is the portion of the graph where the nodes participated in the process of creating that correspondence.

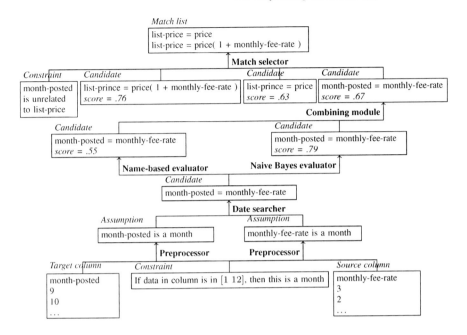

Fig. 9.5. Dependency graph as generated by iMAP [Dhamankar et al., 2004].

9.5.2 Explaining logical reasoning

A complex explanation may be required if users are not familiar with or do not trust the inference engine(s) embedded in a matching system. As the web starts to rely more on information manipulations, instead of simply information retrieval, explanations of embedded manipulation or inference engines become more important. In the current version of S-Match, a propositional satisfiability engine is used (§6.1.19), more precisely, this is the Davis–Putnam–Longemann–Loveland procedure [Davis and Putnam, 1960, Davis et al., 1962] as implemented in JSAT/SAT4J [Le Berre, 2004].

The task of a SAT solver is to find an assignment $\mu \in \{\top, \bot\}$ for atoms of a propositional formula φ such that φ evaluates to *true*. φ is *satisfiable* if and only if $\mu \models \varphi$ for some μ. If μ does not exist, φ is *unsatisfiable*. A *literal* is a propositional atom or its negation. A *clause* is a disjunction of one or more literals. φ is said to be in conjunctive normal form if and only if it is a conjunction of disjunctions of literals. The basic DPLL procedure recursively implements three rules: *unit resolution*, *pure literal* and *split*. We only consider the unit resolution rule to facilitate the presentation.

Let l be a literal and φ a propositional formula in conjunctive normal form. A clause is called a *unit clause* if and only if it has exactly one unassigned literal. *Unit resolution* is an application of *resolution* to a unit clause.

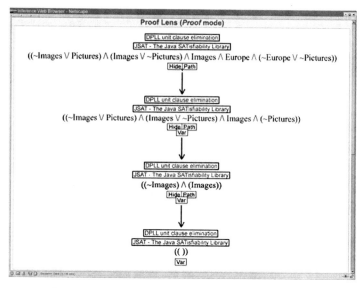

Fig. 9.6. A graphical explanation of the unit clause rule.

$$unit\ resolution : \frac{\varphi \wedge \{l\}}{\varphi[l \mid \top]}$$

Let us consider the propositional formula standing for the problem of testing if the concept at node with label Europe in o is less general than the concept at node with label Pictures in o' in Fig. 9.1. The propositional formula encoding the above stated matching problem is as follows:

((Images ≡ Pictures) ∧ (Europe ≡ Europe)) →

((Images ∧ Europe) → (Europe ∧ Pictures))

Its intuitive reading is 'Assuming that Images and Pictures denote the same concept, is there any situation such that the concept Images of Europe is less general than the concept European Pictures?'. The proof of the fact that this is not the case is shown in Fig. 9.6. Since the DPLL procedure of JSAT/SAT4J only handles conjunctive normal form formulas, in Fig. 9.6, we show the conjunctive normal form of the above formula.

From the explanation in Fig. 9.6, users may learn that the proof of the fact that the concept at node with label Europe in o is less general than the concept at node with label Pictures in o' requires 4 steps and at each proof step (excepting the first one, which is a problem statement) the *unit resolution* rule is applied. Moreover, users may learn the assumptions that are made by JSAT/SAT4J. For example, at the second step, the DPLL procedure assigns the truth value to all instances of the atom Europe, therefore making an assumption that there is a model where what an agent says about Europe is always true. According to the *unit resolution* rule, the atom

Europe should then be deleted from the input sentence, and, hence it does not appear in the sentence of the step 2.

The explanation of Fig. 9.6 represents some technical details (only the less generality test) of the default explanation in Fig. 9.2. This type of explanations is the most verbose. It assumes that, even if the graphical representation of a decision tree is quite intuitive, the matching system users have some background knowledge in logics and SAT. However, if they do not, they have a possibility to learn it by following the publications mentioned in the source metadata information of the DPLL *unit resolution* rule and JSAT, by clicking the *DPLL unit clause elimination* and the *JSAT-The Java SATisfiability Library* buttons, respectively.

9.6 Arguing about correspondences

The goal of argumentation is not strictly to explain the alignments, but to give arguments in favour or against the correspondences. It can have two roles:

- negotiating an alignment between two agents, if they accept each others arguments,
- achieving an alignment through matching. In particular, the multiagent negotiation of alignments can be seen as another aggregation technique (§5.2) between two alignments. [Silva *et al.*, 2005] presents such a system based on quantitative negotiation rather than arguments.

Argumentation allows agents to provide counter-arguments and to choose the arguments depending on their preferences. Contrary to the usual explanation work presented above, each agent can generate its own explanation by assembling arguments.

Let us consider two agents C and P using respectively ontology o and o', expressed in description logic as follows:

$$o = \{\text{Micro-company} = \text{Company} \sqcap \leq_5 \text{employee}\}$$
$$o' = \{\text{SME} = \text{Firm} \sqcap \leq_{10} \text{associate}\}$$

Let us suppose that they have discovered alignment A:

$$A = \{\langle \text{Company}, \text{Firm}, =, .89 \rangle, \quad (\gamma_1)$$
$$\langle \text{employee}, \text{associate}, \sqsubseteq, 1.0 \rangle, \quad (\gamma_2)$$
$$\langle \text{Micro-company}, \text{SME}, \sqsubseteq, .97 \rangle \} \quad (\gamma_3)$$

The three correspondences are denoted, respectively, as γ_1, γ_2 and γ_3. The set of arguments in favour of γ_1 include:

a_1 all the known Company on the one side are Firm on the other side and vice versa;
a_2 the two names Company and Firm are synonyms in WordNet;

The set of arguments in favour of γ_3 include:

a_3 the alignment (without γ_3) plus the two ontologies entail the correspondence;
a_4 all the known micro-companies on the one side are SME on the other side (and not vice versa);

and the counter-arguments include:

a_5 the two names Micro-company and SME are not alike by any string distance, and they are not synonyms in WordNet;
a_6 the only features they share are associate and employee and they have different domains and cardinalities.

In [Laera et al., 2006], the arguments are expressed following the value-based argumentation framework [Bench-Capon, 2003]. They are made of a flag denoting if they are in favour (+) or against (−) the correspondence and the type of method that supports this correspondence (basic methods). A simple way to express these arguments is as follows:

a_1 : ⟨Company, Firm, =, .89, ⟨+, $extensional$⟩⟩
a_2 : ⟨Company, Firm, =, .89, ⟨+, $terminological$⟩⟩
a_3 : ⟨Micro-company, SME, ⊑, .97, ⟨+, $semantic$⟩⟩
a_4 : ⟨Micro-company, SME, ⊑, .97, ⟨+, $extensional$⟩⟩
a_5 : ⟨Micro-company, SME, ⊑, .97, ⟨−, $terminological$⟩⟩
a_6 : ⟨Micro-company, SME, ⊑, .97, ⟨−, $structural$⟩⟩

Such kind of arguments could be delivered by existing basic matchers. Another, more elaborate way to define arguments is to allow correspondences themselves to be justifications. This permits, for instance, to express that the structural similarity of Micro-company and SME depends on the terminological similarity of employee and associate.

The rationale behind these kinds of arguments is that some agents may prefer, or trust, better some techniques than others. For instance, one can imagine that agent C prefers terminological arguments over extensional arguments, extensional arguments over semantic arguments and semantic arguments over structural arguments. This order induces a partial order on the arguments themselves: $a_5 \succ_C a_2$, $a_1 \succ_C a_2$, $a_5 \succ_C a_4$, $a_1 \succ_C a_4$, $a_2 \succ_C a_3$, $a_4 \succ_C a_3$, $a_3 \succ_C a_6$. Similarly, P could have a different preference ordering favouring structural, semantic, terminological and then extensional arguments.

There are logical theories [Dung, 1995, Amgoud et al., 2000, Bench-Capon, 2003] that, given a set of arguments and the preferences of agents, define what is the consensus alignment between both parties. They usually define an admissible argument a with regard to a set of arguments S as an argument to which every counter-argument is attacked by an argument of S. A set S is said conflict-free if no argument of S attacks another argument of S. A maximal conflict free set of arguments acceptable with regard to S is called admissible. Finally, a preferred extension is an admissible set of arguments where there is no other such set that contains arguments preferred to some in the set that are admissible for their preferred arguments in the set. For instance, C will have for preferred extension $\{a_5, a_1, a_2, a_6\}$ and P, in turn, will have $\{a_6, a_5, a_2, a_1\}$. However together, the maximal common subset of arguments between C and P is $\{a_1, a_2, a_5, a_6\}$ which selects the preferred alignment made up of γ_1 and γ_2.

A consensus alignment can also be achieved by a dialogue between the agents during which they exchange arguments. Such a dialogue is presented below. The agent C starts the dialogue by asserting alignment A between the two ontologies o and o' (the agent C is committed to support the alignment A and each correspondence it contains). A possible dialogue between C and P is as follows:

//The agent C is committed to support the alignment
$C-$assert(:content A :reply-with 1)$\rightarrow P$
//The agent P asks to justify the correspondence γ_1 (P does not have counter-argument)
$C \leftarrow$ question(:content γ_1 :reply-with 2) - P
// The agent C justifies the correspondence γ_1 with the arguments a_1 and a_2
$C-$support(:content $a_1, a_2 \vdash^+ \gamma_1$:in-reply-to 2)$\rightarrow P$
//The agent P asks to justify the correspondence γ_3 (P is ready to justify the opposite)
$C \leftarrow$ challenge(:content γ_3 :reply-with 3) - P
// The agent C justifies the correspondence γ_3 with the arguments a_3 and a_4
$C-$support(:content $a_3, a_4 \vdash^+ \gamma_3$:in-reply-to 3)$\rightarrow P$
// The agent P contests the correspondence γ_3 with the counter-arguments a_5 and a_6
$C \leftarrow$contest(:content $a_5, a_6 \vdash^- \gamma_3$:in-reply-to 3) - P
// The agent C retracts the correspondence γ_3
$C-$retract(:content γ_3 :in-reply-to 3)$\rightarrow P$

This results in the selection of the alignment $A' = \{\gamma_1, \gamma_2\}$.

These argumentation techniques have not been used in alignment explanation so far. However, they could be used in interactively explaining to users the arguments in favour or against correspondences. In the argumentation dialogue above, one of the agents can be a human user. The system knowing the preferences of users can provide them with a more adapted arguments.

9.7 Summary

Delivering alignments to users for inspection and revision is an important topic not deeply developed so far. Providing the justifications for correspondences can also be used for helping computer systems like agents to better understand alignments and control matching results.

We have presented the type of raw justifications matchers should supply and the manipulations that explanation systems can perform on these justifications in order to provide an intelligible picture of alignments to users. Some of these manipulations are based on proof presentation techniques, some others are based on a kind of dependency graphs. We have also presented techniques used by agents for exchanging justifications of correspondences and reaching a common agreement.

By using explanations, a matching system can provide users with meaningful prompts and suggestions on further steps towards the production of a desired result. Having understood the alignments returned by a matching system, users can deliberately edit them manually, thereby providing the feedback to the system. Beside explanations, matching systems should provide facilities for users to explore the alternative paths not followed by the system. These systems should enable the users to re-launch the matching process with different parameters in an intermediate state.

10

Processing alignments

In this book we have taken a two steps view on reducing semantic heterogeneity: (i) matching of entities to determine alignment and (ii) processing the alignment according to application needs. In the previous chapters we have discussed various themes related to the first step. In this chapter, in turn, we present how the alignments can be specifically used by the applications, thus focusing on the alignment processing step.

Since this book is devoted to ontology matching, our goal is not to present a complete panorama of the different uses of alignments. This would require another book. Rather we present the broad classes of alignment use and the tools for implementing these usages. Meanwhile, most of the commercially available ontology integration tools focus on automation of the alignment processing, by opposition to matching. They are very often specialised in a particular segment of the matching space. Altova MapForce and Stylus Studio XSLT Mapper are specialised in XML integration. They integrate data from XML sources as well as databases or other structured sources. Microsoft BizTalk Schema Mapper is targeted at the business process and information integration, using the proprietary BizTalk language. Ontoprise SemanticIntegrator offers ontology-based integration of data coming from databases or ontologies. There are unfortunately no scholar references describing these systems in depth and URLs change so often that we refer the reader to ontologymatching.org for accurate and up to date information.

The matching operation itself is not automated within those tools, though they facilitate manual matching by visualising input ontologies (XML, database, flat files formats, etc.) and the correspondences between them. Once the correspondences have been established it is possible to specify, for instance, some data translation operations over the correspondences such as adding, multiplying, and dividing the values of fields in the source document and storing the result in a field in the target document.

We discuss below a minimal set of operations that can be performed from alignments, including ontology merging (§10.1), ontology transformation (§10.2), data translation (§10.3), mediation (§10.4), and reasoning (§10.5). They are presented as operators in the style of [Melnik *et al.*, 2005] (see also Sect. 8.2.1). We provide some

10.1 Ontology merging

Ontology merging is a first natural use of ontology matching. As depicted in Fig. 10.1, it consists of obtaining a new ontology o'' from two matched ontologies o and o' so that the matched entities in o and o' are related as prescribed by the alignment. Merging can be presented as the following operator:

$$\mathsf{Merge}(o, o', A) = o''$$

The ideal property of a merge would be that

$$\mathsf{Merge}(o, o', A) \models o$$
$$\mathsf{Merge}(o, o', A) \models o'$$
$$\mathsf{Merge}(o, o', A) \models \alpha(A)$$

if $\alpha(A)$ is the alignment expressed in the logical language of $\mathsf{Merge}(o, o', A)$, and

$$o, o', A \models \mathsf{Merge}(o, o', A)$$

The former set of assertions means that the merge preserves the consequences of both ontologies and of the relations expressed by the alignment. The latter assertion means that the merge does not entail more consequences than specified by the semantics of alignments (§2.5.4. Of course, this is not restricted to the union of the consequences of o, o' and A.

When the ontologies are expressed in the same language, merging often involves putting the ontologies together and generating bridge or articulation axioms. Merging does not usually require a total alignment: those entities which have no corresponding entity in the other ontology will remain unchanged in the merged ontology.

Ontology merging is especially used when it is necessary to carry out reasoning involving several ontologies. It is also used when editing ontologies in order to create ontologies tailored for a particular application. In such a case, it is most of the time followed by a phase of ontology reengineering, e.g., suppressing unwanted parts from the obtained ontology.

Protégé (§8.3.2) and Rondo (§8.2.1) offer independent operators for ontology merging. OntoMerge (§6.1.17) takes bridge rules expressed in predicate calculus and can merge ontologies in OWL. The Alignment API (§8.2.4) can generate axioms in OWL or SWRL for merging ontologies. Other systems are able to match ontologies and merge them directly: FCA-merge (§6.2.3), SKAT (§6.1.5), DIKE (§6.1.4). OntoBuilder (§6.1.10) uses ontology merging as an internal operation: the system creates an ontology that is mapped to query forms. This ontology is merged with the global ontology so that queries can be directly answered from the global ontology.

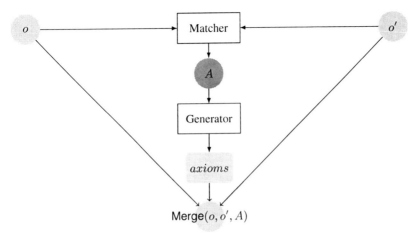

Fig. 10.1. Ontology merging. From two matched ontologies o and o', resulting in alignment A, articulation *axioms* are generated. This allows the creation of a new ontology covering the matched ontologies.

10.2 Ontology transformation

Ontology transformation, from an alignment A between two ontologies o and o', consists of generating an ontology o'' expressing the entities of o with respect to those of o' according to the correspondences in A. It can be denoted as the following operator:

$$\text{Transform}(o, A) = o''$$

Contrary to merging, ontology transformation, and the operators to follow, are oriented. This means that the operation has an identified source and target and from an alignment it is possible to generate two different operations depending on source and target.

Ontology transformation is not well supported by tools. It is useful when one wants to express one ontology with regard to another one. This can be particularly useful for connecting an ontology to a common upper level ontology, for instance, or local schemas to a global schema in data integration.

10.3 Data translation

Data translation, presented in Fig. 10.2, consists of translating instances from entities of ontology o into instances of connected entities of matched ontology o'. This can be expressed by the following operator:

$$\text{Translate}(d, A) = d'$$

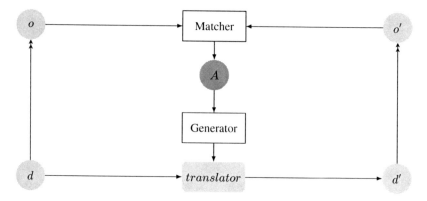

Fig. 10.2. Data translation. From two matched ontologies o and o', resulting in alignment A, a *translator* is generated. This allows the translation of the instance data (d) of the first ontology into instance data (d') for the second one.

Data translation usually involves generating some transformation program from the alignment.

This requires a total alignment if one wants to translate all the extensional information. Non total alignments risk loosing instance information in the translation (this can also be acceptable if one does not want to import all the instance information).

Data translation is used for importing data under another ontology without importing the ontology itself. This is typically what is performed by database views in data integration (§1.2), in multiagent communication for translating messages (§1.5.1), and in semantic web services for translating the flow of information in mediators (§1.4).

Rondo (§8.2.1) provides tools for data translation. The Alignment API (§8.2.4) can generate translations in XSLT or C-OWL. Many tools developed for data integration can generate translators under the form of SQL queries. Drago [Serafini and Tamilin, 2005] is an implementation of C-OWL, which can process alignments expressed in C-OWL for transferring data from one ontology to another one.

Some of the tools reviewed in Chap. 6, provide their output as data translation or process themselves the translation. These include: Clio (§6.3.2), ToMAS (§6.1.15), TransScm (§6.1.3), MapOnto (§6.1.16) and sPLMap (§6.2.12).

10.4 Mediation

In this section, we consider a mediator as an independent software component that is introduced between two other components in order to help them interoperate. There

10.4 Mediation

are many different forms of mediators, including some acting as brokers or dispatchers. We concentrate here on a query mediator which can perform two operations:

$$\text{TransformQuery}(q, A) = q'$$

and

$$\text{Translate}(a', \text{Invert}(A)) = a$$

TransformQuery is a kind of ontology transformation which transforms a query expressed using ontology o into a query expressed with the corresponding entities of a matched ontology o'. The Translate operation performs data translation on the answer of the query if necessary. This process is presented in Fig. 10.3.

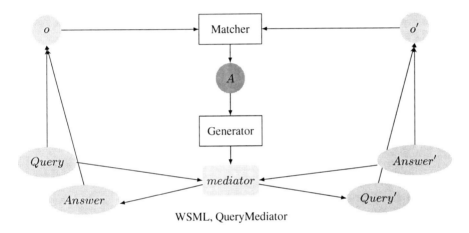

Fig. 10.3. Query mediation. From two matched ontologies o and o', resulting in alignment A, a *mediator* is generated. This allows the transformation of queries expressed with the entities of the first ontology into a query using the corresponding entities of a matched ontology and the translation back of the results from the second ontology to the first one.

Translating the answers requires the possibility of inverting the alignments (Invert operator). The generated functions should be compatible, otherwise the translated answer may not be a valid answer to the initial query. Compatibility can be expressed as follows:

$$\forall e \in o, \text{TransformQuery}(\text{TransformQuery}(e, A), \text{Invert}(A)) \sqsubseteq e$$

Here we use a subsumption relation ($\sqsubseteq$), but it can be replaced by any suitable relation ensuring that the answer is compatible.

However, it is not always necessary to translate the answers, since they can be objects independent from the ontologies, e.g., picture files, strings. Query mediation is mainly used in data integration (§1.2) and peer-to-peer systems (§1.3). When the mediator content is expressed as SQL view definitions, many database systems can process them. The Alignment API (§8.2.4) can behave as a SPARQL query mediator from simple alignments. Some systems directly generate mediators after matching, such as Wise-Integrator (§6.3.7).

10.5 Reasoning

Reasoning consists of using the result of the matching as rules for reasoning with the two matched ontologies. Bridge axioms used for merging can also be viewed as such rules.

$$\mathsf{TransformAsRules}(A) = o$$

Here the set of rules is represented as an ontology o which must be written in an ontology language supporting rules or the expression of bridge axioms.

Typically, any transformation of the alignments under a form suitable for reasoning, such as SWRL (§8.1.4), OWL (§8.1.2) or C-OWL (§8.1.3) can be used by inference engines for these languages, such as Drago [Serafini and Tamilin, 2005] or Pellet [Sirin et al., 2007].

10.6 Towards an alignment service

Throughout this book, ontology matching has been considered as a one-shot private operation. However, there are several reasons why applications using ontology matching could benefit from sharing ontology matching techniques and results:

- *Each application can benefit from more algorithms*: Many different applications have comparable needs. It is thus appropriate to share the solutions to these problems. This is especially true as alignments are quite difficult to provide.
- *Each algorithm can be used in more applications*: Alignments can be used for different purposes and must be expressed as such instead of as bridge axioms, mediators or translation functions.
- *Each individual alignment can be reused by different applications*: There is no magic algorithm for quickly providing a useful alignment. Once high quality alignments have been established – either automatically or manually – it is very important to be able to store, share and reuse them.

For that purpose, it is useful to provide an alignment service able to store, retrieve and manipulate existing alignments as well as to generate new alignments on-the-fly. This kind of service should be shared by the applications using ontologies on the semantic web. They should be seen as a directory or a service by web services, as an agent by agents, as a library in ambient computing applications, etc.

Operations that are necessary in such a service include:

- the ability to store alignments and retrieve them, whether they are provided by automatic means or manually;

- the proper annotation of alignments in order for the clients to evaluate the opportunity to use one of them or to start from it (this starts with the information about the matching algorithms and the justifications for correspondences that can be used in agent argumentation);
- the ability to produce alignments on-the-fly through various algorithms that can be extended and tuned;
- the ability to generate knowledge processors, such as mediators, transformations, translators and rules as well as to run these processors if necessary;
- the possibility to discover similar ontologies and to interact with other such services in order to ask them for operations that the current service cannot provide by itself.

Such a service would require a standardisation support, such as the choice of an alignment format or at least of metadata format (§8.1.5). There have been proposals for providing matching systems and alignment stores that can be considered as servers [Euzenat, 2005, Zhdanova and Shvaiko, 2006], but they need a wider availability (to agents, services, etc.) and achieving a critical mass of users to really be helpful.

10.7 Summary

This chapter only considers in general the issue of alignment processing. Alignments can be used in different ways (merging, transformation, translation, mediation) and there are different languages adapted to each of these ways (SWRL, OWL, C-OWL, XSLT, SQL, etc.). So far there are only a few systems able to generate output in several of these languages. Fortunately, they are independent from matching algorithms. Several matching systems process directly their results in one of these operation, while others deliver alignments. Unfortunately, most often the delivered alignments are in a format that cannot be exploited by other systems and operator generators, thus requiring additional efforts to embed them into the new environments.

Useful alignments are such a scarce resource that storing them in an independent format such as those presented in Chap. 8 is very important. It would allow sharing and processing them in different ways independently form the applications. This would give more freedom to application developers to choose the best suited algorithm and to process alignments adequately.

Finally, the operations to be performed raise constraints against the alignment to be used. As a consequence, matching systems should be able to provide information about the properties the alignments satisfy.

Part V

Conclusions

11

Conclusions

In this book we have attempted at covering ontology matching in its diversity. In particular, we have shown that there are many applications that may need ontology matching (Chap. 1) and that there are different forms of ontologies that may need to be matched (Chap. 2). Ontology matching can take advantage of innumerable basic techniques (Chap. 4) composed and supervised in diverse ways (Chap. 5). The output of matching can be provided according to different representations (Chap. 8) or executable forms (Chap. 10) which may need to be justified (Chap. 9). This, in turn, has led to a profusion of available systems (Chap. 6).

We have provided a systematic view over the resources for helping users, researchers and developers in selecting the system or technique most adapted to their needs. This has been substantiated by identifying application needs (Chap. 1), by classifying matching techniques (Chap. 3) and by proposing an adapted methodology for evaluating matching solutions (Chap. 7). This does not mean that, for any application need, the ideal ontology matching solution is directly provided to readers. Techniques presented in this book can be composed in so many ways that the solution space is open-ended and is far from having thoroughly been explored. There remains a lot of work to be done by researchers to investigate better solutions and by system developers to find the appropriate settings for their applications.

In the remainder of this chapter we overview some general trends in the ontology matching field (§11.1). We present some promising research directions which we believe worth and need further investigations (§11.2). These stem from all parts of the book. We conclude with general remarks (§11.3).

11.1 A brief outlook of the trends in the field

In the past, the ontology matching problem has been addressed in several areas. However, most often this happened in an isolated manner among: (i) database schemas in the world of information integration, (ii) XML schemas and catalogues on the web, (iii) ontologies (axiomatised theories) in artificial intelligence, semantic web, knowledge representation, and (iv) objects and entities in data mining. Technical issues these areas had encountered were rarely addressed from multidisciplinary and cross-community viewpoints.

During the last decade the areas mentioned above have made substantial progress in matching. However, they require other technologies and cross-fertilisation to continue their growth. This was one of the motivations behind this book and such an initiative as Ontology Matching[1], which aims at increasing awareness of the existing matching efforts across the relevant communities and at facilitating the cross-fertilisation between them.

The number and variety of solutions to the ontology matching problem keep growing at a fast pace. Fig. 11.1 shows (approximately) how many works devoted to diverse aspects of matching have been published at various conferences all over the world in recent years[2].

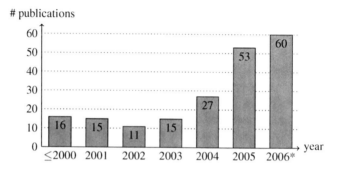

Fig. 11.1. Dynamics of publications devoted to ontology matching. * Value for 2006 is an estimation, since a complete information concerning the previous year usually becomes stable when the current year finishes. The actual value for 2006 on the access date is 48.

In the future, we expect a continuing growth of works on matching due to the constantly increasing interest in solutions for the semantic heterogeneity problem from both academia and industry.

11.2 Future challenges

We provide some directions in which, in our opinion, research on ontology matching should or is likely to evolve. In particular, in this section, we point out current needs that are not addressed by the technology and that will have to be addressed for the field to be considered mature. We also present our opinion on the directions technology is leading to.

---

[1] http://www.OntologyMatching.org
[2] Source: www.OntologyMatching.org, Publications section. Accessed: 02.04.2007.

This section is organised following the structure of the book in terms of chapters. Its description remains at the technical level, while some details at the market level can be found in [Cuel *et al.*, 2006].

11.2.1 Applications

The pressure of applications on ontology matching is tangible. Application developers test the available technology and either adopt it or bypass the heterogeneity problems. In real world, applications are more mixed than what was presented in Chap. 1: they are not pure peer-to-peer or data integration applications. They have a purpose and can provide a more specific background than that which was mentioned.

We can expect that there will be definitely applications which use ontology matching. They will start in niche places with a specific setting rather than presenting a general solution to a global problem. Then, gradually, the proven solutions will start spreading to other applications.

11.2.2 Foundations

Foundations of ontology matching, and particularly the semantics of alignments, deserve additional investigations. Available model-theoretic semantics are sufficiently similar, so they could eventually converge. Recent work on categorical characterisation of ontology matching raised some questions about the statement in categorical terms of expressive alignments which go beyond equivalence [Zimmermann *et al.*, 2006]. Therefore, interesting and useful work could be pursued in this direction.

One can also think of relating matching methods and properties to the resulting alignment. Indeed, it would be useful to know in advance that using a particular method will provide a one-to-one alignment or that the transformation extracted form it preserves some order property.

11.2.3 Basic techniques

As testified by Chap. 4, there is a wealth of basic matching techniques and certainly their number will continue growing. We can expect new types of basic automatic matchers addressing a larger variety and more sophisticated situations than is currently the case. In particular, we can expect:

– methods for matching processes, as opposed to concepts,
– new methods for alignment reuse,
– new techniques for automatic discovery and exploitation of background domain specific knowledge [Giunchiglia *et al.*, 2006c],
– new matchers relying on social networks and web communities [Zhdanova and Shvaiko, 2006],
– techniques able to deal with multilinguality,
– new libraries of matchers or extensions of existing libraries.

The development of new basic matchers may also lead the classifications of Chap. 3 to evolve.

11.2.4 Matching strategies

There is already a lot of creativity deployed in designing matchers, but more will be needed to face ontology matching in its entirety. We briefly outline some of the most important directions:

- One of the important issues to deal with is the proper combination and integration of various categories of matchers. In particular, the integration of semantic (deductive) and inductive techniques is of high interest: is the semantic part used as an alignment amplifier or as a consistency checker? When semantic techniques have given their results, is it worthwhile to use inductive methods from the provided alignment again or not? These questions have yet to find an answer.
- We will also see new approaches to automate the combination of individual matchers and libraries of matchers. Some existing solutions can be found in [Doan et al., 2001, Ehrig et al., 2005], but certainly more results are to come.
- The place of users in semi-automatic matching tools requires attention. In particular, alignments are already complex enough objects that are difficult to display on a computer screen. Enabling natural edition and manipulation of alignments, keeping user involvement lightweight, is still a research issue.
- The work on automatically tuning matcher parameters is also very important because users cannot be expected to find correct parameters by themselves and sometimes the parameters have to be adjusted dynamically [Sayyadian et al., 2005].
- Another important issue consists of delivering partial alignments. In real world cases, it is rare that two ontologies can be exactly matched. In general, ontologies most often only partially overlap. However, matching algorithms tend to maximise this overlap. Developing techniques for deciding when to stop matching entities is an important problem.

11.2.5 Matching systems

More systems will be developed. This should not come as a surprise. With the improvements of tool support for semantic web languages, we expect matching systems to become more mature as well. Beside this, systems will also have to improve on performance issues such as time and memory consumption. We expect the following trends in system improvement:

- Most of the approaches will tend to be increasingly generic, i.e., able to handle multiple input forms of ontologies.
- In contrast to the previous point, we also expect some very good solutions to narrower problems. For example, ontology versions matching is a problem for which current matching technology turns out to work well.

- New types of input, such as plain text and query interfaces from the deep web should enter intensively into practice.
- Approaches will try to suitably handle an increasing number of constructs available from the input, e.g., constraints.
- Different (new) internal representations of the input data, e.g., descriptors of the entries for the learning algorithms, should appear as well.

11.2.6 Evaluation of matching systems

It is necessary to pursue current efforts on extensive evaluation of ontology matching systems. It would be good, however, to have improvements on that topic. In particular, there is an important need for high quality data sets that satisfy all the requirements mentioned in Chap. 7. According to our experience, the number of such data sets is increasing fast.

There is also a need for application specific settings in evaluations, i.e., the possibility to evaluate the quality of alignments within the context of a particular task with evaluation measurements related to the task accomplishment. As mentioned in Chap. 7, we need measures that take the semantics of ontology languages into account, i.e., measures that really match recall with completeness and precision with correctness [Euzenat, 2007]. Finally, when it is required to involve users, an open and important topic is to take this into account in the evaluation.

Beside evaluating systems, it is necessary to be able to help users in choosing the appropriate matcher or to combine the most appropriate matchers for their tasks [Mochol *et al.*, 2006, Huza *et al.*, 2006]. This can be achieved by exploiting the evaluation campaign results and by a better understanding of the problem space of ontology matching. We have tried in this book to provide first steps towards this goal, but a lot remains to be investigated.

11.2.7 Representing alignments

Current alignment formats have advantages and disadvantages. It would be worthwhile to establish one (or two) standard alignment formats for exchanging the alignments. In particular, such formats should have the power to record expressive alignments (not just name mappings) and also to provide more metadata about the alignment, its production process as well as its properties.

In the long term, we also expect substantial progress on the frameworks for integrating different matching systems. In fact, infrastructures, which are able to store and provide alignments to those who need it, are still missing. Such an infrastructure should also match ontologies and process the alignments on specified data. In this context, alignment formats and metadata become crucial. The infrastructure, in order to maximise its usefulness, should be easy to re-configure and be usable through many protocols (adapted to peer-to-peer systems, agents, web services, etc.).

Graphical alignment editors are needed. They should be easy to use for ordinary users. Scalable alignment visualisation techniques should also be developed. Finally, an alignment editor working on a standard format would be very helpful for anyone who plans to use and experiment with ontology matchers.

11.2.8 Explaining alignments

There are only a few matching systems able to provide a justification of their results. In order for matching systems to gain a wider acceptance, it will be necessary that they can provide arguments for their results to users or to other programs that use them. Explanation is thus an important challenge for ontology matching as well as user interfaces in general. For example, a user interface able to help users to efficiently review alignments and to modify them in an interactive manner is needed.

Concerning the interaction of matching systems with other programs, it will be necessary to provide justifications and arguments in a standardised way. This will only happen when more programs take advantage of these justifications. An extension of this paradigm would be for the matching algorithms to assert the properties satisfied by the produced alignments as well as the proofs of these properties in a machine readable form. [Euzenat, 2002, Pinheiro da Silva *et al.*, 2004] have discussed how to take advantage of such proofs.

11.2.9 Processing alignments

Processing alignments according to application needs is the ultimate goal of matching. It has been considered at the end of this book as the last step but would certainly deserve a book on its own.

Currently, many systems are rather monolithic and perform matching and alignment processing at once. We hope to see more modularisation in future. We also expect to see more alignment processors to be developed. Only under these conditions ontology matching can start being used by applications more intensively. Ideally, such processors should be integrated into some infrastructure as mentioned in Sect. 11.2.7.

11.3 Final words

We admit that even if a good progress has been made in the matching field, as such, ontology matching may appear to be virtually impossible. Indeed, for finding the correspondences between concepts, it is necessary to understand their meaning. Besides the general meaning ascribed by model-theoretic semantics, the ultimate meaning of concepts is in the head of the people who developed those concepts and we cannot program a computer to learn it.

However, the same remark leads to the conclusion that communication, even between people, is impossible. We know that human beings achieve communication; they at least, succeed quite often in communicating and sometimes fail. Achieving this communication can be viewed as a continuous task of negotiating the relations between concepts, i.e., arguing about alignments, building new ones, questioning them, etc. Therefore, matching ontologies is an on-going work and further substantial progress in the field can be made by considering it in its dynamics.

A
Legends of figures

We present below the three sets of notations that are used in the pictures of this book.

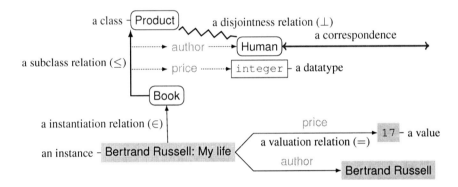

Fig. A.1. Graphic representation of ontologies.

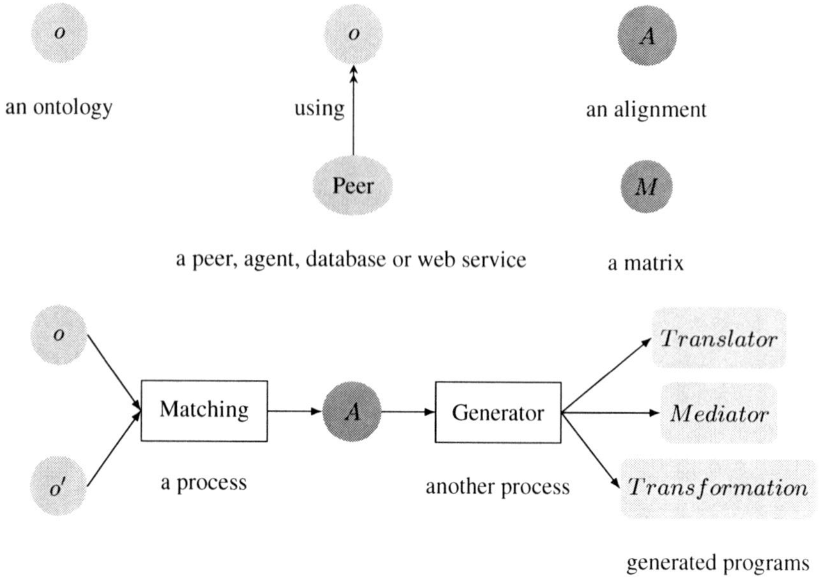

Appendix A: Legends of figures

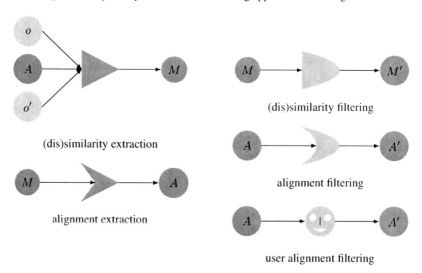

Fig. A.2. Graphic representation of matching application building blocks.

Fig. A.3. Graphic representation of matching system building blocks.

B
Running example

The following two ontologies correspond to those displayed in Fig. 2.7: `culture-shop.owl` is the left-hand side ontology and `library.owl` is the right-hand side one.

B.1 culture-shop.owl

```
<?xml version="1.0" encoding="iso-8859-1"?>
<!DOCTYPE rdf:RDF [
         <!ENTITY xsd "http://www.w3.org/2001/XMLSchema#" >
  <!ENTITY rdf "http://www.w3.org/1999/02/22-rdf-syntax-ns#" >
  <!ENTITY rdfs "http://www.w3.org/2000/01/rdf-schema#" >
  <!ENTITY dc "http://purl.org/dc/elements/1.1/" >
  <!ENTITY owl "http://www.w3.org/2002/07/owl#" >
  <!ENTITY foaf  "http://xmlns.com/foaf/0.1/">
  <!ENTITY ical  "http://www.w3.org/2002/12/cal/ical#">]>

<rdf:RDF
xmlns="http://book.ontologymatching.org/example/culture-shop.owl#"
xml:base="http://book.ontologymatching.org/example/culture-shop.owl#"
xmlns:foaf ="&foaf;"
xmlns:ical ="&ical;"
xmlns:rdf ="&rdf;"
xmlns:xsd ="&xsd;"
xmlns:rdfs ="&rdfs;"
xmlns:owl ="&owl;"
xmlns:dc ="&dc;">

  <!-- ################## ONTOLOGY ################## -->

  <owl:Ontology rdf:about="">
    <dc:creator>Jérôme Euzenat</dc:creator>
    <dc:contributor>Pavel Shvaiko</dc:contributor>
    <dc:description>Fragments of a cultural product shop ontology</dc:description>
    <dc:date>2006/04/12</dc:date>
    <rdfs:label>Culture shop ontology</rdfs:label>
    <rdfs:comment>An example for the Ontology matching book.
    This ontology fragments organises some cultural product
    the way it could be organised for a cultural product
    e-commerce site.</rdfs:comment>
    <owl:versionInfo>
      $Id: culture-shop.owl,v 1.2 2007/03/15 21:05:21 cvs Exp $
    </owl:versionInfo>
```

```xml
</owl:Ontology>

<!-- In OWL-DL all items must be declared -->
<owl:DatatypeProperty rdf:about="&dc;creator" />
<owl:DatatypeProperty rdf:about="&dc;contributor" />
<owl:DatatypeProperty rdf:about="&dc;description" />
<owl:DatatypeProperty rdf:about="&dc;date" />

<!-- ################### CLASSES ################### -->

<owl:Class rdf:ID="Product">
  <rdfs:label xml:lang="en">item</rdfs:label>
  <rdfs:label xml:lang="fr">Marchandise</rdfs:label>
  <rdfs:comment xml:lang="en">The goods which are for sale at our site.
    </rdfs:comment>
  <rdfs:subClassOf>
    <owl:Restriction>
      <owl:onProperty rdf:resource="#price" />
      <owl:cardinality
          rdf:datatype="&xsd;nonNegativeInteger">1</owl:cardinality>
    </owl:Restriction>
  </rdfs:subClassOf>
  <rdfs:subClassOf>
    <owl:Restriction>
      <owl:onProperty rdf:resource="#name" />
      <owl:maxCardinality
          rdf:datatype="&xsd;nonNegativeInteger">1</owl:maxCardinality>
    </owl:Restriction>
  </rdfs:subClassOf>
  <rdfs:subClassOf>
    <owl:Restriction>
      <owl:onProperty rdf:resource="#creator" />
      <owl:maxCardinality
          rdf:datatype="&xsd;nonNegativeInteger">1</owl:maxCardinality>
    </owl:Restriction>
  </rdfs:subClassOf>
  <rdfs:subClassOf>
    <owl:Restriction>
      <owl:onProperty rdf:resource="#id" />
      <owl:maxCardinality
          rdf:datatype="&xsd;nonNegativeInteger">1</owl:maxCardinality>
    </owl:Restriction>
  </rdfs:subClassOf>
  <rdfs:subClassOf>
    <owl:Restriction>
      <owl:onProperty rdf:resource="#topic" />
      <owl:maxCardinality
          rdf:datatype="&xsd;nonNegativeInteger">1</owl:maxCardinality>
    </owl:Restriction>
  </rdfs:subClassOf>
</owl:Class>

<owl:Class rdf:ID="DVD">
  <rdfs:subClassOf rdf:resource="#Product" />
  <rdfs:label xml:lang="en">DVD</rdfs:label>
  <rdfs:comment xml:lang="en"></rdfs:comment>
</owl:Class>

<owl:Class rdf:ID="CD">
  <rdfs:subClassOf rdf:resource="#Product" />
  <rdfs:label xml:lang="en">CD</rdfs:label>
  <rdfs:comment xml:lang="en"></rdfs:comment>
</owl:Class>

<owl:Class rdf:ID="Book">
  <rdfs:subClassOf rdf:resource="#Product" />
  <rdfs:label xml:lang="en">book</rdfs:label>
```

```xml
      <rdfs:comment xml:lang="en">A book.</rdfs:comment>
      <rdfs:subClassOf>
        <owl:Restriction>
          <owl:onProperty rdf:resource="#author" />
          <owl:cardinality
            rdf:datatype="&xsd;nonNegativeInteger">1</owl:cardinality>
        </owl:Restriction>
      </rdfs:subClassOf>
      <rdfs:subClassOf>
        <owl:Restriction>
          <owl:onProperty rdf:resource="#publisher" />
          <owl:cardinality
            rdf:datatype="&xsd;nonNegativeInteger">1</owl:cardinality>
        </owl:Restriction>
      </rdfs:subClassOf>
    </owl:Class>

    <owl:Class rdf:ID="Science">
      <rdfs:subClassOf rdf:resource="#Book" />
      <rdfs:label xml:lang="en">science book</rdfs:label>
      <rdfs:comment xml:lang="en"></rdfs:comment>
    </owl:Class>

    <owl:Class rdf:ID="Textbook">
      <rdfs:subClassOf rdf:resource="#Science" />
      <rdfs:label xml:lang="en">science textbook</rdfs:label>
      <rdfs:comment xml:lang="en">Science book for students.</rdfs:comment>
    </owl:Class>

    <owl:Class rdf:ID="Popular">
      <rdfs:subClassOf rdf:resource="#Science" />
      <rdfs:label xml:lang="en">popular science book</rdfs:label>
      <rdfs:comment xml:lang="en">
        Science book for a wide audience.</rdfs:comment>
    </owl:Class>

    <owl:Class rdf:ID="Pocket">
      <rdfs:subClassOf rdf:resource="#Book" />
      <rdfs:label xml:lang="en">pocket book</rdfs:label>
      <rdfs:comment xml:lang="en">
        Paperback bound books of small size.</rdfs:comment>
    </owl:Class>

    <owl:Class rdf:ID="Children">
      <rdfs:subClassOf rdf:resource="#Book" />
      <rdfs:label xml:lang="en">children book</rdfs:label>
      <rdfs:comment xml:lang="en">Books for children.</rdfs:comment>
    </owl:Class>

    <owl:Class rdf:ID="Person">
      <rdfs:label xml:lang="en">person</rdfs:label>
      <rdfs:comment xml:lang="en">
        @@Developer: should link with FOAF some day</rdfs:comment>
    </owl:Class>

    <owl:Class rdf:ID="Publisher">
      <rdfs:label xml:lang="en">publisher</rdfs:label>
      <rdfs:comment xml:lang="en">A book or music publisher.</rdfs:comment>
    </owl:Class>

    <!-- ################## PROPERTIES ################### -->

    <owl:DatatypeProperty rdf:ID="price">
      <rdfs:domain rdf:resource="#Product" />
      <rdfs:range rdf:resource="&xsd;integer" />
      <rdfs:label xml:lang="en">price</rdfs:label>
      <rdfs:comment xml:lang="en">The list price of a particular item on
```

```xml
    our site. Does not include taxes, shipping or rebates.</rdfs:comment>
</owl:DatatypeProperty>

<owl:DatatypeProperty rdf:ID="firstname">
  <rdfs:domain rdf:resource="#Person" />
  <rdfs:range rdf:resource="&xsd;string" />
  <rdfs:label xml:lang="en">firstname</rdfs:label>
</owl:DatatypeProperty>

<owl:DatatypeProperty rdf:ID="lastname">
  <rdfs:domain rdf:resource="#Person" />
  <rdfs:range rdf:resource="&xsd;string" />
  <rdfs:label xml:lang="en">lastname</rdfs:label>
</owl:DatatypeProperty>

<owl:DatatypeProperty rdf:ID="name">
  <rdfs:domain rdf:resource="#Product" />
  <rdfs:range rdf:resource="&xsd;string" />
  <rdfs:label xml:lang="en">name</rdfs:label>
  <rdfs:comment xml:lang="en">The name identifying an item for the common
    shoppers.</rdfs:comment>
</owl:DatatypeProperty>

<owl:DatatypeProperty rdf:ID="topic">
  <rdfs:domain rdf:resource="#Product" />
  <rdfs:range rdf:resource="&xsd;string" />
  <rdfs:label xml:lang="en">topic</rdfs:label>
  <rdfs:comment xml:lang="en">Some (artistic or cultural) topic under which
    the item could be classified from a customer standpoint.</rdfs:comment>
</owl:DatatypeProperty>

<owl:DatatypeProperty rdf:ID="id">
  <rdfs:domain rdf:resource="#Product" />
  <rdfs:range rdf:resource="&xsd;anyURI" />
  <rdfs:label xml:lang="en">id</rdfs:label>
  <rdfs:comment xml:lang="en">The unique identifier of the item in our infor-
  mation system. This is typically the isbn number for books, the
  doi for electronic documents, etc.</rdfs:comment>
</owl:DatatypeProperty>

<owl:ObjectProperty rdf:ID="creator">
  <rdfs:domain rdf:resource="#Product"/>
  <rdfs:range rdf:resource="#Person" />
  <rdfs:label xml:lang="en">creator</rdfs:label>
  <rdfs:comment xml:lang="en">The human creator of a product.</rdfs:comment>
</owl:ObjectProperty>

<owl:ObjectProperty rdf:ID="author">
  <rdfs:domain rdf:resource="#Book"/>
  <rdfs:range rdf:resource="#Person" />
  <rdfs:label xml:lang="en">author</rdfs:label>
  <rdfs:comment xml:lang="en">The author of a book.</rdfs:comment>
</owl:ObjectProperty>

<owl:ObjectProperty rdf:ID="publisher">
  <rdfs:domain rdf:resource="#Book"/>
  <rdfs:range rdf:resource="#Publisher" />
  <rdfs:label xml:lang="en">publisher</rdfs:label>
  <rdfs:comment xml:lang="en">The publisher of a book.</rdfs:comment>
</owl:ObjectProperty>

<!-- ################## INSTANCES ################## -->

<Popular rdf:about="#a674639524">
  <rdfs:label>Bertrand Russell: My life</rdfs:label>
  <author>
    <Person rdf:about="br">
```

```
            <firstname>Bertrand</firstname>
            <lastname>Russell</lastname>
         </Person>
      </author>
      <publisher>
         <Publisher rdf:about="http://www.routledge.co.uk"/>
      </publisher>
      <name>My life</name>
      <id></id>
      <price rdf:datatype="&xsd;integer">60</price>
   </Popular>

   <Book rdf:about="#a6746390923">
      <rdf:type rdf:resource="#Pocket"/>
      <rdfs:label>Albert Camus: La chute</rdfs:label>
      <author>
         <Person rdf:about="ac">
            <firstname>Albert</firstname>
            <lastname>Camus</lastname>
         </Person>
      </author>
      <publisher>
         <Publisher rdf:about="http://www.gallimard.fr"/>
      </publisher>
      <name>La chute</name>
      <id>http://dx.doi.org/10.1002/prot.999</id>
      <price rdf:datatype="&xsd;integer">9.95</price>
   </Book>

</rdf:RDF>
```

B.2 library.owl

```
<?xml version="1.0" encoding="iso-8859-1"?>
<!DOCTYPE rdf:RDF [
         <!ENTITY xsd "http://www.w3.org/2001/XMLSchema#" >
   <!ENTITY rdf "http://www.w3.org/1999/02/22-rdf-syntax-ns#" >
   <!ENTITY rdfs "http://www.w3.org/2000/01/rdf-schema#" >
   <!ENTITY dc "http://purl.org/dc/elements/1.1/" >
   <!ENTITY owl "http://www.w3.org/2002/07/owl#" >
   <!ENTITY foaf  "http://xmlns.com/foaf/0.1/">
   <!ENTITY ical  "http://www.w3.org/2002/12/cal/ical#">]>

<rdf:RDF
 xmlns="http://book.ontologymatching.org/example/library.owl#"
 xml:base="http://book.ontologymatching.org/example/library.owl#"
 xmlns:foaf ="&foaf;"
 xmlns:ical ="&ical;"
 xmlns:rdf ="&rdf;"
 xmlns:xsd ="&xsd;"
 xmlns:rdfs ="&rdfs;"
 xmlns:owl ="&owl;"
 xmlns:dc ="&dc;">

   <!-- ################### ONTOLOGY ################### -->

   <owl:Ontology rdf:about="">
      <dc:creator>Jérôme Euzenat</dc:creator>
      <dc:contributor>Pavel Shvaiko</dc:contributor>
      <dc:description>Fragments of a library ontology</dc:description>
      <dc:date>2006/04/13</dc:date>
      <rdfs:label>Library ontology</rdfs:label>
      <rdfs:comment>An example for the Ontology matching book. This ontology
         fragment provide a first classification for books.</rdfs:comment>
```

```xml
<owl:versionInfo>
  $Id: library.owl,v 1.3 2007/03/15 21:05:21 cvs Exp $
</owl:versionInfo>
</owl:Ontology>

<!-- In OWL-DL all items must be declared -->
<owl:DatatypeProperty rdf:about="&dc;creator" />
<owl:DatatypeProperty rdf:about="&dc;contributor" />
<owl:DatatypeProperty rdf:about="&dc;description" />
<owl:DatatypeProperty rdf:about="&dc;date" />

<!-- ################## CLASSES ################## -->

<owl:Class rdf:ID="Volume">
  <rdfs:label xml:lang="en">volume</rdfs:label>
  <rdfs:label xml:lang="fr">Volume</rdfs:label>
  <rdfs:comment xml:lang="en">Books referenced in the library.</rdfs:comment>
  <rdfs:subClassOf>
    <owl:Restriction>
      <owl:onProperty rdf:resource="#year" />
      <owl:cardinality
        rdf:datatype="&xsd;nonNegativeInteger">1</owl:cardinality>
    </owl:Restriction>
  </rdfs:subClassOf>
  <rdfs:subClassOf>
    <owl:Restriction>
      <owl:onProperty rdf:resource="#author" />
      <owl:minCardinality
        rdf:datatype="&xsd;nonNegativeInteger">1</owl:minCardinality>
    </owl:Restriction>
  </rdfs:subClassOf>
  <rdfs:subClassOf>
    <owl:Restriction>
      <owl:onProperty rdf:resource="#title" />
      <owl:minCardinality
        rdf:datatype="&xsd;nonNegativeInteger">1</owl:minCardinality>
    </owl:Restriction>
  </rdfs:subClassOf>
  <rdfs:subClassOf>
    <owl:Restriction>
      <owl:onProperty rdf:resource="#isbn" />
      <owl:minCardinality
        rdf:datatype="&xsd;nonNegativeInteger">1</owl:minCardinality>
    </owl:Restriction>
  </rdfs:subClassOf>
</owl:Class>

<owl:Class rdf:ID="Essay">
  <rdfs:subClassOf rdf:resource="#Volume" />
  <rdfs:label xml:lang="en">essay</rdfs:label>
  <rdfs:comment xml:lang="en">A book whose main interest reside in the topic
    considered.</rdfs:comment>
  <rdfs:subClassOf>
    <owl:Restriction>
      <owl:onProperty rdf:resource="#subject" />
      <owl:minCardinality
        rdf:datatype="&xsd;nonNegativeInteger">1</owl:minCardinality>
    </owl:Restriction>
  </rdfs:subClassOf>
</owl:Class>

<owl:Class rdf:ID="LiteraryCritic">
  <rdfs:subClassOf rdf:resource="#Essay" />
  <rdfs:label xml:lang="en">litterary critic</rdfs:label>
  <rdfs:comment xml:lang="en">An essay about Literature.</rdfs:comment>
</owl:Class>
```

B.2 library.owl

```xml
<owl:Class rdf:ID="Politics">
  <rdfs:subClassOf rdf:resource="#Essay" />
  <rdfs:label xml:lang="en">political writings</rdfs:label>
  <rdfs:comment xml:lang="en">An essay about politics.</rdfs:comment>
</owl:Class>

<owl:Class rdf:ID="Biography">
  <rdfs:subClassOf rdf:resource="#Essay" />
  <rdfs:label xml:lang="en">biography</rdfs:label>
  <rdfs:comment xml:lang="en">An essay about a person.</rdfs:comment>
</owl:Class>

<owl:Class rdf:ID="Autobiography">
  <rdfs:subClassOf rdf:resource="#Essay" />
  <rdfs:label xml:lang="en">autobiography</rdfs:label>
  <rdfs:comment xml:lang="en">
    A biography whose author is the subject.</rdfs:comment>
  <rdfs:subClassOf>
    <owl:Restriction>
      <owl:onProperty rdf:resource="#subject" />
      <owl:allValuesFrom rdf:resource="#Human" />
    </owl:Restriction>
  </rdfs:subClassOf>
</owl:Class>

<owl:Class rdf:ID="Literature">
  <rdfs:subClassOf rdf:resource="#Volume" />
  <rdfs:label xml:lang="en">literature</rdfs:label>
  <rdfs:comment xml:lang="en">A volume whose main interest reside in
    the threatment of the topic.</rdfs:comment>
</owl:Class>

<owl:Class rdf:ID="Novel">
  <rdfs:subClassOf rdf:resource="#Literature" />
  <rdfs:label xml:lang="en">novel</rdfs:label>
  <rdfs:comment xml:lang="en">A narative text.</rdfs:comment>
</owl:Class>

<owl:Class rdf:ID="Poetry">
  <rdfs:subClassOf rdf:resource="#Literature" />
  <rdfs:label xml:lang="en">poetry</rdfs:label>
</owl:Class>

<owl:Class rdf:ID="Human">
  <rdfs:label xml:lang="en">human</rdfs:label>
  <rdfs:comment xml:lang="en">A Human being.</rdfs:comment>
</owl:Class>

<owl:Class rdf:ID="Writer">
  <rdfs:subClassOf rdf:resource="#Human" />
  <rdfs:label xml:lang="en">writer</rdfs:label>
  <rdfs:comment xml:lang="en">Someone who authors books.</rdfs:comment>
</owl:Class>

<!-- ################# PROPERTIES ################### -->

<owl:DatatypeProperty rdf:ID="year">
  <rdfs:domain rdf:resource="#Volume" />
  <rdfs:range rdf:resource="&xsd;integer" />
  <rdfs:label xml:lang="en">year</rdfs:label>
  <rdfs:comment xml:lang="en">The year of first publication of this edition
    of the volume.</rdfs:comment>
</owl:DatatypeProperty>

<owl:DatatypeProperty rdf:ID="title">
  <rdfs:domain rdf:resource="#Volume" />
  <rdfs:range rdf:resource="&xsd;string" />
```

284   Appendix B: Running example

```xml
    <rdfs:label xml:lang="en">title</rdfs:label>
    <rdfs:comment xml:lang="en">The title of a volume.</rdfs:comment>
  </owl:DatatypeProperty>

  <owl:DatatypeProperty rdf:ID="isbn">
    <rdfs:domain rdf:resource="#Volume" />
    <rdfs:range rdf:resource="&xsd;integer" />
    <rdfs:label xml:lang="en">year</rdfs:label>
    <rdfs:comment xml:lang="en">
      The International Standard Book Number of a volume.</rdfs:comment>
  </owl:DatatypeProperty>

  <owl:ObjectProperty rdf:ID="author">
    <rdfs:domain rdf:resource="#Volume"/>
    <rdfs:range rdf:resource="#Writer" />
    <rdfs:label xml:lang="en">author</rdfs:label>
    <rdfs:comment xml:lang="en">The author of a volume.</rdfs:comment>
  </owl:ObjectProperty>

  <owl:ObjectProperty rdf:ID="subject">
    <rdfs:domain rdf:resource="#Essay"/>
    <rdfs:range rdf:resource="&owl;Thing" />
    <rdfs:label xml:lang="en">subject</rdfs:label>
    <rdfs:comment xml:lang="en">The subject of an essay.</rdfs:comment>
  </owl:ObjectProperty>

  <owl:DatatypeProperty rdf:ID="name">
    <rdfs:domain rdf:resource="#Human"/>
    <rdfs:range rdf:resource="&xsd;string" />
    <rdfs:label xml:lang="en">name</rdfs:label>
  </owl:DatatypeProperty>

  <!-- ################## INSTANCES ################## -->

  <owl:Thing rdf:about="#a674639524">
    <rdf:type rdf:resource="#Autobiography"/>
    <rdfs:label>"My life" by Bertrand Russell</rdfs:label>
    <author>
      <Writer rdf:about="#br">
        <name>Bertrand Russell</name>
      </Writer>
    </author>
    <isbn>0415189853</isbn>
    <subject rdf:resource="#br"/>
    <year rdf:datatype="&xsd;integer">1969</year>
    <title>My life</title>
  </owl:Thing>

  <Novel rdf:about="#a6746390923">
    <rdfs:label>"La chute" by Albert Camus</rdfs:label>
    <author>
      <Writer rdf:about="#ac">
        <name>Albert Camus</name>
      </Writer>
    </author>
    <isbn>2070360105</isbn>
    <year rdf:datatype="&xsd;integer">1956</year>
    <title>La chute</title>
  </Novel>

</rdf:RDF>
```

B.3 srcalign.rdf

The following alignment in the Alignment format (see §8.1.5) is that of Fig. 2.9. It is considered as a reference alignment.

```
<?xml version='1.0' encoding='utf-8' standalone='no'?>
<!DOCTYPE rdf:RDF [
<!ENTITY xsd "http://www.w3.org/2001/XMLSchema#" >
<!ENTITY ont1 "http://book.ontologymatching.org/example/culture-shop.owl#" >
<!ENTITY ont2 "http://book.ontologymatching.org/example/library.owl#" >]>

<rdf:RDF xmlns='http://knowledgeweb.semanticweb.org/heterogeneity/alignment'
         xml:base='http://knowledgeweb.semanticweb.org/heterogeneity/alignment'
         xmlns:rdf='http://www.w3.org/1999/02/22-rdf-syntax-ns#'
         xmlns:xsd='&xsd;'>
<Alignment>
  <xml>yes</xml>
  <level>0</level>
  <type>**</type>
  <method>Manually generated (Jérôme Euzenat, 2005/04/13)</method>
  <onto1>http://book.ontologymatching.org/examples/culture-shop.owl</onto1>
  <onto2>http://book.ontologymatching.org/examples/library.owl</onto2>
  <uri1>http://book.ontologymatching.org/example/culture-shop.owl</uri1>
  <uri2>http://book.ontologymatching.org/example/library.owl</uri2>
  <map>
    <Cell>
      <entity1 rdf:resource='&ont1;name'/>
      <entity2 rdf:resource='&ont2;title'/>
      <measure rdf:datatype='&xsd;float'>1.0</measure>
      <relation>=</relation>
    </Cell>
  </map>
  <map>
    <Cell>
      <entity1 rdf:resource='&ont1;id'/>
      <entity2 rdf:resource='&ont2;isbn'/>
      <measure rdf:datatype='&xsd;float'>.9</measure>
      <relation>&gt;</relation>
    </Cell>
  </map>
  <map>
    <Cell>
      <entity1 rdf:resource='&ont1;author'/>
      <entity2 rdf:resource='&ont2;author'/>
      <measure rdf:datatype='&xsd;float'>1.0</measure>
      <relation>=</relation>
    </Cell>
  </map>
  <map>
    <Cell>
      <entity1 rdf:resource='&ont1;Person'/>
      <entity2 rdf:resource='&ont2;Human'/>
      <measure rdf:datatype='&xsd;float'>1.0</measure>
      <relation>=</relation>
    </Cell>
  </map>
  <map>
    <Cell>
      <entity1 rdf:resource='&ont1;Science'/>
      <entity2 rdf:resource='&ont2;Essay'/>
      <measure rdf:datatype='&xsd;float'>.8</measure>
      <relation>&lt;</relation>
    </Cell>
  </map>
  <map>
    <Cell>
```

286   Appendix B: Running example

```
        <entity1 rdf:resource='&ont1;Book'/>
        <entity2 rdf:resource='&ont2;Volume'/>
        <measure rdf:datatype='&xsd;float'>.8</measure>
        <relation>&lt;</relation>
      </Cell>
    </map>
</Alignment>
</rdf:RDF>
```

B.4 Alternative alignments for evaluation

Here are the alignments considered in Chap. 7.

B.4.1 refalign.rdf

This is the previous alignment involving only correspondences between classes.

```
<?xml version='1.0' encoding='utf-8' standalone='no'?>
<!DOCTYPE rdf:RDF [
<!ENTITY xsd "http://www.w3.org/2001/XMLSchema#" >
<!ENTITY ont1 "http://book.ontologymatching.org/example/culture-shop.owl#" >
<!ENTITY ont2 "http://book.ontologymatching.org/example/library.owl#" >]>

<rdf:RDF xmlns='http://knowledgeweb.semanticweb.org/heterogeneity/alignment'
         xml:base='http://knowledgeweb.semanticweb.org/heterogeneity/alignment'
         xmlns:rdf='http://www.w3.org/1999/02/22-rdf-syntax-ns#'
         xmlns:xsd='&xsd;'>
<Alignment>
  <xml>yes</xml>
  <level>0</level>
  <type>**</type>
  <method>Manually generated (Jérôme Euzenat, 2005/04/13)</method>
  <onto1>http://book.ontologymatching.org/examples/culture-shop.owl</onto1>
  <onto2>http://book.ontologymatching.org/examples/library.owl</onto2>
  <uri1>http://book.ontologymatching.org/example/culture-shop.owl</uri1>
  <uri2>http://book.ontologymatching.org/example/library.owl</uri2>
  <map>
    <Cell>
      <entity1 rdf:resource='&ont1;Person'/>
      <entity2 rdf:resource='&ont2;Human'/>
      <measure rdf:datatype='&xsd;float'>1.0</measure>
      <relation>=</relation>
    </Cell>
  </map>
  <map>
    <Cell>
      <entity1 rdf:resource='&ont1;Science'/>
      <entity2 rdf:resource='&ont2;Essay'/>
      <measure rdf:datatype='&xsd;float'>.8</measure>
      <relation>&lt;</relation>
    </Cell>
  </map>
  <map>
    <Cell>
      <entity1 rdf:resource='&ont1;Book'/>
      <entity2 rdf:resource='&ont2;Volume'/>
      <measure rdf:datatype='&xsd;float'>.8</measure>
      <relation>&lt;</relation>
    </Cell>
  </map>
</Alignment>
</rdf:RDF>
```

B.4 Alternative alignments for evaluation

B.4.2 nearmiss.rdf

```
<?xml version='1.0' encoding='utf-8' standalone='no'?>
<!DOCTYPE rdf:RDF [
<!ENTITY xsd "http://www.w3.org/2001/XMLSchema#" >
<!ENTITY ont1 "http://book.ontologymatching.org/example/culture-shop.owl#" >
<!ENTITY ont2 "http://book.ontologymatching.org/example/library.owl#" >]>

<rdf:RDF xmlns='http://knowledgeweb.semanticweb.org/heterogeneity/alignment'
         xml:base='http://knowledgeweb.semanticweb.org/heterogeneity/alignment'
         xmlns:rdf='http://www.w3.org/1999/02/22-rdf-syntax-ns#'
         xmlns:xsd='&xsd;'>
<Alignment>
  <xml>yes</xml>
  <level>0</level>
  <type>**</type>
  <method>Manually generated (Jérôme Euzenat, 2005/04/13)</method>
  <onto1>http://book.ontologymatching.org/examples/culture-shop.owl</onto1>
  <onto2>http://book.ontologymatching.org/examples/library.owl</onto2>
  <uri1>http://book.ontologymatching.org/example/culture-shop.owl</uri1>
  <uri2>http://book.ontologymatching.org/example/library.owl</uri2>
  <map>
    <Cell>
      <entity1 rdf:resource='&ont1;Product'/>
      <entity2 rdf:resource='&ont2;Volume'/>
      <measure rdf:datatype='&xsd;float'>1.0</measure>
      <relation>=</relation>
    </Cell>
  </map>
  <map>
    <Cell>
      <entity1 rdf:resource='&ont1;Science'/>
      <entity2 rdf:resource='&ont2;Essay'/>
      <measure rdf:datatype='&xsd;float'>.8</measure>
      <relation>&lt;</relation>
    </Cell>
  </map>
  <map>
    <Cell>
      <entity1 rdf:resource='&ont1;Person'/>
      <entity2 rdf:resource='&ont2;Writer'/>
      <measure rdf:datatype='&xsd;float'>.8</measure>
      <relation>&lt;</relation>
    </Cell>
  </map>
</Alignment>
</rdf:RDF>
```

B.4.3 farone.rdf

```
<?xml version='1.0' encoding='utf-8' standalone='no'?>
<!DOCTYPE rdf:RDF [
<!ENTITY xsd "http://www.w3.org/2001/XMLSchema#" >
<!ENTITY ont1 "http://book.ontologymatching.org/example/culture-shop.owl#" >
<!ENTITY ont2 "http://book.ontologymatching.org/example/library.owl#" >]>

<rdf:RDF xmlns='http://knowledgeweb.semanticweb.org/heterogeneity/alignment'
         xml:base='http://knowledgeweb.semanticweb.org/heterogeneity/alignment'
         xmlns:rdf='http://www.w3.org/1999/02/22-rdf-syntax-ns#'
         xmlns:xsd='&xsd;'>
<Alignment>
  <xml>yes</xml>
  <level>0</level>
  <type>**</type>
  <method>Manually generated (Jérôme Euzenat, 2005/04/13)</method>
```

```
<onto1>http://book.ontologymatching.org/examples/culture-shop.owl</onto1>
<onto2>http://book.ontologymatching.org/examples/library.owl</onto2>
<uri1>http://book.ontologymatching.org/example/culture-shop.owl</uri1>
<uri2>http://book.ontologymatching.org/example/library.owl</uri2>
<map>
  <Cell>
    <entity1 rdf:resource='&ont1;Book'/>
    <entity2 rdf:resource='&ont2;Volume'/>
    <measure rdf:datatype='&xsd;float'>1.0</measure>
    <relation>=</relation>
  </Cell>
</map>
<map>
  <Cell>
    <entity1 rdf:resource='&ont1;Pocket'/>
    <entity2 rdf:resource='&ont2;Essay'/>
    <measure rdf:datatype='&xsd;float'>.8</measure>
    <relation>&lt;</relation>
  </Cell>
</map>
<map>
  <Cell>
    <entity1 rdf:resource='&ont1;Children'/>
    <entity2 rdf:resource='&ont2;Literature'/>
    <measure rdf:datatype='&xsd;float'>.8</measure>
    <relation>&lt;</relation>
  </Cell>
</map>
</Alignment>
</rdf:RDF>
```

C

Exercises

The following exercises cover only the technical sections of this book. They are provided to help readers check their understanding of the presented concepts rather than to make assignments to students. They consist of application of the presented concepts to a pair of ontologies; they are expected not to be difficult. Due to lack of space, these exercises are applied to small size ontologies. However, interested readers may use their own (larger) ontologies instead. It is certainly worthwhile to use available tools for completing these exercises. The solutions to these exercises are provided on the book web site[1].

C.1 Applications

C.1 (Application definition). Consider two university data sources dealing with people. The first one is a database developed from the whole university management standpoint, while the second one represents the standpoint of a particular research laboratory. These data sources are managed by different departments and will continue to evolve independently, however, users would like to access them through a unified interface. Obviously, this could be useful, e.g., for checking both lecture and room availability from a single interface.

1. Provide an architecture for this application. For instance, by drawing diagrams similar to those of Chap. 1.
2. What are the requirements to ontology matching in this application with regard to Table 1.1?

C.2 The matching problem

C.2 (Ontology representation). Let o be a first data source to be integrated. o is described in English as follows:

---
[1] http://book.ontologymatching.org

- People are divided among Students, Faculty and Staff. Faculty is further divided depending on its departments and sub-departments, e.g., Philosophy, Science.
- People are characterised by their firstname and lastname which are strings, id which is a uri and birthdate which is a date.
- Students will attend Courses which are taught by Faculty and Faculty people have an Office as a room.
- Pr. Carla Cipolla is a visiting professor in Computer science and Stefano Zucchini is a PhD Student.

1. Provide a description for o as a folksonomy.
2. Provide a description for o as a directory.
3. Provide a description for o as an XML schema.
4. Provide a description for o as a relational database schema.
5. Provide a description for o as an entity–relationship schema (or UML diagram).
6. Provide a description for o as an ontology.

C.3 (Ontology semantics). Assume that the left hand side ontology of Fig. C.1 is denoted as o.

1. Express o in OWL.
2. Express o as a set of assertions, e.g., Staff $\leq$ People.
3. Provide its semantics.
4. Does $o \models$ Philosophy $\leq$ People and why?
5. Does $o \models$ teaches $\leq$ attends and why?

C.3 Classification

C.4 (Kinds of techniques). Consider the structure of ontologies o and o' of Problem C.2, as illustrated in Fig. C.1. Describe which techniques can be used for matching them (see Fig. 3.1) and explain the choices made.

C.4 Basic techniques

C.5 (Name-based distance computation). Given the ontologies o and o' of Problem C.1 as illustrated in Fig. C.1. Provide the tables (similar to those of Example 4.14 or Example 4.25) for class matching with the following techniques:

1. String distances between all the labels occuring in Fig. C.1:
   - substring similarity;
   - 3-gram similarity;
   - edit distance;
   - Jaro–Winckler measure.
2. Linguistic distances between all the labels by using the last version of WordNet:
   - cosynonymy similarity;

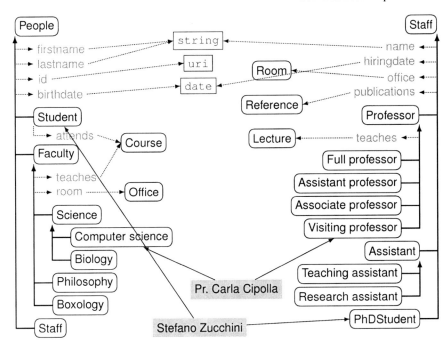

Fig. C.1. Two ontologies to be matched.

- gloss overlap;
- Wu–Palmer similarity.

C.6 (Extensional distance computation). Consider the two ontologies o and o' of Problem C.1 as illustrated in Fig. C.1. Assume that the two following tables specify data instances for o and o', respectively.

Class	firstname	lastname	id	birthdate
Computer Science	G.	Cetriolo	445	04/07/1978
Biology	P.	Pomodoro	1678	01/08/1972
Computer Science	C.	Cipolla	1998	13/06/1977
Philosophy	P.	Carciofo	128	03/09/1982
Student	C.	Fragola	1664	12/12/1985
Biology	A.	Verdura	88	07/09/1981
Student	S.	Zucchini	1178	16/04/1987
Computer Science	F.	di Guava		23/02/1966
Staff	C.	Melocoton	178	14/01/1962

Class	name	hiring date	office
Assistant professor	Giancarlo Cetriolo	2/9/2004	B45
Associate professor	Paola Pomodoro	23/2/2002	B45
Visiting professor	Carla Cipolla	4/12/2007	C17
Assistant professor	Paolo Carciofo		C18
Full professor	Federico diGuava	7/7/1999	B12
Visiting professor	Pierluiggi Pomodoro	4/12/2007	C18
PhDStudent	Mario Staggioni	2/9/2006	B47
PhDStudent	Stefano Zucchini	17/10/2005	B47
PhDStudent	Domenica Melanzana	15/9/2006	

1. Identify which extensional techniques can be used and why.
2. Design a distance for strings which can compare names with and without abbreviated first name.
3. Use the previous string distance for computing a similarity between instances.
4. Use the substring similarity between lastname and name for identifying instances.
5. Starting with each of the previously computed measures, use the single linkage measure on instances for comparing classes of the two ontologies.

C.5 Strategies

C.7 (Measure aggregation). Consider the distances between ontology entities in o and o' given by (i) edit distance computed on their names (see Problem C.5), and (ii) the distance computed with the single linkage measure applied to the substring distance (see Problem C.6).

1. Compute their aggregation with the $\max(x + y - 1, 0)$ triangular norm.
2. Compute their aggregation with the weighted product, where $2/3$ is the weight for the former distance and $1/3$ is the weight for the latter one.
3. Compute their aggregation with the weighted sum where $2/3$ is the weight for the former distance and $1/3$ is the weight for the latter one.
4. Compute their aggregation with the ordered weighted average where $2/3$ is the weight for the former distance and $1/3$ is the weight for the latter one.

C.8 (Thresholds). Assume that the similarity between entities of ontologies o and o' of Problem C.2 is expressed by the following similarity table:

	People	Student	Faculty	Science	Philosophy	Boxology	Staff	Course	Office
Staff	.56	.65	.33	.64	.12	.11	.63	.22	.13
Professor	.62	.36	.60	.40	.44	.32	.55	.21	.36
Assistant	.40	.44	.58	.62	.46	.33	.43	.32	.22
PhDStudent	.64	.92	.45	.60	.65	.52	.55	.33	.34
Room	.12	.20	.20	.18	.10	.12	.09	.11	.62
Reference	.23	.06	.18	.25	.26	.28	.22	.17	.23
Lecture	.15	.16	.26	.23	.34	.12	.14	.20	.16

Provide the set of correspondences resulting from the application of a .6 threshold, specifically:

1. with a hard threshold of .6;
2. with a delta threshold of .6;
3. with a proportional threshold of .6;
4. with a percentage threshold of .6.

C.9 (Alignment extraction). Consider the two ontologies o and o' of Problem C.1 as illustrated in Fig. C.1. Assume that the similarity between their entities is expressed by the similarity table of Problem C.8.

1. Extract an alignment based on the similarity with the help of the greedy algorithm;
2. Extract an alignment based on the similarity as a stable marriage;
3. Extract an alignment based on the similarity as the maximum weight graph matching.

C.10 (Composing matchers). Consider the application described in Problem C.1.

1. Provide an architecture of a matching system suitable to match these ontologies (use the answer identified to Problem C.4 for guiding your choice).
2. Compute the alignment with this architecture.

C.6 Evaluation of matching systems

C.11 (Precision and recall computation). Let R as described in Fig. C.2 be the reference alignment between the two ontologies o and o' of Problem C.1 as illustrated in Fig. C.1. Consider the three alignments (A_1, A_2, A_3) as follows.

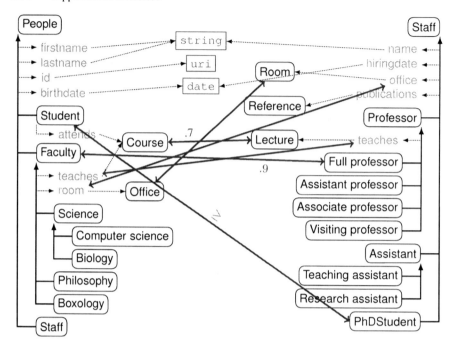

Fig. C.2. Reference alignment R between the ontologies of Fig. C.1. Correspondences are expressed by arrows. By default their relation is $=$ and their confidence value is 1.0; otherwise, these are mentioned near the arrows.

| room = Room | Office = office | Staff = Staff | |
| firstname $\leq$ name | Student $\geq$ PhDStudent | teaches = teaches | (A_1) |

| Professor $\geq$ Faculty | Office = Room | room = office | |
| People = Staff | Course = Lecture | teaches = teaches | (A_2) |

| Faculty = Full professor | Student = PhDStudent | Staff = Assistant | |
| hiringdate = birthdate | firstname = name | teaching = teaching | (A_3) |

1. Compute the precision of alignments A_1, A_2 and A_3;
2. Compute the recall of alignments A_1, A_2 and A_3;
3. Compute the F-measure of alignments A_1, A_2 and A_3;
4. Compute the fallout of alignments A_1, A_2 and A_3;
5. Compute the overall of alignments A_1, A_2 and A_3;
6. Compute the Hamming distance between alignments A_1, A_2 and A_3 and R.

C.12 (Application oriented evaluation). Let R as described in Fig. C.2 be the reference alignment between the two ontologies o and o' of Problem C.1 as illustrated in Fig. C.1. Apply the aggregation technique based on weighted harmonic mean of Sect. 7.4.1 with the criterion identified for Problem C.1 on alignments A_1, A_2 and A_3 from Problem C.11.

C.7 Representing alignments

C.13 (Representation generation). Consider R as described in Fig. C.2.

1. Express it in OWL;
2. Express it in C-OWL;
3. Express it in SWRL;
4. Express it in the SEKT mapping language;
5. Express it in SKOS;
6. Discuss the advantages and limitations of these formalisms.

C.8 Explaining alignments

C.14 (Alignment explanation). Given the architecture and the solution devised for Problem C.10, provide a process trace (see for example Fig. 9.5) for:

1. Several correctly identified correspondence (true positives);
2. Several incorrectly identified correspondence (false positives).
3. For both cases provide a natural language intuitive explanation (similar to what can be achieved by reading Fig. 9.2 or Fig. 9.3) of why the correspondences have been correctly and incorrectly identified.

C.9 Processing alignments

C.15 (Merging ontologies). Let R be the alignment described in Fig. C.2.

1. Describe the merge between o and o' according to R using the OWL import capability;
2. Describe it as one integrated OWL ontology.

C.16 (Data translation). Assume that we want to transform the data instances from ontology o into instances of o'. Consider the data instances in the first table of Problem C.6.

1. Provide their translation with regard to the reference alignment R described in Fig. C.2.
2. Develop a program able to perform this translation.

C.17 (Mediation). Assume that we want that each of the sources of Problem C.1 can query the other one. It is thus necessary to transform the queries and the returned answers.

1. Given the query 'SELECT ?x.room WHERE Faculty(?x)' expressed with respect to ontology o, transform it, in a query expressed with regard to ontology o', according to the reference alignment R described in Fig. C.2.
2. Apply the transformed query to the data instances of the second table of Problem C.6.
3. Develop a mediator able to perform the query transformations and answer translation depending on the reference alignment R described in Fig. C.2.

References

Pages in which a reference is cited are mentioned within square brackets.

[Aberer et al., 2004a] Karl Aberer, Tiziana Catarci, Philippe Cudré-Mauroux, Tharam Dillon, Stephan Grimm, Mohand-Saïd Hacid, Arantza Illarramendi, Mustafa Jarrar, Vipul Kashyap, Massimo Mecella, Eduardo Mena, Erich Neuhold, Aris Ouksel, Thomas Risse, Monica Scannapieco, Félix Saltor, Luca De Santis, Stefano Spaccapietra, Steffen Staab, Rudi Studer, and Olga De Troyer. Emergent semantics systems. In *Proc. 1st International Conference on Semantics of a Networked World (ICSNW)*, volume 3226 of *Lecture notes in computer science*, pages 14–43, Paris (FR), 2004. [18]

[Aberer et al., 2004b] Karl Aberer, Philippe Cudré-Mauroux, Aris Ouksel, Tiziana Catarci, Mohand-Saïd Hacid, Arantza Illarramendi, Vipul Kashyap, Massimo Mecella, Eduardo Mena, Erich Neuhold, Olga De Troyer, Thomas Risse, Monica Scannapieco, Félix Saltor, Luca de Santis, Stefano Spaccapietra, Steffen Staab, and Rudi Studer. Emergent semantics principles and issues. In *Proc. 9th International Conference on Database Systems for Advanced Applications (DASFAA)*, volume 2973 of *Lecture notes in computer science*, pages 25–38, Jeju Island (KR), 2004. [18]

[Agrawal and Srikant, 2001] Rakesh Agrawal and Ramakrishnan Srikant. On integrating catalogs. In *Proc. 10th International World Wide Web Conference (WWW)*, pages 603–612, Hong Kong (CN), 2001. [11]

[Alagic and Bernstein, 2001] Suad Alagic and Philip Bernstein. A model theory for generic schema management. In *Proc. 8th International Workshop on Database Programming Languages (DBPL)*, volume 2397 of *Lecture notes in computer science*, pages 228–246, Frascati (IT), 2001. Springer. [51]

[Aleksovski et al., 2006] Zharko Aleksovski, Michel Klein, Warner ten Kate, and Frank van Harmelen. Matching unstructured vocabularies using a background ontology. In *Proc. 15th International Conference on Knowledge Engineering and Knowledge Management (EKAW)*, volume 4248 of *Lecture notes in computer science*, pages 182–197, Praha (CZ), 2006. [68, 111]

[Amgoud et al., 2000] Leïla Amgoud, Simon Parsons, and Nicolas Maudet. Arguments, dialogue and negotiation. In *Proc. 14th European Conference on Artificial Intelligence (ECAI)*, pages 338–342, Berlin (DE), 2000. [256]

[An et al., 2005a] Yuan An, Alexander Borgida, and John Mylopoulos. Constructing complex semantic mappings between XML data and ontologies. In *Proc. 4th International Se-

mantic Web Conference (ISWC), volume 3729 of *Lecture notes in computer science*, pages 6–20, Galway (IE), 2005. [163]

[An *et al.*, 2005b] Yuan An, Alexander Borgida, and John Mylopoulos. Inferring complex semantic mappings between relational tables and ontologies from simple correspondences. In *Proc. 4th International Conference on Ontologies, Databases and Applications of Semantics (ODBASE)*, volume 3761 of *Lecture notes in computer science*, pages 1152–1169, Agia Napa (CY), 2005. [163]

[An *et al.*, 2006] Yuan An, Alexander Borgida, and John Mylopoulos. Discovering the semantics of relational tables through mappings. *Journal on Data Semantics*, VII:1–32, 2006. [163]

[Arens *et al.*, 1996] Yigal Arens, Chun-Nan Hsu, and Craig Knoblock. Query processing in the SIMS information mediator. In Austin Tate, editor, *Readings in agents*, pages 82–90. AAAI press, Menlo Park (CA US), 1996. [153]

[Ashpole *et al.*, 2005] Benjamin Ashpole, Marc Ehrig, Jérôme Euzenat, and Heiner Stuckenschmidt, editors. *Proc. K-CAP Workshop on Integrating Ontologies*, Banff (CA), October 2005. [196]

[Atzeni *et al.*, 2005] Paolo Atzeni, Paolo Cappellari, and Philip Bernstein. Modelgen: Model independent schema translation. In *Proc. 21st International Conference on Data Engineering (ICDE)*, pages 1111–1112, Tokyo (JP), 2005. [236]

[Atzeni *et al.*, 2006] Paolo Atzeni, Paolo Cappellari, and Philip Bernstein. Model-independent schema and data translation. In *Proc. 10th Conference on Extending Database Technology (EDBT)*, volume 3896 of *Lecture notes in computer science*, pages 368–385, München (DE), 2006. [236]

[Aumüller *et al.*, 2005] David Aumüller, Hong-Hai Do, Sabine Maßmann, and Erhard Rahm. Schema and ontology matching with COMA++. In *Proc. 24th International Conference on Management of Data (SIGMOD), Software Demonstration*, pages 906–908, Baltimore (MD US), 2005. [68]

[Avesani *et al.*, 2005] Paolo Avesani, Fausto Giunchiglia, and Mikalai Yatskevich. A large scale taxonomy mapping evaluation. In *Proc. 4th International Semantic Web Conference (ISWC)*, volume 3729 of *Lecture notes in computer science*, pages 67–81, Galway (IE), 2005. [202, 203]

[Bach and Dieng-Kuntz, 2005] Thanh-Le Bach and Rose Dieng-Kuntz. Measuring similarity of elements in OWL ontologies. In *Proc. AAAI Workshop on Contexts and Ontologies (C&O)*, pages 96–99, Pittsburgh (PA US), 2005. [167]

[Bach *et al.*, 2004] Than-Le Bach, Rose Dieng-Kuntz, and Fabien Gandon. On ontology matching problems (for building a corporate semantic web in a multi-communities organization). In *Proc. 6th International Conference on Enterprise Information Systems (ICEIS)*, pages 236–243, Porto (PT), 2004. [99, 167]

[Bailin and Truszkowski, 2002] Sidney Bailin and Walt Truszkowski. Ontology negotiation: How agents can really get to know each other. In *Proc. 1st International Workshop on Radical Agent Concepts (WRAC)*, volume 2564 of *Lecture notes in computer science*, pages 320–334, McLean (VA US), 2002. [21]

[Barthélemy and Guénoche, 1992] Jean-Pierre Barthélemy and Alain Guénoche. *Trees and proximity representations*. John Wiley & Sons, Chichester (UK), 1992. [101]

[Barwise and Seligman, 1997] Jon Barwise and Jerry Seligman. *Information flow: the logic of distributed systems*, volume 44 of *Cambridge tracts in theoretical computer science*. Cambridge University Press, Cambridge (UK), 1997. [177]

[Batini *et al.*, 1986] Carlo Batini, Maurizio Lenzerini, and Shamkant Navathe. A comparative analysis of methodologies for database schema integration. *ACM Computing Surveys*, 18(4):323–364, 1986. [11, 13, 40, 61]

[Bechhofer *et al.*, 2003] Sean Bechhofer, Raphael Volz, and Phillip Lord. Cooking the semantic web with the OWL API. In *Proc. 2nd International Semantic Web Conference (ISWC)*, volume 2870 of *Lecture notes in computer science*, pages 659–675, Sanibel Island (FL US), 2003. [239]

[Bench-Capon and Malcolm, 1999] Trevor Bench-Capon and Grant Malcolm. Formalising ontologies and their relations. In *Proc. 16th International Conference on Database and Expert Systems Applications (DEXA)*, volume 1677 of *Lecture notes in computer science*, pages 250–259, 1999. [51]

[Bench-Capon, 2003] Trevor Bench-Capon. Persuasion in practical argument using value-based argumentation frameworks. *Journal of Logic and Computation*, 13(3):429–448, 2003. [256]

[Benerecetti *et al.*, 2000] Massimo Benerecetti, Paolo Bouquet, and Chiara Ghidini. Contextual reasoning distilled. *Journal of Experimental and Theoretical Artificial Intelligence*, 12(3):279–305, July 2000. [41]

[Benerecetti *et al.*, 2001] Massimo Benerecetti, Paolo Bouquet, and Chiara Ghidini. On the dimensions of context dependence: partiality, approximation, and perspective. In *Proc. 3rd International and Interdisciplinary Conference on Modeling and Using Context (CONTEXT)*, volume 2116 of *Lecture notes in computer science*, pages 59–72, Dundee (UK), 2001. [41]

[Beneventano *et al.*, 1998] Domenico Beneventano, Sonia Bergamaschi, Stefano Lodi, and Claudio Sartori. Consistency checking in complex object database schemata with integrity constraints. *IEEE Transactions on Knowledge and Data Engineering*, 10(4):576–598, 1998. [157]

[Bergamaschi *et al.*, 1998] Sonia Bergamaschi, Domenico Beneventano, Silvana Castano, and Maurizio Vincini. MOMIS: An intelligent system for the integration of semistructured and structured data. Technical Report T3-R07, Università di Modena e Reggio Emilia, Modena (IT), 1998. [157]

[Bergamaschi *et al.*, 1999] Sonia Bergamaschi, Silvana Castano, and Maurizio Vincini. Semantic integration of semistructured and structured data sources. *ACM SIGMOD Record*, 28(1):54–59, 1999. [157]

[Berge, 1970] Claude Berge. *Graphes et hypergraphes*. Dunod, Paris (FR), 1970. [148]

[Berlin and Motro, 2002] Jacob Berlin and Amihai Motro. Database schema matching using machine learning with feature selection. In *Proc. 14th International Conference on Advanced Information Systems Engineering (CAiSE)*, volume 2348 of *Lecture notes in computer science*, pages 452–466, Toronto (CA), 2002. [172]

[Berners-Lee *et al.*, 2001] Tim Berners-Lee, James Hendler, and Ora Lassila. The semantic web. *Scientific American*, 284(5):34–43, May 2001. [70]

[Bernstein and Rahm, 2000] Philip Bernstein and Erhard Rahm. Data warehouse scenarios for model management. In *Proc. 19 International Conference on Conceptual Modeling (ER)*, volume 1920 of *Lecture notes in computer science*, pages 1–15, Salt Lake City (UT US), 2000. [11, 14]

[Bernstein *et al.*, 2000] Philip Bernstein, Alon Halevy, and Rachel Pottinger. A vision of management of complex models. *ACM SIGMOD Record*, 29(4):55–63, 2000. [44, 235]

[Bernstein *et al.*, 2002] Philip Bernstein, Fausto Giunchiglia, A. Kementsietsidis, John Mylopoulos, Luciano Serafini, and Ilya Zaihrayeu. Data management for peer-to-peer computing: A vision. In *Proc. 5th International Workshop on the Web and Databases (WebDB)*, Madison (WI US), 2002. [16, 17]

[Bernstein *et al.*, 2006] Philip Bernstein, Sergey Melnik, and John Churchill. Incremental schema matching. In *Proc. 32nd International Conference on Very Large Data Bases (VLDB)*, pages 1167–1170, Seoul (KR), 2006. [144]

[Bilke and Naumann, 2005] Alexander Bilke and Felix Naumann. Schema matching using duplicates. In *Proc. 21st International Conference on Data Engineering (ICDE)*, pages 69–80, Tokyo (JP), 2005. [174]

[Birkes and Dodge, 2001] David Birkes and Yadolah Dodge. *Alternative Methods of Regression*. John Wiley & Sons, Inc., 2001. [140]

[Biron and Malhotra (ed.), 2004] Paul Biron and Ashok Malhotra (ed.). XML schema part 2: Datatypes. Recommendation, W3C, 2004. [95]

[Bisson, 1992] Gilles Bisson. Learning in FOL with similarity measure. In *Proc. 10th National Conference on Artificial Intelligence (AAAI)*, pages 82–87, San-Jose (CA US), 1992. [130]

[Booch et al., 1998] Grady Booch, James Rumbaugh, and Ivar Jacobson. *The Unified Modeling Language user guide*. Addison-Wesley, Reading (MA US), 1998. [35]

[Borgida and Serafini, 2003] Alexander Borgida and Luciano Serafini. Distributed description logics: Assimilating information from peer sources. *Journal on Data Semantics*, I:153–184, 2003. [52]

[Bouquet and Serafini, 2003] Paolo Bouquet and Luciano Serafini. On the difference between bridge rules and lifting axioms. In *Proc. 4th International and Interdisciplinary Conference on Modeling and Using Context (CONTEXT)*, volume 2680 of *Lecture notes in computer science*, pages 80–93, Stanford (CA US), 2003. [113]

[Bouquet et al., 2003a] Paolo Bouquet, Fausto Giunchiglia, Frank van Harmelen, Luciano Serafini, and Heiner Stuckenschmidt. C-OWL – contextualizing ontologies. In *Proc. 2nd International Semantic Web Conference (ISWC)*, volume 2870 of *Lecture notes in computer science*, pages 164–179, Sanibel Island (FL US), 2003. [223]

[Bouquet et al., 2003b] Paolo Bouquet, Bernardo Magnini, Luciano Serafini, and Stefano Zanobini. A SAT-based algorithm for context matching. In *Proc. 4th International and Interdisciplinary Conference on Modeling and Using Context (CONTEXT)*, volume 2680 of *Lecture notes in computer science*, pages 66–79, Stanford (CA US), 2003. [164]

[Bouquet et al., 2003c] Paolo Bouquet, Luciano Serafini, and Stefano Zanobini. Semantic coordination: A new approach and an application. In *Proc. 2nd International Semantic Web Conference (ISWC)*, volume 2870 of *Lecture notes in computer science*, pages 130–145, Sanibel Island (FL US), 2003. [11, 13, 63, 164]

[Bouquet et al., 2004a] Paolo Bouquet, Marc Ehrig, Jérôme Euzenat, Enrico Franconi, Pascal Hitzler, Markus Krötzsch, Luciano Serafini, Giorgos Stamou, York Sure, and Sergio Tessaris. Specification of a common framework for characterizing alignment. Deliverable D2.2.1, Knowledge web NoE, 2004. [41, 42, 44, 45, 55]

[Bouquet et al., 2004b] Paolo Bouquet, Fausto Giunchiglia, Frank van Harmelen, Luciano Serafini, and Heiner Stuckenschmidt. Contextualizing ontologies. *Journal of Web Semantics*, 1(1):325–343, 2004. [223]

[Bouquet et al., 2006] Paolo Bouquet, Luciano Serafini, Stefano Zanobini, and Simone Sceffer. Bootstrapping semantics on the web: meaning elicitation from schemas. In *Proc. 15th International World Wide Web Conference (WWW)*, pages 505–512, Edinburgh (UK), 2006. [114, 164]

[Bourigault and Jacquemin, 1999] Didier Bourigault and Christian Jacquemin. Term extraction + term clustering: an integrated platform for computer-aided terminology. In *Proc. European Chapter of the Association for Computational Linguistics (EACL)*, pages 15–22, Bergen (NO), 1999. [85]

[Boyd et al., 2004] Michael Boyd, Sasivimol Kittivoravitkul, Charalambos Lazanitis, Peter McBrien, and Nikos Rizopoulos. AutoMed: A BAV data integration system for heterogeneous data sources. In *Proc. 16th International Conference on Advanced Information*

Systems Engineering (CAiSE), volume 3084 of *Lecture notes in computer science*, pages 82–97, Riga (LV), 2004. [153]
[Breiman, 1996] Leo Breiman. Stacked regressions. *Machine Learning*, 24(1):49–64, 1996. [140]
[Breitbart, 1990] Yuri Breitbart. Multidatabase interoperability. *ACM SIGMOD Record*, 19(3):53–60, 1990. [40]
[Brill, 1992] Eric Brill. A simple rule-based part of speech tagger. In *Proc. 3rd Conference on Applied Natural Language Processing (ANLC)*, pages 152–155, Trento (IT), 1992. [84]
[Brodie *et al.*, 1984] Michael Brodie, John Mylopoulos, and Joachim Schmidt. *On conceptual modeling*. Springer, New York (NY US), 1984. [1, 35]
[Budanitsky and Hirst, 2006] Alexander Budanitsky and Graeme Hirst. Evaluating WordNet-based measures of lexical semantic relatedness. *Computational Linguistics*, 32(1):13–47, 2006. [90, 91]
[Bussler *et al.*, 2002] Christoph Bussler, Dieter Fensel, and Alexander Mädche. A conceptual architecture for semantic web enabled web services. *ACM SIGMOD Record*, 31(4):24–29, 2002. [19]
[Calvanese *et al.*, 2002a] Diego Calvanese, Giuseppe De Giacomo, and Maurizio Lenzerini. Description logics for information integration. In Anthonis Kakas and Fariba Sadri, editors, *Computational logic: logic programming and beyond, essays in honour of Robert A. Kowalski*, volume 2408 of *Lecture notes in computer science*, pages 41–60. Springer, Heidelberg (DE), 2002. [153]
[Calvanese *et al.*, 2002b] Diego Calvanese, Giuseppe De Giacomo, and Maurizio Lenzerini. A framework for ontology integration. In Isabel Cruz, Stefan Decker, Jérôme Euzenat, and Deborah McGuinness, editors, *The emerging semantic web*, pages 201–214. IOS Press, Amsterdam (NL), 2002. [52, 228]
[Calvanese *et al.*, 2004] Diego Calvanese, Giuseppe De Giacomo, Maurizio Lenzerini, and Riccardo Rosati. Logical foundations of peer-to-peer data integration. In *Proc. 23rd Symposium on Principles of Database Systems (PODS)*, pages 241–251, Paris (FR), 2004. [52]
[Castano *et al.*, 2000] Silvana Castano, Valeria De Antonellis, and Sabrina De Capitani di Vimercati. Global viewing of heterogeneous data sources. *IEEE Transactions on Knowledge and Data Engineering*, 13(2):277–297, 2000. [157]
[Castano *et al.*, 2005] Silvana Castano, Alfio Ferrara, and Stefano Montanelli. Dynamic knowledge discovery in open, distributed and multi-ontology systems: Techniques and applications. In David Taniar and Johanna Rahayu, editors, *Web semantics and ontology*, chapter 8, pages 226–258. Idea Group Publishing, Hershey (PA US), 2005. [157]
[Castano *et al.*, 2006] Silvana Castano, Alfio Ferrara, and Stefano Montanelli. Matching ontologies in open networked systems: Techniques and applications. *Journal on Data Semantics*, V:25–63, 2006. [157]
[Castro *et al.*, 2004] Raúl García Castro, Diana Maynard, Doug Foxvog, Holger Wache, and Rafael González-Cabero. Specification of a methodology, general criteria, and benchmark suites for benchmarking ontology tools. Deliverable D2.1.4, Knowledge web NoE, 2004. [194]
[Cerbah and Euzenat, 2001] Farid Cerbah and Jérôme Euzenat. Traceability between models and texts through terminology. *Data and Knowledge Engineering*, 38(1):31–43, 2001. [85]
[Chalupsky, 2000] Hans Chalupsky. OntoMorph: a translation system for symbolic knowledge. In *Proc. 7th International Conference on the Principles of Knowledge Representation and Reasoning (KR)*, pages 471–482, Breckenridge (CO US), 2000. [41, 42]
[Chan *et al.*, 1996] Lois Mai Chan, John Comaromi, Joan Mitchell, and Mohinder Satija. *Dewey decimal classifcation: a practical guide*. OCLC Forest Press, Dublin (OH US), 1996. [32]

[Chang et al., 2004] Kevin Chang, Bin He, Chengkai Li, Mitesh Patel, and Zhen Zhang. Structured databases on the web: observations and implications. *SIGMOD Record*, 33(3):61–70, 2004. [24]

[Chang et al., 2005] Kevin Chang, Bin He, and Zhen Zhang. Toward large scale integration: Building a metaquerier over databases on the web. In *Proc. 2nd Biennial Conference on Innovative Data Systems Research (CIDR)*, pages 44–55, Asilomar (CA US), 2005. [168]

[Chawathe et al., 1994] Sudarshan Chawathe, Hector Garcia-Molina, Joachim Hammer, Kelly Ireland, Yannis Papakonstantinou, Jeffrey Ullman, and Jennifer Widom. The TSIMMIS project: Integration of heterogeneous information sources. In *Proc. 16th Meeting of the Information Processing Society of Japan (IPSJ)*, pages 7–18, Tokyo (JP), 1994. [11, 153]

[Chen, 1976] Peter Chen. The entity-relationship model–toward a unified view of data. *ACM Transactions on Database Systems*, 1(1):9–36, 1976. [35]

[Clark and DeRose (ed.), 2001] James Clark and Steve DeRose (ed.). XML path language (XPath) version 1.0. Recommendation, W3C, 2001. [221]

[Clifton et al., 1997] Chris Clifton, Ed Hausman, and Arnon Rosenthal. Experience with a combined approach to attribute matching across heterogeneous databases. In *Proc. 7th IFIP Conference on Database Semantics*, pages 428–453, Leysin (CH), 1997. [26, 154]

[Cohen and Hirsh, 1998] William Cohen and Haym Hirsh. Joins that generalize: text classification using WHIRL. In *Proc. 4th International Conference on Knowledge Discovery and Data Mining (KDD)*, pages 169–173, New York (NY US), 1998. [135]

[Cohen et al., 2003a] William Cohen, Pradeep Ravikumar, and Stephen Fienberg. A comparison of string distance metrics for name-matching tasks. In *Proc. IJCAI Workshop on Information Integration on the Web (IIWeb)*, pages 73–78, Acapulco (MX), 2003. [174]

[Cohen et al., 2003b] William Cohen, Pradeep Ravikumar, and Stephen Fienberg. A comparison of string metrics for matching names and records. In *Proc. KDD Workshop on Data Cleaning and Object Consolidation*, pages 73–78, Washington (DC US), 2003. [76, 83]

[Cohen, 1998] William Cohen. Integration of heterogeneous databases without common domains using queries based on textual similarity. In *Proc. 17th International Conference on Management of Data (SIGMOD)*, pages 201–212, Seattle (WA US), 1998. [135]

[Corcho, 2004] Oscar Corcho. *A declarative approach to ontology translation with knowledge preservation*. PhD thesis, Universidad Politécnica de Madrid, Madrid (ES), 2004. [41]

[Coutaz et al., 2005] Joelle Coutaz, James Crowley, Simon Dobson, and David Garlan. Context is key. *Communications of the ACM*, 48(3):49–53, 2005. [22]

[Cramer, 2000] Erhard Cramer. Probability measures with given marginals and conditionals: i-projections and conditional iterative proportional fitting. *Statistics and Decisions*, 18(3):311–329, 2000. [168]

[Cuel et al., 2006] Roberta Cuel, Alain Léger, Fausto Giunchiglia, Pavel Shvaiko, Anna Zhdanova, Diana Maynard, Jérôme Euzenat, Ying Ding, Luigi Lucchese, York Sure, Arthur Stutt, Martin Dzbor, Enrico Motta, Michele Pasin, Ina O'Murchu, and John Breslin. Technology roadmap. Deliverable D1.4.1, Knowledge web NoE, 2006. [271]

[da Silva, 2004] Nuno Alexandre Pinto da Silva. *Multi-dimensional service-oriented ontology mapping*. PhD thesis, Universidade de Trás-os-Montes e Alto Douro, Villa Real (PT), 2004. [221, 238]

[Davis and Putnam, 1960] Martin Davis and Hilary Putnam. A computing procedure for quantification theory. *Journal of the ACM*, 7(3):201–215, 1960. [246, 253]

[Davis et al., 1962] Martin Davis, George Longemann, and Donald Loveland. A machine program for theorem proving. *Communications of the ACM*, 5(7):394–397, 1962. [246, 253]

[de Bruijn et al., 2004] Jos de Bruijn, Douglas Foxvog, and Kerstin Zimmerman. Ontology mediation patterns library. Deliverable D4.3.1, SEKT, 2004. [229]

[Dean and Schreiber (eds.), 2004] Mike Dean and Guus Schreiber (eds.). OWL web ontology language reference. Recommendation, W3C, February 2004. [36]

[Deerwester et al., 1990] Scott Deerwester, Susan Dumais, George Furnas, Thomas Landauer, and Richard Harshman. Indexing by latent semantic analysis. *Journal of the American Society for Information Science*, 41(6):391–407, 1990. [81]

[Dhamankar et al., 2004] Robin Dhamankar, Yoonkyong Lee, An-Hai Doan, Alon Halevy, and Pedro Domingos. iMAP: Discovering complex semantic matches between database schemas. In *Proc. 23rd International Conference on Management of Data (SIGMOD)*, pages 383–394, Paris (FR), 2004. [171, 245, 248, 250, 253]

[Dieng and Hug, 1998] Rose Dieng and Stefan Hug. Comparison of "personal ontologies" represented through conceptual graphs. In *Proc. 13th European Conference on Artificial Intelligence (ECAI)*, pages 341–345, Brighton (UK), 1998. [103]

[Do and Rahm, 2002] Hong-Hai Do and Erhard Rahm. COMA – a system for flexible combination of schema matching approaches. In *Proc. 28th International Conference on Very Large Data Bases (VLDB)*, pages 610–621, Hong Kong (CN), 2002. [68, 104, 145, 161, 236]

[Do et al., 2002] Hong-Hai Do, Sergei Melnik, and Erhard Rahm. Comparison of schema matching evaluations. In *Proc. Workshop on Web, Web-Services, and Database Systems*, volume 2593 of *Lecture notes in computer science*, pages 221–237, Erfurt (DE), 2002. [153, 198, 205]

[Do, 2005] Hong-Hai Do. *Schema matching and mapping-based data integration*. PhD thesis, University of Leipzig, Leipzig (DE), 2005. [71, 236, 237, 238]

[Doan and Halevy, 2005] An-Hai Doan and Alon Halevy. Semantic integration research in the database community: A brief survey. *AI Magazine*, 26(1):83–94, 2005. Special issue on Semantic integration. [71, 153]

[Doan et al., 2001] An-Hai Doan, Pedro Domingos, and Alon Halevy. Reconciling schemas of disparate data sources: A machine-learning approach. In *Proc. 20th International Conference on Management of Data (SIGMOD)*, pages 509–520, Santa Barbara (CA US), 2001. [171, 272]

[Doan et al., 2003] An-Hai Doan, Pedro Domingos, and Alon Halevy. Learning to match the schemas of data sources: A multistrategy approach. *Machine Learning*, 50(3):279–301, 2003. [133, 135, 139]

[Doan et al., 2004] An-Hai Doan, Jayant Madhavan, Pedro Domingos, and Alon Halevy. Ontollogy matching: a machine learning approach. In Steffen Staab and Rudi Studer, editors, *Handbook on ontologies*, chapter 18, pages 385–404. Springer Verlag, Berlin (DE), 2004. [171]

[Domingos and Pazzani, 1996] Pedro Domingos and Michael Pazzani. Beyond independence: Conditions for the optimality of the simple Bayesian classifier. In *Proc. 13th International Conference on Machine Learning (ICML)*, pages 105–112, Bari (IT), 1996. [133, 134]

[Dou et al., 2005] Dejing Dou, Drew McDermott, and Peishen Qi. Ontology translation on the semantic web. *Journal on Data Semantics*, II:35–57, 2005. [163]

[Dragut and Lawrence, 2004] Eduard Dragut and Ramon Lawrence. Composing mappings between schemas using a reference ontology. In *Proc. 3rd International Conference on Ontologies, DataBases, and Applications of Semantics (ODBASE)*, volume 3290 of *Lecture notes in computer science*, pages 783–800, Larnaca (CY), 2004. [238]

[Draper et al., 2001] Denise Draper, Alon Halevy, and Daniel Weld. The nimble integration engine. In *Proc. 20th International Conference on Management of Data SIGMOD*, pages 567–568, Santa Barbara (CA US), 2001. [11]

[Dung, 1995] Phan Minh Dung. On the acceptability of arguments and its fundamental role in nonmonotonic reasoning, logic programming and n-person games. *Artificial Intelligence*, 77(2):321–358, 1995. [256]

[Dzbor et al., 2003] Martin Dzbor, John Domingue, and Enrico Motta. Magpie – towards a semantic web browser. In *Proc. 2nd International Semantic Web Conference (ISWC)*, volume 2870 of *Lecture notes in computer science*, pages 690–705, Sanibel Island (FL US), 2003. [22]

[Dzbor et al., 2004] Martin Dzbor, Enrico Motta, and John Domingue. Opening up Magpie via semantic services. In *Proc. 3rd International Semantic Web Conference (ISWC)*, volume 3298 of *Lecture notes in computer science*, pages 635–649, Hiroshima (JP), 2004. [22]

[Ehrig and Euzenat, 2005] Marc Ehrig and Jérôme Euzenat. Relaxed precision and recall for ontology matching. In *Proc. K-CAP Workshop on Integrating Ontologies*, pages 25–32, Banff (CA), 2005. [210, 211]

[Ehrig and Staab, 2004] Marc Ehrig and Steffen Staab. QOM – quick ontology mapping. In *Proc. 3rd International Semantic Web Conference (ISWC)*, volume 3298 of *Lecture notes in computer science*, pages 683–697, Hiroshima (JP), 2004. [178, 179, 212]

[Ehrig and Sure, 2004] Marc Ehrig and York Sure. Ontology mapping – an integrated approach. In *Proc. 1st European Semantic Web Symposium (ESWS)*, volume 3053 of *Lecture notes in computer science*, pages 76–91, Hersounisous (GR), May 2004. [62, 98, 103, 145, 146, 178]

[Ehrig et al., 2005] Marc Ehrig, Steffen Staab, and York Sure. Bootstrapping ontology alignment methods with APFEL. In *Proc. 4th International Semantic Web Conference (ISWC)*, volume 3729 of *Lecture notes in computer science*, pages 186–200, Galway (IE), 2005. [136, 138, 184, 240, 272]

[Ehrig, 2007] Marc Ehrig. *Ontology alignment: bridging the semantic gap*. Semantic web and beyond: computing for human experience. Springer, New-York (NY US), 2007. [70, 100, 185, 208, 214, 240, 241]

[Elfeky et al., 2002] Mohamed Elfeky, Ahmed Elmagarmid, and Vassilios Verykios. Tailor: A record linkage tool box. In *Proc. 18th International Conference on Data Engineering (ICDE)*, pages 17–28, San Jose (CA US), 2002. [107]

[Elmagarmid et al., 1999] Ahmed Elmagarmid, Marek Rusinkiewicz, and Amith Sheth, editors. *Management of heterogeneous and autonomous database systems*. Morgan Kaufmann, San Francisco (CA US), 1999. [13]

[Embley et al., 2004] David Embley, Li Xu, and Yihong Ding. Automatic direct and indirect schema mapping: Experiences and lessons learned. *ACM SIGMOD Record*, 33(4):14–19, 2004. [179]

[Euzenat and Stuckenschmidt, 2003] Jérôme Euzenat and Heiner Stuckenschmidt. The 'family of languages' approach to semantic interoperability. In Borys Omelayenko and Michel Klein, editors, *Knowledge transformation for the semantic web*, pages 49–63. IOS press, Amsterdam (NL), 2003. [41]

[Euzenat and Valtchev, 2004] Jérôme Euzenat and Petko Valtchev. Similarity-based ontology alignment in OWL-lite. In *Proc. 15th European Conference on Artificial Intelligence (ECAI)*, pages 333–337, Valencia (ES), 2004. [97, 99, 100, 102, 129, 181, 240]

[Euzenat et al., 2004a] Jérôme Euzenat, Thanh Le Bach, Jesús Barrasa, Paolo Bouquet, Jan De Bo, Rose Dieng-Kuntz, Marc Ehrig, Manfred Hauswirth, Mustafa Jarrar, Rubén Lara, Diana Maynard, Amedeo Napoli, Giorgos Stamou, Heiner Stuckenschmidt, Pavel Shvaiko, Sergio Tessaris, Sven Van Acker, and Ilya Zaihrayeu. State of the art on ontology alignment. Deliverable D2.2.3, Knowledge web NoE, 2004. [85, 99, 100]

[Euzenat *et al.*, 2004b] Jérôme Euzenat, Marc Ehrig, and Raúl García Castro. Specification of a benchmarking methodology for alignment techniques. Deliverable D2.2.2, Knowledge web NoE, 2004. [196]

[Euzenat *et al.*, 2005a] Jérôme Euzenat, Loredana Laera, Valentina Tamma, and Alexandre Viollet. Negotiation/argumentation techniques among agents complying to different ontologies. Deliverable 2.3.7, Knowledge web NoE, 2005. [21]

[Euzenat *et al.*, 2005b] Jérôme Euzenat, Heiner Stuckenschmidt, and Mikalai Yatskevich. Introduction to the ontology alignment evaluation 2005. In *Proc. K-CAP Workshop on Integrating Ontologies*, pages 61–71, Banff (CA), 2005. [196, 202, 209]

[Euzenat, 1994] Jérôme Euzenat. Brief overview of T-tree: the Tropes taxonomy building tool. In *Proc. 4th ASIS SIG/CR Workshop on Classification Research*, pages 69–87, Columbus (OH US), 1994. [169]

[Euzenat, 2001] Jérôme Euzenat. Towards a principled approach to semantic interoperability. In *Proc. IJCAI Workshop on Ontologies and Information Sharing*, pages 19–25, Seattle (WA US), 2001. [41, 42]

[Euzenat, 2002] Jérôme Euzenat. An infrastructure for formally ensuring interoperability in a heterogeneous semantic web. In Isabel Cruz, Stefan Decker, Jérôme Euzenat, and Deborah McGuinness, editors, *The emerging semantic web*, pages 245–260. IOS press, Amsterdam (NL), 2002. [274]

[Euzenat, 2003] Jérôme Euzenat. Towards composing and benchmarking ontology alignments. In *Proc. ISWC Workshop on Semantic Integration*, pages 165–166, Sanibel Island (FL US), 2003. [49, 196, 226]

[Euzenat, 2004] Jérôme Euzenat. An API for ontology alignment. In *Proc. 3rd International Semantic Web Conference (ISWC)*, volume 3298 of *Lecture notes in computer science*, pages 698–712, Hiroshima (JP), 2004. [45, 226, 239]

[Euzenat, 2005] Jérôme Euzenat. Alignment infrastructure for ontology mediation and other applications. In *Proc. International Workshop on Mediation in Semantic Web Services (MEDIATE)*, pages 81–95, Amsterdam (NL), 2005. [265]

[Euzenat, 2007] Jérôme Euzenat. Semantic precision and recall for ontology alignment evaluation. In *Proc. 20th International Joint Conference on Artificial Intelligence (IJCAI)*, pages 248–253, Hyderabad (IN), 2007. [113, 212, 273]

[Everitt, 1993] Brian Everitt. *Cluster analysis*. New York Press, 1993. [162]

[Fagin *et al.*, 1995] Ronald Fagin, Joseph Halpern, Yoram Moses, and Moshe Vardi. *Reasoning about knowledge*. The MIT press, Cambridge (MA US), 1995. [52, 54, 55]

[Fagin *et al.*, 2004] Ronald Fagin, Phokion Kolaitis, Lucian Popa, and Wang Chiew Tan. Composing schema mappings: Second-order dependencies to the rescue. In *Proc. 23rd Symposium on Principles of Database systems (PODS)*, pages 83–94, Paris (FR), 2004. [177]

[Fellbaum, 1998] Christiane Fellbaum. *WordNet: an electronic lexical database*. The MIT Press, Cambridge (MA US), 1998. [87]

[Fellegi and Sunter, 1969] Ivan Fellegi and Alan Sunter. A theory for record linkage. *Journal of the American Statistical Association*, 64(328):1183–1210, 1969. [107]

[Fensel *et al.*, 2007] Dieter Fensel, Holger Lausen, Axel Polleres, Jos de Bruijn, Michael Stollberg, Dumitru Roman, and John Domingue. *Enabling semantic web services: the web service modeling ontology*. Springer, Heidelberg (DE), 2007. [19]

[Fensel, 2004] Dieter Fensel. *Ontologies: a silver bullet for knowledge management and electronic commerce*. Springer, Heidelberg (DE), 2nd edition, 2004. [1]

[FIPA0037, 2002] FIPA0037. FIPA ACL communicative act library specification. Technical report, FIPA, 2002. http://www.fipa.org/specs/fipa00037. [21]

[FIPA0061, 2002] FIPA0061. FIPA ACL message structure specification, 2002. http://www.fipa.org/specs/fipa00061. [21]

[Fleiss, 1973] Joseph Fleiss. *Statistical methods for rates and proportions*. John Wiley & Sons, Chichester (UK), 1973. [173]

[Fowler et al., 1999] Jerry Fowler, Brad Perry, Marian Nodine, and Bruce Bargmeyer. Agent-based semantic interoperability in InfoSleuth. *ACM SIGMOD Record*, 28(1):60–67, 1999. [153]

[Franconi et al., 2003] Enrico Franconi, Gabriel Kuper, Andrei Lopatenko, and Luciano Serafini. A robust logical and computational characterisation of peer-to-peer database systems. In *Proc. VLDB International Workshop on Databases, Information Systems and Peer-to-Peer Computing (DBISP2P)*, pages 64–76, Berlin (DE), 2003. [54, 55]

[Gal et al., 2005a] Avigdor Gal, Ateret Anaby-Tavor, Alberto Trombetta, and Danilo Montesi. A framework for modeling and evaluating automatic semantic reconciliation. *The VLDB Journal*, 14(1):50–67, 2005. [46, 125]

[Gal et al., 2005b] Avigdor Gal, Giovanni Modica, Hassan Jamil, and Ami Eyal. Automatic ontology matching using application semantics. *AI Magazine*, 26(1):21–32, 2005. [160]

[Gal, 2006] Avigdor Gal. Managing uncertainty in schema matching with top-k schema mappings. *Journal on Data Semantics*, VI:90–114, 2006. [160]

[Gale and Shapley, 1962] David Gale and Lloyd Stowell Shapley. College admissions and the stability of marriage. *American Mathematical Monthly*, 69(1):5–15, 1962. [148]

[Gangemi et al., 2003] Aldo Gangemi, Nicola Guarino, Claudio Masolo, and Alessandro Oltramari. Sweetening WordNet with DOLCE. *AI Magazine*, 24(3):13–24, 2003. [66, 68, 111]

[Gangemi, 2004] Aldo Gangemi. Restructuring semi-structured terminologies for ontology building: a realistic case study in fishery information systems. Deliverable D16, WonderWeb, 2004. [111]

[Ganter and Wille, 1999] Bernhard Ganter and Rudolf Wille. *Formal concept analysis: mathematical foundations*. Springer Verlag, Berlin (DE), 1999. [106]

[Garey and Johnson, 1979] Michael Garey and David Johnson. *Computers and intractability: a guide to the theory of NP-completeness*. W. H. Freeman & Co., New York (NY US), 1979. [98]

[Ghidini and Giunchiglia, 2001] Chiara Ghidini and Fausto Giunchiglia. Local models semantics, or contextual reasoning = locality + compatibility. *Artificial Intelligence*, 127(2):221–259, 2001. [56]

[Ghidini and Giunchiglia, 2004] Chiara Ghidini and Fausto Giunchiglia. A semantics for abstraction. In *Proc. 15th European Conference on Artificial Intelligence (ECAI)*, pages 343–347, Valencia (ES), 2004. [41]

[Ghidini and Serafini, 1998] Chiara Ghidini and Luciano Serafini. Distributed first order logics. In *Proc. 2nd Conference on Frontiers of Combining Systems (FroCoS)*, pages 121–139, Amsterdam (NL), 1998. [52, 54]

[Giunchiglia and Shvaiko, 2003a] Fausto Giunchiglia and Pavel Shvaiko. Semantic matching. *The Knowledge Engineering Review*, 18(3):265–280, 2003. [62, 71, 113, 164]

[Giunchiglia and Shvaiko, 2003b] Fausto Giunchiglia and Pavel Shvaiko. Semantic matching. In *Proc. IJCAI Workshop on Ontologies and Distributed Systems*, pages 139–146, Acapulco (MX), 2003. [227]

[Giunchiglia and Yatskevich, 2004] Fausto Giunchiglia and Mikalai Yatskevich. Element level semantic matching. In *Proc. ISWC Meaning Coordination and Negotiation Workshop*, pages 37–48, Hiroshima (JP), 2004. [91, 164]

[Giunchiglia and Zaihrayeu, 2002] Fausto Giunchiglia and Ilya Zaihrayeu. Making peer databases interact – a vision for an architecture supporting data coordination. In *Proc. 6th International Workshop on Cooperative Information Agents (CIA)*, pages 18–35, Madrid (ES), 2002. [17]

[Giunchiglia et al., 2004] Fausto Giunchiglia, Pavel Shvaiko, and Mikalai Yatskevich. S-Match: an algorithm and an implementation of semantic matching. In *Proc. 1st European Semantic Web Symposium (ESWS)*, volume 3053 of *Lecture notes in computer science*, pages 61–75, Hersounisous (GR), 10-12 May 2004. [63, 88, 113, 164]

[Giunchiglia et al., 2005a] Fausto Giunchiglia, Pavel Shvaiko, and Mikalai Yatskevich. Semantic schema matching. In *Proc. 13rd International Conference on Cooperative Information Systems (CoopIS)*, volume 3761 of *Lecture notes in computer science*, pages 347–365, Agia Napa (CY), 2005. [11, 164]

[Giunchiglia et al., 2005b] Fausto Giunchiglia, Mikalai Yatskevich, and Enrico Giunchiglia. Efficient semantic matching. In *Proc. 2nd European Semantic Web Conference (ESWC)*, volume 3532 of *Lecture notes in computer science*, pages 272–289, Hersounisous (GR), 2005. [165]

[Giunchiglia et al., 2006a] Fausto Giunchiglia, Maurizio Marchese, and Ilya Zaihrayeu. Encoding classifications into lightweight ontologies. In *Proc. 3rd European Semantic Web Conference (ESWC)*, volume 4011 of *Lecture notes in computer science*, pages 80–94, Budva (ME), 2006. [31]

[Giunchiglia et al., 2006b] Fausto Giunchiglia, Fiona McNeill, and Mikalai Yatskevich. Web service composition via semantic matching of interaction specifications. Technical Report DIT-06-080, University of Trento, 2006. [19]

[Giunchiglia et al., 2006c] Fausto Giunchiglia, Pavel Shvaiko, and Mikalai Yatskevich. Discovering missing background knowledge in ontology matching. In *Proc. 16th European Conference on Artificial Intelligence (ECAI)*, pages 382–386, Riva del Garda (IT), 2006. [91, 110, 164, 271]

[Giunchiglia et al., 2007] Fausto Giunchiglia, Mikalai Yatskevich, and Pavel Shvaiko. Semantic matching: Algorithms and implementation. *Journal on Data Semantics*, IX, 2007. [164]

[Goasdoué et al., 2000] François Goasdoué, Véronique Lattes, and Marie-Christine Rousset. The use of CARIN language and algorithms for information integration: The PICSEL system. *International Journal of Cooperative Information Systems*, 9(4):383–401, 2000. [153]

[Goh, 1997] Cheng-Hian Goh. Representing and reasoning about semantic conflicts in heterogeneous information sources. PhD thesis, MIT, Cambridge (MA US), 1997. [41]

[Good, 1965] Irving John Good. *The estimation of probabilities: an essay on modern Bayesian methods*. Classics series. The MIT press, Cambridge (MA US), 1965. [133]

[Gotoh, 1981] Osamu Gotoh. An improved algorithm for matching biological sequences. *Journal of Molecular Biology*, 162(3):705–708, 1981. [79]

[Haas et al., 2005] Laura Haas, Mauricio Hernández, Howard Ho, Lucian Popa, and Mary Roth. Clio grows up: from research prototype to industrial tool. In *Proc. 24th International Conference on Management of Data (SIGMOD)*, pages 805–810, Baltimore (MD US), 2005. [177]

[Haase et al., 2004] Peter Haase, Björn Schnizler, Jeen Broekstra, Marc Ehrig, Frank van Harmelen, Maarteen Menken, Peter Mika, Michal Plechawski, Pawel Pyszlak, Ronny Siebes, Steffen Staab, and Christoph Tempich. Bibster – a semantics-based bibliographic peer-to-peer system. *Journal of Web Semantics*, 2(1):99–103, 2004. [17]

[Hájek, 1998] Petr Hájek. *The metamathematics of fuzzy logic*. Kluwer, Dordrecht (NL), 1998. [121]

[Halevy et al., 2005] Alon Halevy, Naveen Ashish, Dina Bitton, Michael Carey, Denise Draper, Jeff Pollock, Arnon Rosenthal, and Vishal Sikka. Enterprise information integration: successes, challenges and controversies. In *Proc. 24th International Conference on Management of Data (SIGMOD)*, pages 778–787, Baltimore (MD US), 2005. [11, 14]

[Hameed et al., 2004] Adil Hameed, Alun Preece, and Derek Sleeman. Ontology reconciliation. In Steffen Staab and Rudi Studer, editors, *Handbook on ontologies*, chapter 12, pages 231–250. Springer Verlag, Berlin (DE), 2004. [41, 43]

[Hamming, 1950] Richard Hamming. Error detecting and error correcting codes. Technical Report 2, Bell System Technical Journal, 1950. [77]

[Hausdorff, 1914] Felix Hausdorff. *Grundzüge der Mengenlehre*. Verlag Veit, Leipzig (DE), 1914. [109]

[He and Chang, 2003] Bin He and Kevin Chang. Statistical schema matching across web query interfaces. In *Proc. 22nd International Conference on Management of Data (SIGMOD)*, pages 217–228, San Diego (CA US), 2003. [175]

[He and Chang, 2006] Bin He and Kevin Chang. Automatic complex schema matching across web query interfaces: A correlation mining approach. *ACM Transactions on Database Systems*, 31(1):1–45, 2006. [168, 199]

[He et al., 2004] Hai He, Weiyi Meng, Clement Yu, and Zonghuan Wu. Automatic integration of web search interfaces with WISE-Integrator. *The VLDB Journal*, 13(3):256–273, 2004. [180]

[He et al., 2005] Hai He, Weiyi Meng, Clement Yu, and Zonghuan Wu. WISE-Integrator: A system for extracting and integrating complex web search interfaces of the deep web. In *Proc. 31st International Conference on Very Large Data Bases (VLDB)*, pages 1314–1317, Trondheim (NO), 2005. [180]

[Hitzler et al., 2005] Pascal Hitzler, Jérôme Euzenat, Markus Krötzsch, Luciano Serafini, Heiner Stuckenschmidt, Holger Wache, and Antoine Zimmermann. Integrated view and comparison of alignment semantics. Deliverable 2.2.5, Knowledge web NoE, 2005. [51]

[Horrocks et al., 2004] Ian Horrocks, Peter Patel-Schneider, Harold Boley, Said Tabet, Benjamin Grosof, and Mike Dean. SWRL: a semantic web rule language combining OWL and RuleML, 2004. http://www.w3.org/Submission/SWRL/. [225, 229]

[Hovy, 1998] Eduard Hovy. Combining and standardizing large-scale, practical ontologies for machine translation and other uses. In *Proc. 1st International Conference on Language Resources and Evaluation (LREC)*, pages 535–542, Granada (ES), 1998. [154]

[Hu et al., 2005] Wei Hu, Ningsheng Jian, Yuzhong Qu, and Qanbing Wang. GMO: A graph matching for ontologies. In *Proc. K-CAP Workshop on Integrating Ontologies*, pages 43–50, Banff (CA), 2005. [182]

[Hull, 1997] Richard Hull. Managing semantic heterogeneity in databases: a theoretical prospective. In *Proc. 16th Symposium on Principles of Database Systems (PODS)*, pages 51–61, Tucson (AZ US), 1997. [41]

[Huza et al., 2006] Mirella Huza, Mounira Harzallah, and Francky Trichet. OntoMas: a tutoring system dedicated to ontology matching. In *Proc. 1st ISWC International Workshop on Ontology Matching (OM)*, pages 228–323, Athens (GA US), 2006. [273]

[Ichise et al., 2003] Ryutaro Ichise, Hideaki Takeda, and Shinichi Honiden. Integrating multiple internet directories by instance-based learning. In *Proc. 18th International Joint Conference on Artificial Intelligence (IJCAI)*, pages 22–30, Acapulco (MX), 2003. [11, 173]

[Ichise et al., 2004] Ryutaro Ichise, Masahiro Hamasaki, and Hideaki Takeda. Discovering relationships among catalogs. In *Proc. 7th International Conference on Discovery Science*, volume 3245 of *Lecture notes in computer science*, pages 371–379, Padova (IT), 2004. [173]

[Ide and Véronis, 1998] Nancy Ide and Jean Véronis. Word Sense Disambiguation: the state of the art. *Computational Linguistics*, 24(1):1–40, 1998. [87]

[Ives *et al.*, 2004] Zachary Ives, Alon Halevy, Peter Mork, and Igor Tatarinov. Piazza: mediation and integration infrastructure for semantic web data. *Journal of Web Semantics*, 1(2):155–175, 2004. [17]

[Jaccard, 1901] Paul Jaccard. Distribution de la flore alpine dans le bassin des dranses et dans quelques régions voisines. *Bulletin de la société vaudoise des sciences naturelles*, 37:241–272, 1901. [106]

[Jacquemin and Tzoukermann, 1999] Christian Jacquemin and Évelyne Tzoukermann. NLP for term variant extraction: synergy between morphology, lexicon and syntax. In Tomek Strzalkowski, editor, *Language information retrieval*, pages 25–74. Kluwer, Boston (MA, US), 1999. [85]

[Jaro, 1976] Matthew Jaro. UNIMATCH: A record linkage system: User's manual. Technical report, U.S. Bureau of the Census, Washington (DC US), 1976. [80]

[Jaro, 1989] Matthew Jaro. Advances in record-linkage methodology as applied to matching the 1985 census of Tampa, Florida. *Journal of the American Statistical Association*, 84(406):414–420, 1989. [80]

[Jeffrey, 1983] Richard Jeffrey. *The logic of decisions*. University of Chicago Press, Chicago (IL US), 1983. [168]

[Jian *et al.*, 2005] Ningsheng Jian, Wei Hu, Gong Cheng, and Yuzhong Qu. Falcon-AO: Aligning ontologies with Falcon. In *Proc. K-CAP Workshop on Integrating Ontologies*, pages 87–93, Banff (CA), 2005. [182]

[Kalfoglou and Schorlemmer, 2003a] Yannis Kalfoglou and Marco Schorlemmer. IF-Map: an ontology mapping method based on information flow theory. *Journal on Data Semantics*, I:98–127, 2003. [177]

[Kalfoglou and Schorlemmer, 2003b] Yannis Kalfoglou and Marco Schorlemmer. Ontology mapping: the state of the art. *The Knowledge Engineering Review*, 18(1):1–31, 2003. [42, 44, 51, 61, 153]

[Kang and Naughton, 2003] Jaewoo Kang and Jeffrey Naughton. On schema matching with opaque column names and data values. In *Proc. 22nd International Conference on Management of Data (SIGMOD)*, pages 205–216, San Diego (CA US), 2003. [64, 173]

[Kashyap and Sheth, 1996] Vipul Kashyap and Amit Sheth. Semantic and schematic similarities between database objects: a context-based approach. *The VLDB Journal*, 5(4):276–304, 1996. [61]

[Kashyap and Sheth, 1998] Vipul Kashyap and Amit Sheth. Semantic heterogeneity in global information systems: The role of metadata, context and ontologies. In Michael Papazoglou and Gunter Schlageter, editors, *Cooperative information systems*, pages 139–178. Academic Press, New York (NY US), 1998. [41]

[Kensche *et al.*, 2005] David Kensche, Christoph Quix, Mohamed Amine Chatti, and Matthias Jarke. GeRoMe: A generic role based metamodel for model management. In *Proc. 4th International Conference on Ontologies, DataBases, and Applications of Semantics (ODBASE)*, volume 3761 of *Lecture notes in computer science*, pages 1206–1224, Agia Napa (CY), 2005. [236]

[Kim and Seo, 1991] Won Kim and Jungyun Seo. Classifying schematic and data heterogeneity in multidatabase systems. *IEEE Computer*, 24(12):12–18, 1991. [40]

[Kim *et al.*, 2005] Jaehong Kim, Minsu Jang, Young-Guk Ha, Joo-Chan Sohn, and Sang-Jo Lee. MoA: OWL ontology merging and alignment tool for the semantic web. In *Proc. 18th International Conference on Industrial and Engineering Applications of Artificial Intelligence and Expert Systems (IEA/AIE)*, volume 3533 of *Lecture notes in computer science*, pages 722–731, Bari (IT), 2005. [166]

[Klein, 2001] Michel Klein. Combining and relating ontologies: an analysis of problems and solutions. In *Proc. IJCAI Workshop on Ontologies and Information Sharing*, Seattle (WA US), 2001. [41, 42]

[Kohonen, 2001] Teuvo Kohonen. *Self-organizing maps*. Springer, Berlin (DE), 2001. [136]

[Kotis and Vouros, 2004] Konstantinos Kotis and George Vouros. HCONE approach to ontology merging. In *Proc. 1st European Semantic Web Symposium (ESWS)*, volume 3053 of *Lecture notes in computer science*, pages 137–151, Hersounisous (GR), 2004. [165]

[Kotis *et al.*, 2006] Konstantinos Kotis, George Vouros, and Konstantinos Stergiou. Towards automatic merging of domain ontologies: The HCONE-merge approach. *Journal of Web Semantics*, 4(1):60–79, 2006. [165]

[Lacher and Groh, 2001] Martin Lacher and Georg Groh. Facilitating the exchange of explicit knowledge through ontology mappings. In *Proc. 14th International Florida Artificial Intelligence Research Society Conference (FLAIRS)*, pages 305–309, Key West (FL US), 2001. [170]

[Laera *et al.*, 2006] Loredana Laera, Valentina Tamma, Jérôme Euzenat, Trevor Bench-Capon, and Terry Payne. Reaching agreement over ontology alignments. In *Proc. 5th International semantic web Conference (ISWC)*, volume 4273 of *Lecture notes in computer science*, pages 371–384, Athens (GA US), 2006. [21, 245, 248, 256]

[Lambrix and Edberg, 2003] Patrick Lambrix and Anna Edberg. Evaluation of ontology merging tools in bioinformatics. In *Proc. Pacific Symposium on Biocomputing*, pages 589–600, Kauai (HA US), 2003. [202]

[Langlais *et al.*, 1998] Philippe Langlais, Jean Véronis, and Michel Simard. Methods and practical issues in evaluating alignment techniques. In *Proc. 17th International Conference on Computational Linguistics (CoLing)*, pages 711–717, Montréal (CA), 1998. [210]

[Larson *et al.*, 1989] James Larson, Shamkant Navathe, and Ramez Elmasri. A theory of attributed equivalence in databases with application to schema integration. *IEEE Transactions on Software Engineering*, 15(4):449–463, 1989. [61, 105]

[Le Berre, 2004] Daniel Le Berre. SAT4J: A satisfiability library for Java. http://www.sat4j.org/, 2004. [253]

[Leacock *et al.*, 1998] Claudia Leacock, Martin Chodorow, and George Miller. Using corpus statistics and WordNet relations for sense identification. *Computational Linguistics*, 24(1):1–40, 1998. [101]

[Lee *et al.*, 2002] Mong Li Lee, Liang Huai Yang, Wynne Hsu, and Xia Yang. XClust: clustering XML schemas for effective integration. In *Proc. 11th International Conference on Information and Knowledge Management (CIKM)*, pages 292–299, McLean (VA US), 2002. [96, 162]

[Léger *et al.*, 2005] Alain Léger, Lyndon Nixon, and Pavel Shvaiko. On identifying knowledge processing requirements. In *Proc. 4th International Semantic Web Conference (ISWC)*, volume 3729 of *Lecture notes in computer science*, pages 928–943, Galway (IE), 2005. [247]

[Lenat and Guha, 1990] Douglas Lenat and Ramanathan Guha. *Building large knowledge-based systems*. Addison Wesley, Reading (MA US), 1990. [66, 68, 111]

[Lenzerini, 2002] Maurizio Lenzerini. Data integration: A theoretical perspective. In *Proc. 21st Symposium on Principles of Database Systems (PODS)*, pages 233–246, Madison (WI US), 2002. [15, 16, 18, 44, 176]

[Leone *et al.*, 2005] Nicola Leone, Gianluigi Greco, Giovambattista Ianni, Vincenzino Lio, Giorgio Terracina, Thomas Eiter, Wolfgang Faber, Michael Fink, Georg Gottlob, Riccardo Rosati, Domenico Lembo, Maurizio Lenzerini, Marco Ruzzi, Edyta Kalka, Bartosz Nowicki, and Witold Staniszkis. The INFOMIX system for advanced integration of incomplete

and inconsistent data. In *Proc. 24th International Conference on Management of Data (SIGMOD)*, pages 915–917, Baltimore (MD US), 2005. [153]
[Lerner, 2000] Barbara Staudt Lerner. A model for compound type changes encountered in schema evolution. *ACM Transactions on Database Systems*, 25(1):83–127, 2000. [158]
[Lesk, 1986] Michael Lesk. Automatic sense disambiguation using machine readable dictionaries: how to tell a pine cone from an ice cream cone. In *Proc. 5th Annual International Conference on Systems Documentation (SIGDOC)*, pages 24–26, Toronto (CA), 1986. [87, 90]
[Levenshtein, 1965] Vladimir Levenshtein. Binary codes capable of correcting deletions, insertions, and reversals. *Doklady akademii nauk SSSR*, 163(4):845–848, 1965. In Russian. English Translation in Soviet Physics Doklady, 10(8) p. 707–710, 1966. [79]
[Li and Clifton, 1994] Wen-Syan Li and Chris Clifton. Semantic integration in heterogeneous databases using neural networks. In *Proc. 10th International Conference on Very Large Data Bases (VLDB)*, pages 1–12, Santiago (CL), 1994. [108, 136, 176]
[Li and Clifton, 2000] Wen-Syan Li and Chris Clifton. SEMINT: a tool for identifying attribute correspondences in heterogeneous databases using neural networks. *Data and Knowledge Engineering*, 33(1):49–84, 2000. [176]
[Li et al., 2006] Yi Li, Juanzi Li, Duo Zhang, and Jie Tang. Result of ontology alignment with RiMOM at OAEI-06. In *Proc. 1st ISWC International Workshop on Ontology Matching (OM)*, pages 181–190, Athens (GA US), 2006. [182]
[Lim et al., 1993] Ee-Peng Lim, Jaideep Srivastava, Satya Prabhakar, and James Richardson. Entity identification in database integration. In *Proc. 9th International Conference on Data Engineering (ICDE)*, pages 294–301, Wien (AT), 1993. [107]
[Lin, 1998] Dekang Lin. An information-theoretic definition of similarity. In *Proc. 15th International Conference of Machine Learning (ICML)*, pages 296–304, Madison (WI US), 1998. [90]
[Lopez et al., 2005] Vanessa Lopez, Michele Pasin, and Enrico Motta. AquaLog: An ontology-portable question answering system for the semantic web. In *Proc. 2nd European Semantic Web Conference (ESWC)*, volume 3532 of *Lecture notes in computer science*, pages 546–562, Hersounisous (GR), 2005. [23]
[Lopez et al., 2006] Vanessa Lopez, Enrico Motta, and Victoria Uren. PowerAqua: Fishing the semantic web. In York Sure and John Domingue, editors, *Proc. 3rd European Semantic Web Conference (ESWC)*, volume 4011 of *Lecture notes in computer science*, pages 393–410, Budva (ME), 2006. [23]
[Lovász and Plummer, 1986] László Lovász and Michael Plummer. *Matching theory*. North-Holland, Amsterdam (NL), 1986. [148]
[Lovins, 1968] Julie Beth Lovins. Development of a stemming algorithm. *Mechanical Translation and Computational Linguistics*, 11(1):22–31, 1968. [85]
[Mädche and Staab, 2002] Alexander Mädche and Steffen Staab. Measuring similarity between ontologies. In *Proc. 13th International Conference on Knowledge Engineering and Knowledge Management (EKAW)*, volume 2473 of *Lecture notes in computer science*, pages 251–263, Siguenza (ES), 2002. [99, 104]
[Mädche and Zacharias, 2002] Alexander Mädche and Valentin Zacharias. Clustering ontology-based metadata in the semantic web. In *Proc. 6th European Conference on Principles and Practice of Knowledge Discovery in Databases (PKDD)*, pages 348–360, Helsinki (FI), 2002. [102]
[Mädche et al., 2002] Alexander Mädche, Boris Motik, Nuno Silva, and Raphael Volz. MAFRA – a mapping framework for distributed ontologies. In *Proc. 13th International Conference on Knowledge Engineering and Knowledge Management (EKAW)*, volume 2473 of *Lecture notes in computer science*, pages 235–250, Siguenza (ES), 2002. [221, 222, 238]

[Madhavan *et al.*, 2001] Jayant Madhavan, Philip Bernstein, and Erhard Rahm. Generic schema matching with Cupid. In *Proc. 27th International Conference on Very Large Data Bases (VLDB)*, pages 48–58, Roma (IT), 2001. [104, 160]

[Madhavan *et al.*, 2002] Jayant Madhavan, Philip Bernstein, Pedro Domingos, and Alon Halevy. Representing and reasoning about mappings between domain models. In *Proc. 18th National Conference on Artificial Intelligence (AAAI)*, pages 122–133, Edmonton (CA), 2002. [35, 235]

[Madhavan *et al.*, 2005] Jayant Madhavan, Philip Bernstein, An-Hai Doan, and Alon Halevy. Corpus-based schema matching. In *Proc. 21st International Conference on Data Engineering (ICDE)*, pages 57–68, Tokyo (JP), 2005. [183]

[Masolo *et al.*, 2003] Claudio Masolo, Stefano Borgo, Aldo Gangemi, Nicola Guarino, and Alessandro Oltramari. Ontology library. Deliverable D18, Wonderweb, 2003. [228]

[Maynard and Ananiadou, 2001] Diana Maynard and Sophia Ananiadou. Term extraction using a similarity-based approach. In Didier Bourigault, Christian Jacquemin, and Marie-Claude Lhomme, editors, *Recent advances in computational terminology*, pages 261–278. John Benjamins, Amsterdam (NL), 2001. [84, 85]

[Maynard, 1999] Diana Maynard. *Term Recognition Using Combined Knowledge Sources*. PhD thesis, Department of Computing and Mathematics, Manchester Metropolitan University, Manchester (UK), 1999. [85]

[McCallum and Nigam, 1998] Andrew McCallum and Kamal Nigam. A comparison of event models for naive Bayes text classification. In *Proc. AAAI Workshop on Learning for Text Categorization*, pages 41–48, Madison (WI US), 1998. [133]

[McDermott and Dou, 2002] Drew McDermott and Dejing Dou. Representing disjunction and quantifiers in RDF. In *Proc 1st International Semantic Web Conference (ISWC)*, volume 2342 of *Lecture notes in computer science*, pages 250–263, Chia Laguna (IT), 2002. [163]

[McGuinness and Pinheiro da Silva, 2003] Deborah McGuinness and Paulo Pinheiro da Silva. Infrastructure for web explanations. In *Proc. 2nd International Semantic Web Conference (ISWC)*, volume 2870 of *Lecture notes in computer science*, pages 113–129, Sanibel Island (FL US), 2003. [248]

[McGuinness and Pinheiro da Silva, 2004] Deborah McGuinness and Paulo Pinheiro da Silva. Explaining answers from the semantic web: The Inference Web approach. *Journal of Web Semantics*, 1(4):397–413, 2004. [245, 246, 247]

[McGuinness *et al.*, 2000] Deborah McGuinness, Richard Fikes, James Rice, and Steve Wilder. An environment for merging and testing large ontologies. In *Proc. 7th International Conference on the Principles of Knowledge Representation and Reasoning (KR)*, pages 483–493, Breckenridge (CO US), 2000. [241]

[McNeill, 2006] Fiona McNeill. *Dynamic ontology refinement*. PhD thesis, University of Edinburgh, Edinburgh (UK), 2006. [24]

[MDC, 1999] Open information model, version 1.0. http://mdcinfo/oim/oim10.html, 1999. [161]

[Medjahed and Bouguettaya, 2005] Brahim Medjahed and Athman Bouguettaya. A multi-level composability model for semantic web services. *IEEE Transactions on Knowledge and Data Engineering*, 17(7):954–968, 2005. [19]

[Meilicke *et al.*, 2006] Christian Meilicke, Heiner Stuckenschmidt, and Andrei Tamilin. Improving automatically created mappings using logical reasoning. In *Proc. 1st ISWC International Workshop on Ontology Matching (OM)*, pages 61–72, Athens (GA US), 2006. [115]

[Melnik *et al.*, 2002] Sergey Melnik, Hector Garcia-Molina, and Erhard Rahm. Similarity flooding: a versatile graph matching algorithm. In *Proc. 18th International Conference on Data Engineering (ICDE)*, pages 117–128, San Jose (CA US), 2002. [127, 161, 207, 211]

[Melnik et al., 2003a] Sergey Melnik, Erhard Rahm, and Philip Bernstein. Developing metadata-intensive applications with Rondo. *Journal of Web Semantics*, 1(1):47-74, 2003. [236]

[Melnik et al., 2003b] Sergey Melnik, Erhard Rahm, and Philip Bernstein. Rondo: A programming platform for model management. In *Proc. 22nd International Conference on Management of Data (SIGMOD)*, pages 193-204, San Diego (CA US), 2003. [236]

[Melnik et al., 2005] Sergey Melnik, Philip Bernstein, Alon Halevy, and Erhard Rahm. Supporting executable mappings in model management. In *Proc. 24th International Conference on Management of Data (SIGMOD)*, pages 167-178, Baltimore (MD US), 2005. [128, 235, 236, 259, 260]

[Melnik, 2004] Sergey Melnik. *Generic Model Management Concepts and Algorithms*. Springer, Heidelberg (DE), 2004. [235]

[Melton (ed.), 2003] Jim Melton (ed.). Information technology — database languages — SQL. ISO standard ISO/CEI 9075:2003, ISO, 2003. [36]

[Mena et al., 1996] Eduardo Mena, Vipul Kashyap, Amit Sheth, and Arantza Illarramendi. Observer: An approach for query processing in global information systems based on interoperability between pre-existing ontologies. In *Proc. 4th International Conference on Cooperative Information Systems (CoopIS)*, pages 14-25, Brussels (BE), 1996. [24, 153]

[Miles and Brickley, 2005a] Alistair Miles and Dan Brickley. SKOS core guide. Note, W3C, 2005. [231]

[Miles and Brickley, 2005b] Alistair Miles and Dan Brickley. SKOS core vocabulary. Note, W3C, 2005. [231]

[Miller et al., 2000] Renée Miller, Laura Haas, and Mauricio Hernández. Schema mapping as query discovery. In *Proc. 26th International Conference on Very Large Data Bases (VLDB)*, pages 77-88, Cairo (EG), 2000. [177]

[Miller et al., 2001] Renée Miller, Mauricio Hernández, Laura Haas, Lingling Yan, Howard Ho, Ronald Fagin, and Lucian Popa. The Clio project: managing heterogeneity. *ACM SIGMOD Record*, 30(1):78-83, 2001. [177]

[Miller, 1995] George Miller. WordNet: A lexical database for english. *Communications of the ACM*, 38(11):39-41, 1995. [66, 86, 87]

[Milo and Zohar, 1998] Tova Milo and Sagit Zohar. Using schema matching to simplify heterogeneous data translation. In *Proc. 24th International Conference on Very Large Data Bases (VLDB)*, pages 122-133, New York (NY US), 1998. [155]

[Mitra and Wiederhold, 2002] Prasenjit Mitra and Gio Wiederhold. Resolving terminological heterogeneity in ontologies. In *Proc. ECAI Workshop on Ontologies and Semantic Interoperability*, pages 45-50, Lyon (FR), 2002. [156]

[Mitra et al., 1999] Prasenjit Mitra, Gio Wiederhold, and Jan Jannink. Semi-automatic integration of knowledge sources. In *Proc. 2nd International Conference on Information Fusion*, pages 572-581, Sunnyvale (CA US), 1999. [156]

[Mitra et al., 2000] Prasenjit Mitra, Gio Wiederhold, and Martin Kersten. A graph-oriented model for articulation of ontology interdependencies. In *Proc. 8th Conference on Extending Database Technology (EDBT)*, volume 1777 of *Lecture notes in computer science*, pages 86-100, Praha (CZ), 2000. [156]

[Mitra et al., 2005] Prasenjit Mitra, Natalya Noy, and Anuj Jaiswal. Ontology mapping discovery with uncertainty. In *Proc. 4th International Semantic Web Conference (ISWC)*, volume 3729 of *Lecture notes in computer science*, pages 537-547, Galway (IE), 2005. [141, 142, 168]

[Mocan et al., 2006] Adrian Mocan, Emilia Cimpian, and Mick Kerrigan. Formal model for ontology mapping creation. In *Proc. 5th International Semantic Web Conference (ISWC)*, volume 4273 of *Lecture notes in computer science*, pages 459–472, Athens (GA US), 2006. [243]

[Mochol et al., 2006] Malgorzata Mochol, Anja Jentzsch, and Jérôme Euzenat. Applying an analytic method for matching approach selection. In *Proc. 1st ISWC International Workshop on Ontology Matching (OM)*, pages 37–48, Athens (GA US), 2006. [273]

[Modica et al., 2001] Giovanni Modica, Avigdor Gal, and Hasan Jamil. The use of machine-generated ontologies in dynamic information seeking. In *Proc. 9th International Conference on Cooperative Information Systems (CoopIS)*, volume 2172 of *Lecture notes in computer science*, pages 433–448, Trento (IT), 2001. [159]

[Monge and Elkan, 1997] Alvaro Monge and Charles Elkan. An efficient domain-independent algorithm for detecting approximately duplicate database records. In *Proc. SIGMOD Workshop on Data Mining and Knowledge Discovery*, Tucson (AZ US), 1997. [79]

[Munkres, 1957] James Munkres. Algorithms for the assignment and transportation problems. *SIAM Journal on Applied Mathematics*, 5(1):32–38, 1957. [148]

[Naumann et al., 2002] Felix Naumann, Ching-Tien Ho, Xuqing Tian, Laura Haas, and Nimrod Megiddo. Attribute classification using feature analysis. In *Proc. 18th International Conference on Data Engineering (ICDE)*, page 271, San Jose (CA US), 2002. [177]

[Navathe and Buneman, 1986] Shamkant Navathe and Peter Buneman. Integrating user views in database design. *IEEE Computer*, 19(1):50–62, 1986. [97]

[Needleman and Wunsch, 1970] Saul Needleman and Christian Wunsch. A general method applicable to the search for similarities in the amino acid sequence of two proteins. *Journal of Molecular Biology*, 48(3):443–453, 1970. [79]

[Nejdl et al., 2002] Wolfgang Nejdl, Boris Wolf, Changtao Qu, Stefan Decker, Michael Sintek, Ambjörn Naeve, Mikael Nilsson, Matthias Palmér, and Tore Risch. Edutella: A P2P networking infrastructure based on RDF. In *Proc. 11th International World Wide Web Conference (WWW)*, pages 604–615, Honolulu (HA US), 2002. [17]

[Niles and Pease, 2001] Ian Niles and Adam Pease. Towards a standard upper ontology. In *Proc. 2nd International Conference on Formal Ontology in Information Systems (FOIS)*, pages 2–9, Ogunquit (ME US), 2001. [66, 68, 111]

[Nodine et al., 2000] Marian Nodine, Jerry Fowler, Tomasz Ksiezyk, Brad Perry, Malcolm Taylor, and Amy Unruh. Active information gathering in infosleuth. *International Journal of Cooperative Information Systems*, 9(1-2):3–28, 2000. [153]

[Nottelmann and Straccia, 2005] Henrik Nottelmann and Umberto Straccia. sPLMap: A probabilistic approach to schema matching. In *Proc. 27th European Conference on Information Retrieval Research (ECIR)*, pages 81–95, Santiago de Compostela (ES), 2005. [176]

[Nottelmann and Straccia, 2006] Henrik Nottelmann and Umberto Straccia. A probabilistic, logic-based framework for automated web directory alignment. In Zongmin Ma, editor, *Soft computing in ontologies and the semantic web*, volume 204 of *Studies in fuzziness and soft computing*, pages 47–77. Springer Verlag, 2006. [176]

[Noy and Klein, 2004] Natalya Noy and Michel Klein. Ontology evolution: Not the same as schema evolution. *Knowledge and Information Systems*, 6(4):428–440, 2004. [11, 42]

[Noy and Musen, 1999] Natalya Noy and Marc Musen. SMART: Automated support for ontology merging and alignment. In *Proc. 12th Workshop on Knowledge Acquisition, Modeling and Management (KAW)*, Banff (CA), 1999. [159]

[Noy and Musen, 2000] Natalya Noy and Mark Musen. PROMPT: Algorithm and tool for automated ontology merging and alignment. In *Proc. 17th National Conference of Artificial Intelligence (AAAI)*, pages 450–455, Austin (TX US), 2000. [10, 242]

[Noy and Musen, 2001] Natalya Noy and Mark Musen. Anchor-PROMPT: Using non-local context for semantic matching. In *Proc. IJCAI Workshop on Ontologies and Information Sharing*, pages 63–70, Seattle (WA US), 2001. [159, 242]

[Noy and Musen, 2002a] Natalya Noy and Mark Musen. Evaluating ontology-mapping tools: requirements and experience. In *Proc. 1st EKAW Workshop on Evaluation of Ontology Tools (EON)*, pages 1–14, Siguenza (ES), 2002. [196, 198]

[Noy and Musen, 2002b] Natalya Noy and Mark Musen. PromptDiff: A fixed-point algorithm for comparing ontology versions. In *Proc. 18th National Conference on Artificial Intelligence (AAAI)*, pages 744–750, Edmonton (CA), 2002. [11, 159, 242]

[Noy and Musen, 2003] Natalya Noy and Marc Musen. The PROMPT suite: interactive tools for ontology merging and mapping. *International Journal of Human-Computer Studies*, 59(6):983–1024, 2003. [159, 242]

[Noy and Musen, 2004] Natalya Noy and Mark Musen. Ontology versioning in an ontology management framework. *IEEE Intelligent Systems*, 19(4):6–13, 2004. [11]

[Noy, 2004a] Natalya Noy. Semantic integration: A survey of ontology-based approaches. *ACM SIGMOD Record*, 33(4):65–70, 2004. [153]

[Noy, 2004b] Natalya Noy. Tools for mapping and merging ontologies. In Steffen Staab and Rudi Studer, editors, *Handbook on ontologies*, chapter 18, pages 365–384. Springer Verlag, Berlin (DE), 2004. [242]

[Oberle *et al.*, 2004] Daniel Oberle, Raphael Volz, Steffen Staab, and Boris Motik. An extensible ontology software environment. In Steffen Staab and Rudi Studer, editors, *Handbook on ontologies*, chapter 15, pages 299–319. Springer Verlag, Berlin (DE), 2004. [241, 243]

[Oundhakar *et al.*, 2005] Swapna Oundhakar, Kunal Verma, Kaarthik Sivashanugam, Amit Sheth, and John Miller. Discovery of web services in a multi-ontology and federated registry environment. *International Journal of Web Services Research*, 2(3):1–32, 2005. [19]

[Özsu and Valduriez, 1999] Tamer Özsu and Patrick Valduriez. *Principles of distributed database systems*. Prentice Hall, Englewood Cliffs (NJ US), 2nd edition, 1999. [13]

[Palopoli *et al.*, 1998] Luigi Palopoli, Domenico Saccà, and Domenico Ursino. An automatic techniques for detecting type conflicts in database schemes. In *Proc. 7th International Conference on Information and Knowledge Management (CIKM)*, pages 306–313, Bethesda (ML US), 1998. [155]

[Palopoli *et al.*, 2000] Luigi Palopoli, Luigi Pontieri, Giorgio Terracina, and Domenico Ursino. Intensional and extensional integration and abstraction of heterogeneous databases. *Data and Knowledge Engineering*, 35(3):201–237, 2000. [155]

[Palopoli *et al.*, 2003a] Luigi Palopoli, Domenico Saccá, Giorgio Terracina, and Domenico Ursino. Uniform techniques for deriving similarities of objects and subschemes in heterogeneous databases. *IEEE Transactions on Knowledge and Data Engineering*, 15(2):271–294, 2003. [155]

[Palopoli *et al.*, 2003b] Luigi Palopoli, Giorgio Terracina, and Domenico Ursino. DIKE: a system supporting the semi-automatic construction of cooperative information systems from heterogeneous databases. *Software–Practice and Experience*, 33(9):847–884, 2003. [155]

[Pan *et al.*, 2005] Rong Pan, Zhongli Ding, Yang Yu, and Yun Peng. A Bayesian network approach to ontology mapping. In *Proc. 3rd International Semantic Web Conference (ISWC)*, volume 3298 of *Lecture notes in computer science*, pages 563–577, Hiroshima (JP), 2005. [141, 167]

[Paolucci *et al.*, 2002] Massimo Paolucci, Takahiro Kawamura, Terry Payne, and Katia Sycara. Semantic matching of web services capabilities. In *Proc. 1st International Semantic Web Conference (ISWC)*, volume 2342 of *Lecture notes in computer science*, pages 333–347, Chia Laguna (IT), 2002. [19]

[Parent and Spaccapietra, 1998] Christine Parent and Stefano Spaccapietra. Issues and approaches of database integration. *Communications of the ACM*, 41(5):166–178, 1998. [11, 13, 61]

[Parent and Spaccapietra, 2000] Christine Parent and Stefano Spaccapietra. Database integration: the key to data interoperability. In Mike Papazoglou, Stefano Spaccapietra, and Zahir Tari, editors, *Object-oriented data modeling*, chapter 9, pages 221–253. The MIT Press, Cambridge (MA US), 2000. [115, 153]

[Pedersen et al., 2004] Ted Pedersen, Siddharth Patwardhan, and Jason Michelizzi. WordNet::Similarity – measuring the relatedness of concepts. In *Proc. 19th National Conference on Artificial Intelligence (AAAI)*, pages 1024–1025, San Jose (CA US), 2004. [91]

[Pinheiro da Silva et al., 2004] Paulo Pinheiro da Silva, Deborah McGuinness, and Richard Fikes. A proof markup language for semantic web services. Technical Report TR KSL-04-01, Stanford University, 2004. [274]

[Pinheiro da Silva et al., 2006] Paulo Pinheiro da Silva, Deborah McGuinness, and Richard Fikes. A proof markup language for semantic web services. *Information Systems*, 31(4):381–395, 2006. [248]

[Porter, 1980] Martin Porter. An algorithm for suffix stripping. *Program*, 14(3):130–137, 1980. [85]

[Preece et al., 2000] Alun Preece, Kit-Ying Hui, Alex Gray, Philippe Marti, Trevor Bench-Capon, Dean Jones, and Zhan Cui. The KRAFT architecture for knowledge fusion and transformation. *Knowledge-Based Systems*, 13(2-3):113–120, 2000. [153]

[Prud'hommeaux and Seaborne (ed.), 2007] Eric Prud'hommeaux and Andrew Seaborne (ed.). SPARQL query language for RDF. Working draft, W3C, 2007. [45]

[Qu et al., 2006] Yuzhong Qu, Wei Hu, and Gong Chen. Constructing virtual documents for ontology matching. In *Proc. 15th International World Wide Web Conference (WWW)*, pages 23–31, Edinburgh (UK), 2006. [81, 182]

[Quinlan, 1993] John Ross Quinlan. *C4.5: Programs for machine learning*. Morgan Kaufmann Publishers, Menlo Park (CA US), 1993. [138]

[Rahm and Bernstein, 2001] Erhard Rahm and Philip Bernstein. A survey of approaches to automatic schema matching. *The VLDB Journal*, 10(4):334–350, 2001. [61, 63, 64, 68, 71, 92, 153]

[Rahm et al., 2004] Erhard Rahm, Hong-Hai Do, and Sabine Maßmann. Matching large XML schemas. *ACM SIGMOD Record*, 33(4):26–31, 2004. [68, 69]

[Resnik, 1995] Philipp Resnik. Using information content to evaluate semantic similarity in a taxonomy. In *Proc. 14th International Joint Conference on Artificial Intelligence (IJCAI)*, pages 448–453, Montréal (CA), 1995. [90]

[Resnik, 1999] Phillip Resnik. Semantic similarity in a taxonomy: an information-based measure and its application to problems of ambiguity in natural language. *Journal of Artificial Intelligence Research*, 11:95–130, 1999. [90]

[Robertson and Jones, 1976] Stephen Robertson and Karen Spärck Jones. Relevance weighting of search terms. *Journal of the American Society for Information Science*, 27(3):129–146, 1976. [81]

[Robertson et al., 2006] Dave Robertson, Fausto Giunchiglia, Frank van Harmelen, Maurizio Marchese, Marta Sabou, Marco Schorlemmer, Nigel Shadbolt, Ronnie Siebes, Carles Sierra, Chris Walton, Srinandan Dasmahapatra, Dave Dupplaw, Paul Lewis, Mikalai Yatskevich, Spyros Kotoulas, Adrian Perreau de Pinninck, and Antonis Loizou. Open knowledge semantic webs through peer-to-peer interaction. Technical Report DIT-06-034, University of Trento, 2006. [19]

[Roddick, 1995] John Roddick. A survey of schema versioning issues for database systems. *Information and Software Technology*, 37(7):383–393, 1995. [11]

[Roman et al., 2004] Dumitru Roman, Holger Lausen, and Uwe Keller. Web service modeling ontology standard (WSMO-standard). Working Draft D2v0.2, WSMO, 2004. [19, 229]

[Rousset et al., 2006] Marie-Christine Rousset, Philippe Adjiman, Philippe Chatalic, François Goasdoué, and Laurent Simon. Somewhere in the semantic web. In *Proc. 32nd International Conference on Current Trends in Theory and Practice of Computer Science (SofSem)*, volume 3831 of *Lecture notes in computer science*, pages 84–99, Merin (CZ), 2006. [16, 17]

[Russell and Norvig, 1995] Stuart Russell and Peter Norvig. *Artificial intelligence: a modern approach*. Prentice Hall, Englewood Cliffs (NJ US), 1995. [141, 172]

[Sabou et al., 2006a] Marta Sabou, Mathieu d'Aquin, and Enrico Motta. Using the semantic web as background knowledge for ontology mapping. In *Proc. 1st ISWC International Workshop on Ontology Matching (OM)*, pages 1–12, Athens (GA US), 2006. [112]

[Sabou et al., 2006b] Marta Sabou, Vanessa Lopez, and Enrico Motta. Ontology selection for the real semantic web: How to cover the Queen birthday dinner? In *Proc. 15th International Conference on Knowledge Engineering and Knowledge Management (EKAW)*, volume 4248 of *Lecture notes in computer science*, pages 96–111, Praha (CZ), 2006. [23]

[Saint-Onge, 1995] David Saint-Onge. Detecting and correcting malapropisms with lexical chains. Master's thesis, University of Toronto, Toronto (CA), 1995. [92]

[Salton and McGill, 1983] Gerard Salton and Michael McGill. *Introduction to modern information retrieval*. McGraw-Hill, New York (NY US), 1983. [81]

[Salton, 1971] Gerard Salton. *The SMART retrieval system: experiments in automatic information processing*. Prentice Hall, Englewood Cliffs (NJ US), 1971. [81]

[Sayyadian et al., 2005] Mayssam Sayyadian, Yoonkyong Lee, An-Hai Doan, and Arnon Rosenthal. Tuning schema matching software using synthetic scenarios. In *Proc. 31st International Conference on Very Large Data Bases (VLDB)*, pages 994–1005, Trondheim (NO), 2005. [185, 272]

[Scharffe, 2005] François Scharffe. Mapping and merging tool design. Deliverable D7.2, Ontology Management Working Group, 2005. [229]

[Schuh, 1999] Randall Schuh. *Biological systematics: principles and applications*. Cornell University Press, Ithaca (NY US), 1999. [32]

[Schulten et al., 2001] Ellen Schulten, Hans Akkermans, Guy Botquin, Martin Dorr, Nicola Guarino, Nelson Lopes, and Norman Sadeh. Call for participants: The e-commerce product classification challenge. *IEEE Intelligent Systems*, 16(4):86–c3, 2001. [14]

[Serafini and Tamilin, 2005] Luciano Serafini and Andrei Tamilin. DRAGO: Distributed reasoning architecture for the semantic web. In *Proc. 2nd European Semantic Web Conference (ESWC)*, volume 3532 of *Lecture notes in computer science*, pages 361–376, Hersounisous (GR), May 2005. [262, 264]

[Serafini et al., 2005] Luciano Serafini, Heiner Stuckenschmidt, and Holger Wache. A formal investigation of mapping language for terminological knowledge. In *Proc. 19th International Joint Conference on Artificial Intelligence (IJCAI)*, pages 576–581, Edinburgh (UK), 2005. [220]

[Sheth and Larson, 1990] Amit Sheth and James Larson. Federated database systems for managing distributed, heterogeneous, and autonomous databases. *ACM Computing Surveys*, 22(3):183–236, 1990. [11, 13, 40]

[Sheth et al., 1988] Amit Sheth, James Larson, Aloysius Cornelio, and Shamkant Navathe. A tool for integrating conceptual schemas and user views. In *Proc. 4th International Conference on Data Engineering (ICDE)*, pages 176–183, Los Angeles (CA US), 1988. [105]

[Shvaiko and Euzenat, 2005] Pavel Shvaiko and Jérôme Euzenat. A survey of schema-based matching approaches. *Journal on Data Semantics*, IV:146–171, 2005. [61, 64, 65, 153]

[Shvaiko et al., 2005] Pavel Shvaiko, Fausto Giunchiglia, Paulo Pinheiro da Silva, and Deborah McGuinness. Web explanations for semantic heterogeneity discovery. In *Proc. 2nd European Semantic Web Conference (ESWC)*, volume 3532 of *Lecture notes in computer science*, pages 303–317, Hersounisous (GR), May 2005. [164, 245, 248]

[Shvaiko et al., 2006a] Pavel Shvaiko, Jérôme Euzenat, Natalya Noy, Heiner Stuckenschmidt, Richard Benjamins, and Michael Uschold, editors. *Proc. 1st ISWC International Workshop on Ontology Matching (OM)*, Athens (GA US), 2006. [196]

[Shvaiko et al., 2006b] Pavel Shvaiko, Fausto Giunchiglia, Marco Schorlemmer, Fiona McNeill, Alan Bundy, Maurizio Marchese, Mikalai Yatskevich, Ilya Zaihrayeu, Bo Ho, Vanessa Lopez, Marta Sabou, Joaqín Abian, Ronny Siebes, and Spyros Kotoulas. Dynamic ontology matching: a survey. Deliverable 3.1, OpenKnowledge STREP, 2006. [17]

[Shvaiko, 2004] Pavel Shvaiko. Iterative schema-based semantic matching. Technical Report DIT-04-020, University of Trento (IT), 2004. [114]

[Shvaiko, 2006] Pavel Shvaiko. *Iterative Schema-based Semantic Matching*. PhD thesis, International Doctorate School in Information and Communication Technology, University of Trento, Trento (IT), November 2006. [113]

[Silva et al., 2005] Nuno Silva, Paulo Maio, and João Rocha. An approach to ontology mapping negotiation. In *Proc. K-CAP Workshop on Integrating Ontologies*, pages 54–60, Banff (CA), 2005. [255]

[Sirin et al., 2007] Evren Sirin, Bijan Parsia, Bernardo Cuenca Grau, Aditya Kalyanpur, and Yarden Katz. Pellet: a practical OWL-DL reasoner. *Journal of Web Semantics*, 5, 2007. To appear. [164, 264]

[Smith and Waterman, 1981] Temple Smith and Michael Waterman. Identification of common molecular subsequences. *Journal of Molecular Biology*, 147(1):195–197, 1981. [79]

[Smith et al., 2004] Mike Smith, Christopher Welty, and Deborah McGuinness (eds.). OWL web ontology language guide. Recommendation, W3C, February 10 2004. [36]

[Smolka, 1992] Gerd Smolka. Feature constraints logics for unification grammars. *Journal of Logic Programming*, 12(1):324–343, 1992. [226]

[Sotnykova et al., 2005] Anastasiya Sotnykova, Christèle Vangenot, Nadine Cullot, Nacéra Bennacer, and Marie-Aude Aufaure. Semantic mappings in description logics for spatio-temporal database schema integration. *Journal on Data Semantics*, III:143–167, 2005. [115]

[Spaccapietra and Parent, 1991] Stefano Spaccapietra and Christine Parent. Conflicts and correspondence assertions in interoperable databases. *SIGMOD Record*, 20(4):49–54, 1991. [11]

[Staab and Stuckenschmidt, 2006] Steffen Staab and Heiner Stuckenschmidt, editors. *Semantic web and peer-to-peer*. Springer, Heidelberg (DE), 2006. [16]

[Staab and Studer, 2004] Steffen Staab and Rudi Studer. *Handbook on ontologies*. International handbooks on information systems. Springer Verlag, Berlin (DE), 2004. [36]

[Stoilos et al., 2005] Georgos Stoilos, Giorgos Stamou, and Stefanos Kollias. A string metric for ontology alignment. In *Proc. 4th International Semantic Web Conference (ISWC)*, volume 3729 of *Lecture notes in computer science*, pages 624–637, Galway (IE), 2005. [80]

[Straccia and Troncy, 2005] Umberto Straccia and Raphaël Troncy. oMAP: Combining classifiers for aligning automatically OWL ontologies. In *Proc. 6th International Conference on Web Information Systems Engineering (WISE)*, pages 133–147, New York (NY US), 2005. [179, 240]

[Straccia and Troncy, 2006] Umberto Straccia and Raphaël Troncy. Towards distributed information retrieval in the semantic web: Query reformulation using the oMAP framework. In *Proc. 3rd European Semantic Web Conference (ESWC)*, volume 4011 of *Lecture notes in computer science*, pages 378–392, Budva (ME), 2006. [179]

[Stumme and Mädche, 2001] Gerd Stumme and Alexander Mädche. FCA-Merge: Bottom-up merging of ontologies. In *Proc. 17th International Joint Conference on Artificial Intelligence (IJCAI)*, pages 225–234, Seattle (WA US), 2001. [170]

[Su *et al.*, 2006] Weifeng Su, Jiying Wang, and Frederick Lochovsky. Holistic schema matching for web query interfaces. In *Proc. 10th Conference on Extending Database Technology (EDBT)*, volume 3896 of *Lecture notes in computer science*, pages 77–94, München (DE), 2006. [199]

[Sun and Lin, 2001] Aixin Sun and Ee-Peng Lin. Hierarchical text classification and evaluation. In *Proc. 1st International Conference on Data Mining (ICDM)*, pages 521–528, San Jose (CA), 2001. [210]

[Sure *et al.*, 2004] York Sure, Oscar Corcho, Jérôme Euzenat, and Todd Hughes, editors. *Proc. 3rd ISWC Workshop on Evaluation of Ontology-based tools (EON)*, Hiroshima (JP), 2004. [196]

[Tang *et al.*, 2006] Jie Tang, Juanzi Li, Bangyong Liang, Xiaotong Huang, Yi Li, and Kehong Wang. Using Bayesian decision for ontology mapping. *Journal of Web Semantics*, 4(1):243–262, 2006. [182]

[Ting and Witten, 1999] Kai Ming Ting and Ian Witten. Issues in stacked generalization. *Journal of Artificial Intelligence Research*, 10:271–289, 1999. [139, 141]

[Tsarkov and Horrocks, 2006] Dmitry Tsarkov and Ian Horrocks. FaCT++ description logic reasoner: system description. In *Proc. 3rd International Joint Conference on Automated Reasoning (IJCAR)*, volume 4130 of *Lecture notes in computer science*, pages 292–297, Seattle (WA US), 2006. Springer. [164]

[Tu and Yu, 2005] Kewei Tu and Yong Yu. CMC: Combining multiple schema-matching strategies based on credibility prediction. In *Proc. 10th International Conference on Database Systems for Advanced Applications (DASFAA)*, volume 3453 of *Lecture notes in computer science*, pages 888–893, Beijing (CN), 2005. [238]

[Tverski, 1977] Amos Tverski. Features of similarity. *Psychological Review*, 84(2):327–352, 1977. [74]

[Uschold and Gruninger, 2004] Mike Uschold and Michael Gruninger. Ontologies and semantics for seamless connectivity. *ACM SIGMOD Record*, 33(4):58–64, 2004. [30]

[Uschold, 2005] Mike Uschold. Achieving semantic interoperability using RDF and OWL - v4, 2005. http://lists.w3.org/Archives/Public/public-swbp-wg/2005Sep/att-0027/SemanticII-v4.htm. [222]

[Valtchev and Euzenat, 1997] Petko Valtchev and Jérôme Euzenat. Dissimilarity measure for collections of objects and values. In *Proc. 2nd Symposium on Intelligent Data Analysis (IDA)*, volume 1280 of *Lecture notes in computer science*, pages 259–272, London (UK), 1997. [94, 100]

[Valtchev, 1999] Petko Valtchev. *Construction automatique de taxonomies pour l'aide à la représentation de connaissances par objets*. Thèse d'informatique, Université Grenoble 1, Grenoble (FR), 1999. [83, 94, 95, 96, 102, 109, 110, 124]

[van Eijk *et al.*, 2001] Rogier van Eijk, Frank de Boer, Wiebe van de Hoek, and John-Jules Meyer. On dynamically generated ontology translators in agent communication. *International Journal of Intelligent Systems*, 16(5):587–607, 2001. [21]

[van Rijsbergen, 1975] Cornelis Joost (Keith) van Rijsbergen. *Information retrieval*. Butterworths, London (UK), 1975. http://www.dcs.gla.ac.uk/Keith/Preface.html. [205]

[Velegrakis *et al.*, 2003] Yannis Velegrakis, Renée Miller, and Lucian Popa. Mapping adaptation under evolving schemas. In *Proc. 29th International Conference on Very Large Data Bases (VLDB)*, pages 584–595, Berlin (DE), 2003. [162]

[Velegrakis *et al.*, 2004a] Yannis Velegrakis, Renée Miller, and Lucian Popa. Preserving mapping consistency under schema changes. *The VLDB Journal*, 13(3):274–293, 2004. [162]

[Velegrakis *et al.*, 2004b] Yannis Velegrakis, Renée Miller, Lucian Popa, and John Mylopoulos. ToMAS: A system for adapting mappings while schemas evolve. In *Proc. 20th International Conference on Data Engineering (ICDE)*, page 862, Boston (MA US), 2004. [162]

[Visser *et al.*, 1998] Pepijn Visser, Dean Jones, Trevor Bench-Capon, and Michael Shave. Assessing heterogeneity by classifying ontology mismatches. In *Proc. 1st International Conference on Formal Ontology in Information Systems (FOIS)*, pages 148–162, Trento (IT), 1998. [41]

[Vouros and Kotis, 2005] George Vouros and Konstantinos Kotis. Extending HCONE-merge by approximating the intended interpretations of concepts iteratively. In *Proc. 2nd European Semantic Web Conference (ESWC)*, volume 3532 of *Lecture notes in computer science*, pages 198–210, Hersounisous (GR), May 2005. [165]

[Wache *et al.*, 2001] Holger Wache, Thomas Voegele, Ubbo Visser, Heiner Stuckenschmidt, Gerhard Schuster, Holger Neumann, and Sebastian Hübner. Ontology-based integration of information – a survey of existing approaches. In *Proc. IJCAI Workshop on Ontologies and Information Sharing*, pages 108–117, Seattle (WA US), 2001. [11, 41, 61]

[Wang and Gasser, 2002] Jun Wang and Les Gasser. Mutual online ontology alignment. In *Proc. AAMAS Workshop on Ontologies in Agent Systems (OAS)*, Bologna (IT), 2002. [21]

[Wang *et al.*, 2004] Jiying Wang, Ji-Rong Wen, Frederick Lochovsky, and Wei-Ying Ma. Instance-based schema matching for web databases by domain-specific query probing. In *Proc. 30th International Conference on Very Large Data Bases (VLDB)*, pages 408–419, Toronto (CA), 2004. [174]

[Wiesman *et al.*, 2002] Floris Wiesman, Nico Roos, and Paul Vogt. Automatic ontology mapping for agent communication. In *Proc. 1st International joint Conference on Autonomous agents and multiagent systems (AAMAS)*, pages 563–564, Bologna (IT), 2002. [21]

[Winkler, 1999] William Winkler. The state of record linkage and current research problems. Technical Report 99/04, Statistics of Income Division, Internal Revenue Service Publication, 1999. [80]

[Wolpert, 1992] David Wolpert. Stacked generalization. *Neural Networks*, 5(2):241–259, 1992. [139]

[Wooldridge, 2000] Mike Wooldridge. *Reasoning about rational agents*. The MIT press, Cambridge (MA US), 2000. [52]

[Wu and Palmer, 1994] Zhibiao Wu and Martha Palmer. Verb semantics and lexical selection. In *Proc. 32nd Annual Meeting of the Association for Computational Linguistics (ACL)*, pages 133–138, Las Cruces (NM US), 1994. [101]

[Xu and Embley, 2003] Li Xu and David Embley. Discovering direct and indirect matches for schema elements. In *Proc. 8th International Conference on Database Systems for Advanced Applications (DASFAA)*, pages 39–46, Kyoto (JP), 2003. [138, 179]

[Yager, 1988] Ronald Yager. On ordered weighted averaging aggregation operators in multicriteria decision making. *IEEE Transactions on System, Man and Cybernetics*, 18(1):183–190, 1988. [126]

[Zaihrayeu, 2006] Ilya Zaihrayeu. *Towards Peer-to-Peer Information Management Systems*. PhD thesis, International Doctorate School in Information and Communication Technology, University of Trento, Trento (IT), March 2006. [16, 17]

[Zanobini, 2006] Stefano Zanobini. *Semantic coordination: the model and an application to schema matching*. PhD thesis, International Doctorate School in Information and Communication Technology, University of Trento, Trento (IT), March 2006. [71]

[Zhang et al., 2004] Songmao Zhang, Peter Mork, and Olivier Bodenreider. Lessons learned from aligning two representations of anatomy. In *Proc. 13th Internation Conference on the Principles of Knowledge Representation and Reasoning Conference (KR)*, pages 555–560, Whistler (CA), 2004. [202]

[Zhdanova and Shvaiko, 2006] Anna Zhdanova and Pavel Shvaiko. Community-driven ontology matching. In *Proc. 3rd European Semantic Web Conference (ESWC)*, volume 4011 of *Lecture notes in computer science*, pages 34–49, Budva (ME), 2006. [240, 265, 271]

[Zhdanova et al., 2005] Anna Zhdanova, Reto Krummenacher, Jan Henke, and Dieter Fensel. Community-driven ontology management: DERI case study. In *Proc. 4th International Conference on Web Intelligence (WI)*, pages 73–79, Compiegne (FR), 2005. [18]

[Zimmermann and Euzenat, 2006] Antoine Zimmermann and Jérôme Euzenat. Three semantics for distributed systems and their relations with alignment composition. In *Proc. 5th International Semantic Web Conference (ISWC)*, volume 4273 of *Lecture notes in computer science*, pages 16–29, Athens (GA US), 2006. [52]

[Zimmermann et al., 2006] Antoine Zimmermann, Markus Krötzsch, Jérôme Euzenat, and Pascal Hitzler. Formalizing ontology alignment and its operations with category theory. In *Proc. 4th International Conference on Formal Ontology in Information Systems (FOIS)*, pages 277–288, Baltimore (MD US), 2006. [44, 51, 271]

[Zohar, 1997] Sagit Zohar. Schema-based data translation. Master's thesis, Tel-Aviv University, Tel-Aviv (IL), 1997. [155]

Index

A searchable index can be found online at http://book.ontologymatching.org.

$=$ (property assignment), *39*
C (ontology classes), *39*
D (domain of interpretation), *39*
$F(\cdot,\cdot)$ (fallout), *207*
$H(\cdot,\cdot)$ (Hamming distance), *204*
I (ontology individuals), *39*
$I(\cdot)$ (interpretation function), *39*
$M_\alpha(\cdot,\cdot)$ (F-measure), *207*
$O(\cdot,\cdot)$ (overall), *208*
$P(\cdot,\cdot)$ (precision), *206*
$Q_L(\cdot)$ (ontology entities), *46*
R (ontology relations), *39*
R (reference alignment), *204*
$R(\cdot,\cdot)$ (recall), *206*
T (ontology types), *39*
V (ontology values), *39*
$W(\cdot,\cdot)$ (weighted harmonic mean), *215*
$\Delta(\cdot,\cdot)$ (linkage measures), *109*
Γ (equalising functions), *52*
Λ (set of alignments), *54*
$\mathcal{M}$ (set of models), *40*
Ω (set of ontologies), *54*
Σ (synonym resource), *87*
Θ (correspondence relations), *46*
Ξ (confidence structure), *46*
α (alignment), *50*
$\perp$ (ontology exclusion), *39*
δ (dissimilarity or distance), *74*
ϵ
  empty string, *75*
  iteration threshold, *130*
$\gamma(\cdot)$ (equalising function), *52*
$\in$ (ontology instantiation), *39*
$\leq$ (ontology specialisation), *39*
$\models$
  alignment satisfaction, *53*
  alignment validity, *53*
  correspondence satisfaction, *53*
  ontology entailment, *40*
  satisfiability of a formula, *39*
μ (merge), *51*
$\omega(\cdot,\cdot)$ (alignment proximity), *210*
π (probability), *90*
σ (similarity), *73*

accuracy (matching -), 207
agent, 9, 19–22, 25, 264, 273
  cognitive, 20
  communication, 26
  language, 21
  reactive, 20
aggregation
  fuzzy, *125*, 125
  similarity -, 121, 240
AGROVOC, 203
algebra, 235
Alignment
  API, 83, 122, 179, 196, 229, 239–240, 243, 260, 262, 264
  format, 196, 226–229, 234, 239, 241
alignment, 2, 42, 47, 45–56, 219
  bijective -, 49
  completeness, *206*
  completion, 199
  correctness, *205*
  edition, 272, 273
  evaluation, 240
  evolution, 16, 163
  extraction, 119, 144–149, 240
  greedy -, *148*

extractor, *145*
filter, *145*
format, 273
infrastructure, 273
initial -, 143
injectivity, *49*, 110, 145, 148, 200
maximal cardinality -, 148
metadata, 265
multiplicity, 49, 199, 200
one-to-one -, *49*, 62, 148, 157, 170, 171, 174, 179, 184, 185, 187, 200, 271
reference -, 204–208, 210
relation, 45
interpretation, 53
reuse, 66, *68*, 271
reversible -, *49*
satisfaction, 53
satisfiability, *53*, 112
semantics, 51–56
service, 264–265
sharing, 264
structure, 45–49
surjective -, *49*
total -, *49*, 145, 148, 200, 260, 262
update, 199
validity, *53*
ambient computing, 22, 264
Anchor-Prompt, 103, 159
anchoring, *111*, 112
antonym, *86*
APFEL, 144, 184–185, 192, 240
application-specific evaluation, *197*, 211, 213, 273
approximate algorithm, *62*
approximation (ontology -), *50*
argumentation, 19, 248, 274
array, 97
Artemis, 62, 157, 188, 190
ArtGen, 156, 157
articulation axiom, *see* bridge axiom

ASCO, 167, 188, 191
associativity, *121*
authoritativeness, 246
Automatch, 172–173, 189, 191
AutoMed, 153
autonomy
  design -, 16, 17
  participation -, 16
  total -, 16, 17
average, 126, 185
  linkage, *109*
  ordered weighted -, *126*, 126
  weighted -, *125*, 124–125, 140, 179, 214
axiom (bridge -), 10, *43*, 163, 260, 264, 265

background
  knowledge, 110
  ontology, 110, 111
bag, 97
  of words, *80*, 81, 83, 167, 182
Bayesian
  classification, 71, 252
  learning, 133–135, 139, 171–173, 176, 177, 179, 183, 184, 187
  network, 141–142, 167, 168
Bayesian classification, 251
BayesOWL, 188, 191
benchmark, *194*
  competence -, *197*
  suite, *194*
best match, 147–149
BibSter, 17
bijective alignment, *49*
BizTalk schema mapper, 259
blank normalisation, *76*
BN mapping, 167–168
boundary condition, *121*
bounded path matcher, 103
boundedness (evaluation measure -), *210*
bridge

axiom, 10, 25, *43*, 163, 260, 264, 265
  concept - (in MAFRA), 221
  property - (in MAFRA), 221
  rule, 163, 221, 223
  semantic - (in MAFRA), 221
Brown corpus, 90
browsing (semantic web -), 22–24
built-in composition, *143*

C-OWL, 223–225, 234, 262, 264, 265
CAIMAN, 170, 189, 191
cardinality, 67
  compatibility, 96, 162
  maximal -, 148
  property -, 92, 93, 95–97
case normalisation, *76*, 167
catalogue integration, 2, 9, 11, 13–14
categorical characterisation, 271
Chebichev distance, *123*
Chimaera, 241–242
City-blocks distance, *see* Manhattan distance
class, 37
  exclusion, 38
  specialisation, 38
classification, *31*, 31–32, 164, 170, 173, 179, 191
  of matching approaches, 63–72, 153, 154, 187
Clio, 162, 163, 177, 189, 191, 262
COMA, 62, 161, 178, 188, 190, 236–238
  COMA++, 143, 161, 188, 190, 236–238
combination (matcher -), *see* composition
communication (agent -), 20–22
commutativity, *121*

comparison evaluation, *197*
compatibility
  cardinality -, 96
  datatype, 96
  datatype -, *94*
  transformation -, 263
competence benchmark, *197*
completeness (alignment -), 206
completion (alignment -), 199
composition, 63
  built-in -, *143*
  opportunistic -, *143*
  parallel -, *119*, 179
    heterogeneous -, *120*
    homogeneous -, *120*
  sequential -, 117, 156, 160, 167, 173, 179
  user-driven -, *143*
  web service -, 19–20, 43, 235
compound similarity, 121
computing
  ambient -, 22
  pervasive -, *see* ambient computing
concatenation of strings, *74*
concept
  bridge (in MAFRA), 221
  lattice, 106
conceptual model, 35
conceptualisation mismatch, 41
conditional probability table, 141, 168
confidence, 62
  degree, *46*
  structure, *46*
consequence (ontology -), 40
constraint-based technique, 63, 66, 67, 92
context, 110
contextualising, *111*
continuity
  evaluation -, *194*
  property, *125*

convergence (fixed point algorithm -), 128, 130, 132
corpus, 90
  -based similarity, 90
  Brown, 90
Corpus-based matching, 183–184
correctness (alignment -), 205
correspondence, 42, *46*
  analysis, 81
  graded -, 62
  justified -, 245
  satisfaction, *53*
cosine similarity, *81*, 84
cosynonymy similarity, 89
count, 226
cover (graph -), 148
coverage, *41*, 51
CtxMatch, 164, 187, 188, 191
Cupid, 62, 160–161, 184, 188, 190
Cyc, 66, 68, 111

DAML+OIL, 221
data
  analysis technique, 66, 70
  integration, 9, 11, 14–17, 25, 52, 118, 175, 234, 235, 240, 262, 263, 271
  set (evaluation -), 194, 198–203
  transformation, 262
  translation, 2, 25, 26, 43, *43*, 222, 235, *261*, 261–265
  translator, *see* data translation
  value, *37*
  warehouse, 9, 11
database
  federated -, 13
  schema, 29, 32–33, 93, 154, 162, 163, 172–174, 176, 177, 184, 187, 235

  matching, 2
datatype, 33, *37*, 62, 93, 97, 170
  compatibility, *94*, 96
DCM, 168–169, 188, 191
decision tree, 135, 137–139, 180
  learning, 138, 185
deep web matching, 24, 273
definiteness (property), *74*
degree
  confidence -, *46*
  of completeness, 206
  of correctness, 205
DELTA, 154, 187, 190
delta threshold, *146*
dependency graph, *252*, 252
description logic, 112, 226
  technique, 70, 114–115, 164
design autonomy, 16, 17
diacritic suppression, 76
Dice coefficient, *81*, 84
dictionary, 84, *86*
digit suppression, 76
DIKE, 155–156, 188, 190, 260
directory, *31*, 31–32
disambiguation (word sense -), 87, 92
disjointness, 63
dissemination (evaluation -), *194*
dissimilarity, *73*, 148
  Leacock–Chodorow -, 92, 101
distance, *74*
  aggregation, *see* similarity aggregation
  Chebichev -, *123*
  City-block -, *see* Manhattan
  edit -, *78*, 159, 161, 165, 180, 184, 185, 207, 209
  Euclidean -, 81, 96, 123, *123*, 174
  Hamming -
    on alignments, 204

on multisets, *81*
on sets, 81, *105*, 123
on strings, 77
Hausdorff -, *109*
Levenshtein -, *79*, 122
Manhattan -, 81, 123, *123*, 124
Minkowski -, 81, 123, *123*
multidimentional -, 122–125
n-gram -, 161, 165, 184, 185
Needleman–Wunch -, *79*
on sequences, 83
path -, *82*
relative-size -, *95*
tree -, 101
distributed
database, 13
knowledge, 55
system, *54*
model, *54*
document frequency (inverse -), *82*
DOLCE, 66, 68, 111
domain
of interpretation, 39, 52
property -, 93, 94, 104
specific ontology (technique based on -), 66, 68
DPLL procedure, 246, 253–255
Drago, 262, 264
DTD, 33, 96, 162, 229
Dublin core, 71
Dumas, 174, 189, 191
DWQ, 153

e-commerce, 13–16
edge count similarity, *see* structural topological dissimilarity
edit
distance, *78*, 159, 161, 165, 180, 184, 185, 207, 209
edition

alignment -, 272, 273
ontology -, 10
effort-based
precision and recall, 211
effort-based precision and recall, 211
element-based technique, 63, 64, *64*, 93
elementary matchers, 63
emergent semantics, 18–19
empty
phrase, 91
word, 91
engineering (ontology -), 9–11
entailment, 50
enterprise information integration, *see* data integration
entity
–relationship model, 35
-relationship model, 62, 154, 155
interpretation, 45
language, *45*, 46–48
ontology -, 37–39
EON, 196, 200
equalising function, *52*
error minimisation, 144
eTuner, 185–186, 192
Euclidean distance, 81, 84, 96, 123, *123*, 174
evaluation, 193–216
application-specific -, *197*, 211, 213
comparison, *197*
type, 196–198
evolution
alignment -, 16, 163
ontology -, 10–11, 241
exact algorithm, *62*, 64
exclusion, *38*
exclusivity, *148*
executability, 234
exhaustivity, *63*
explanation, 274
explicitation mismatch, 41
expressiveness, 233
extendibility, 233

extensional technique, 66, 105–110, 169, 170
external
resource, 199
structure-based technique, 92, 98–105
technique, 62, 64, *64*
extraction (alignment -), 144–149, 240
extractor (alignment -), *145*
extrinsic linguistic technique, 67, 86–91

F-logic, 199
F-measure, 186, 207, *207*, 210, 215
FaCT, 164
Falcon-AO, 143, 181–182, 189, 192, 229
fallout, *207*
false
negative, *205*, 206
positive, 87, 205, *205*
FCA, *see* formal concepts analysis
FCA-merge, 170, 189, 191, 260
feature path equations, 226
federated database, 13
feedback (relevance -), 144
filter
alignment -, *145*
similarity -, *145*
FIPA, 21
fixed point, 55, 127
computation, 126–133, 161, 181, 187
FMA, 68, 202
FOAF, 10
FOAM, 178–179, 184, 229, 240–241, 243
folksonomy, 29–31, 231
formal concept analysis, 106, 170
full linkage, *109*
fuzzy aggregation, *125*, 125

Galen, 202
Galois

connection, 106
lattice, *see* concept lattice
GAV, *see* global-as-view
Gene ontology, 202
GeRoMe, 236
GLAV, *see* global-local-as-view
global
  -as-view, 15, 157, 228
  -local-as-view, 15, 16, 176
  knowledge, 54
  maximal - similarity, 148
gloss, *87*, 90, 91, 231
  overlap, *90*, 92
GLUE, 62, 171, 189, 191
gold standard, *see* reference alignment
Gotoh distance, 79, 84
graded correspondence, 62
granularity, *41*, 51
  matcher -, *64*
graph
  -based technique, 66
  cover, 148
  dependency -, 252
  matching, 127, 148
    maximum weight -, 148, 174
    minimum weight -, 148, 174
greedy alignment extraction, *148*

H-Match, 143, 157–158, 188, 190
Hamming distance
  on alignments, *204*
  on multisets, *81*
  on sets, 81, *105*, 123
  on strings, *77*
hard threshold, *145*
harmonic mean, 207, 215
  weighted -, *215*
Hausdorff distance, *109*
HCONE, 165–166, 188, 191, 229
heterogeneity, 1, 9, 271
  conceptual -, *41*

language -, 198
pragmatic -, *42*
semantic -, *41*
semiotic -, *42*
syntactic -, *41*
terminological -, *41*
heterogeneous parallel composition, *120*
homogeneity, *63*
homogeneous parallel composition, *120*
homonym, 75, *86*, 156
Horn clause, 225, 226
Hungarian method, 148
hypernym, 84, *86*, 87–89, 165
hyponym, 67, 84, *86*, 89, 156, 165

I3CON, 196
idempotency, *125*
IF-Map, 177–178, 189, 192
Illinois Semantic Integration Archive, 203
iMAP, 171–172, 189, 191, 248, 250–252
import (ontology -), 10
individual, *37*
Inference Web, 248
InfoMix, 153
information
  -theoretic similarity, 90, 92
  integration, 9, 11–16, 23
  retrieval, 179
InfoSleuth, 153
infrastructure (alignment -), 273
initial alignment, 143
injectivity (alignment -), *49*, 110, 145, 148, 200
input, 64
  dimensions, *62*
  kind of -, *66*
instance, *see* individual
  -based technique, 62, 63
instantiation, *38*
instrinsic linguistic technique, 84–85

integration
  catalogue -, 9, 11, 13–14
  data -, 9, 11, 14–17, 25, 234, 235, 240, 262, 263, 271
  information -, 9, 11–16, 23
  ontology -, 9, *43*, 235
  schema -, 9, 11, 13, 25
integrity constraint, 97
intelligibility (evaluation -), *194*
internal structure-based technique, *64*, 66, 92–98, 107, 159
interoperability, *see* heterogeneity
interpretation
  domain of -, 39
  entity -, 45
  of alignment relations, *53*
  ontology -, *39*, 45
intrinsic linguistic technique, 67
inverse document frequency, *82*
iPrompt, 242
ISWC, 196

Jaccard similarity, 84, 90, 97, 102, *106*, 167
Jaro measure, *80*, 84
Jaro–Winkler measure, *80*, 84, 167
justification, 245
justified correspondence, 245

K-Cap, 196
k-nearest neighbours, *see* nearest neighbours
KAON2, 241, 243
key, 93, *93*, 107
KIF, 178, 199
kind of input, *66*
knob, 185
knowledge base, *36*
Knowledge web, 5
Kraft, 153

language
  -based technique, 66, 67, 83–92, 181
  independence, 233
  ontology -, 36–40
  query -, 45
largest common directed subgraph, *see* maximum common directed subgraph problem
latent semantic indexing, *81*, 166
LAV, *see* local-as-view
Leacock–Chodorow dissimilarity, 92, 101
learning, 133–141, 273
lemmatisation, *85*, 180
Levenshtein distance, 79, 84, 122, 167
lexicon, *86*, 91
  multilingual -, *86*
  semantico-syntactic -, *86*
library of matchers, 272
linguistic
  technique, 63, 66, 93, 157–160, 167, 180, 182
  based on - resource, 66, 67, 156, 157, 160–162, 164, 165, 188
link stripping, *76*
linkage dissimilarity
  average -, *109*
  full -, *109*
  single -, *109*, 167
list, 97
local
  -as-view, 15, 16, 228
  knowledge, 54
logical mismatch, *41*
LSD, 71, 170–171, 184, 189, 191

MAFRA, 221–222, 234, 238–239
Magpie, 22

Manhattan distance, 81, 84, 123, *123*, 124
MapForce, 259
MapOnto, 163, 188, 191, 262
mapping, *42*, 138
  in model management, 235
  rule, *43*, 138
marriage (stable -), 147–149
match-based similarity, *109*
matcher composition, 117–121
matching, *42*
  accuracy, 207
  coefficient, *81*, 84
  graph, 127
  memory consumption, 212
  multiple -, *44*, 47, 48
  process, *44*, 44, 271
  scalability, 212
  schema -, 92
  speed, 212, 216
  usability, 213
matrix, 74, 119
maximality
  evaluation measure -, *210*
  of similarity measures, *73*
maximum
  common directed subgraph problem, *98*
  weight matching, 148, 174
mean (harmonic -), 207
measure
  Jaro -, *80*
  Jaro–Winkler -, *80*
  Smoa -, *80*
mediation, 27, *43*
  query -, 222, *263*, 262–265
mediator, 10, *43*, 234, *see* mediation
  web service -, 229
memory consumption, 212, 272
mereologic structure, 99, 103

merging (ontology -), 2, 10, 25–27, 43, *43*, 51, 112, 159, 160, 163, 165, 166, 170, 180, 222, 226, 240, 242, *260*, 260–261
meronym, *87*
meronymy, 158
message translation, 26
metadata (alignment -), 265
method
  composition, 117–121
  dynamic -, 142–144
  learning, 133–141
  metric, *74*
  minimality, *74*
minimisation (error -), 144
minimum
  cost maximum flow, 173
  weight matching, 148, 174
Minkowski distance, 81, 123, *123*
mismatch, *see* heterogeneity
  conceptualisation -, 41
  explicitation -, 41
  logical -, *41*
mixed technique, 62
MoA, 166, 188, 191
Moda, 236
modal logic satisfiability, 114
model
  -based technique, *see* semantic technique
  -theoretic semantics, 64, 271
  conceptual -, 35
  entity–relationship -, 35, 154
  entity-relationship -, 155
  in model management, 235
  in Rondo, 236
  management, 235–238
  of a distributed system, *54*
  of aligned ontologies, *54*
  ontology -, *40*

in a distributed system, 55
ModelGen, 236
Monge–Elkan distance, 79, 84, 167
monotonicity, *121*
  increasing, *125*
morphism, 50
  in Rondo, *236*
morphological
  analysis, 85
  normalisation, 160, 199
multi
  -linguality, 271
  -response linear regression, 139
  -set, *80*, 81, 97
  dimentional distance, 122–125
multialignment, *47*
multiple matching, 47, 48, 199
multiplicity, 67
  alignment -, 49, 199, 200
  property -, 92, 93, 95, 96
  similarity, 97

n-gram, 84
  distance, 161, 165, 184, 185
  similarity, *78*
NAL, 203
name-based technique, 66, 74–93, 178, 252
natural language processing, 67
nearest neighbours, 135, 176, 183, 189
Needleman–Wunch distance, *79*, 84
negative
  false -, *205*, 206
  true -, *205*
neighbours (nearest -), 135, 176, 183, 189
neural network, 135–137, 139, 177, 185
NIST, 195, 196
noise, 207

NOM, 178–179, 189, 192, 240
norm (triangular -), *121*, 121–122, 125, 128
normalisation, 160, 179, 180
  measure, 74
  morphological -, 160, 199
  string -, 76–77
    blank -, *76*
    case -, *76*
    diacritic -, *76*
    digit -, *76*
    link stripping -, *76*
    punctuation -, *76*

OAEI, 195–196, 202, 203, 209, 216, 229, 240
object, *see* individual
  -oriented model, 62
  categorical -, 50
  identifier, 236
Observer, 153
occurence
  of a character in a string, 74
  of a substring in a string, 75
OID, 236
OLA, 129, 181, 187, 189, 192, 229, 240
  algorithm, 129–133
oMap, 179, 189, 192, 229, 240
OMEN, 168, 188, 191
one-to-one alignment, *49*, 62, 148, 157, 170, 171, 174, 179, 184, 185, 187, 200, 271
ONION, 156–157
onto function, *49*
OntoBuilder, 159–160, 188, 190, 260
ontology, 29, 36
  approximation, *50*
  background -, 110, 111
  consequence, *40*
  edition, 10, 260
  engineering, 9–11, 25
  entity, 37–39

language, *45*, 46–48
  evolution, 10–11, 241
  import, 10
  integration, 9, *43*, 235
  interpretation, *39*, 45
  language, 36–40, 199
  merging, 2, 10, 25–27, 43, *43*, 51, 159, 160, 163, 165, 166, 170, 180, 222, 226, 240, 242, *260*, 260–261
  model, *40*
  in a distributed system, 55
  peer-to-peer -, 17–18
  reconciliation, *43*
  satisfiability of a formula, *39*
  semantics, 39–40
  syntax, *39*
  transformation, 25, *43*, 226, 234, *261*, 261, 263, 265
  translation, *43*, 163, 242
  upper level -, 261
  version, 10–11, *43*
Ontology Alignment Evaluation Initiative, *see* OEAI
ontologymatching.org, 203
OntoMerge, 163, 187, 188, 191, 260
opportunistic composition, *143*
ordered
  weighted average, *126*
ordered weighted average, 126
oriented precision and recall, 211
output dimension, 62
overall, 207, *207*, 211
overlap, 84
  proximity, *210*
OWL, 17, 36–40, 45, 62, 92, 94, 96, 98, 100, 130, 157, 159, 163, 166, 179, 181, 187,

190, 191, 199, 202,
221–223, 225, 226,
229, 234, 240, 242,
260, 264, 265
-DLP, 241
C-OWL, 223–224, 262,
264
SWRL, 225–226, 264

P2P, *see* peer-to-peer
parallel composition, 179
parameter, 143, 144, 184,
185, 199
tuning, 272
part-of-speech tagger, 84, 91
partially ordered synonym
resource, 87, *87*, 90
participation autonomy, 16
path, 221
bounded - matcher, 103
distance, *82*
peer-to-peer, 16–19, 25,
157, 263
ontology, 17–18
system, 9, 271, 273
semantic -, 16
Pellet, 164, 264
percentage threshold, *146*
PerMIS, 196
perspective, *41*, 51
pervasive computing, *see*
ambient computing
Picsel, 153
PML, 248
polysemy, *75*
positive
false -, 87, 205, *205*
true -, 86, 205, *205*, 206
positiveness, *73*
evaluation measure -, *210*
pre-similarity, *74*, 78
precision, 205, *205*,
208–212, 215
/recall curve, 208
effort-based -, 211
generalised -, 208–212
oriented -, 211
relaxed -, *210*
semantic -, 212

symmetric -, 211
process
dimension, 62
matching, 271
trace, 247
processors, 274
product
classification, 14
weighted -, *121*, 214
Prolog, 178
Prompt, 10, 185, 242–243
Anchor-Prompt, 159, 188,
242
iPrompt, 242
PromptDiff, 159, 242,
243
PromptFactor, 242, 243
proof markup language, 248
property, 92
bridge (in MAFRA), 221
proportional threshold, *146*
propositional
satisfiability, 113
technique, 113–114
Protégé, 10, 143, 242, 260
proximity (overlap -), *210*
punctuation
normalisation, 167
suppression, *76*
purpose independence, 234

QOM, 62, 178–179, 185,
189, 192, 240
quality (evaluation -), *194*
query
answering, 2, 25, 43, 241
language, 45
mediation, 222, *263*,
262–265
transformation, 25, 262

range (property -), 92, 94,
104
RDF, 17, 62, 178, 182, 226,
229, 233, 240
/XML, 229, 239
schema, 17, 159, 167,
191, 199, 231, 233,
238

RDFS, *see* RDF Schema
reasoning, 260, *264*, 264
recall, 205, *206*, 208–212,
215
effort-based -, 211
generalised -, 208–212
oriented -, 211
relaxed -, *210*
semantic -, 212
symeric -, 211
reconciliation (ontology -),
43
reference alignment,
204–208, 210
relation, *37*
alignment, *45*
relational
database model, 62
structure-based technique,
66, 92, 98–105,
156–158, 161, 167,
180–182, 188
relative size distance, *95*
relaxed
precision, *210*
recall, *210*
relevance feedback, 144,
185
repository of structure, 66,
69
Resnik similarity, *90*, 92
resource, 199
consumption, 212
reuse (alignment -), 66, *68*
reuse-oriented matcher, 71
reversible alignment, *49*
RiMOM, 182–183, 189, 192
role-value map, 226
Rondo, 127, 143, 146, *236*,
235–237, 260, 262
rule
bridge -, 163, 221, 223
mapping -, 43, 138
RuleML, 225

S-Match, 62, 164–166, 187,
188, 248–255
SAT, 70, *see* propositional
satisfiability

modal, *see* modal logic
satisfiability
  solver, 113, 164, 246, 249, 253
SAT4J, 253, 254
satisfaction
  of alignment, *53*
  of correspondence, *53*
satisfiability
  alignment -, *53*, 112
  by an ontology, *39*
  modal logic -, 112, 114
  propositional logic -, 112, 113, 255
saturation, *63*
SBI&NB, 173, 189, 191
scalability, 212
schema
  -based technique, 62, 63
  database -, 29, 32–33, 154, 162, 163, 172–174, 176, 177, 184, 187, 235
  integration, 9, 11, 13, 25, 43
  matching, 92
  RDF -, 17, 159, 167, 191, 231, 233, 238
  UML, 29
  XML -, 17, 33, 94, 162–164, 171, 177, 179, 190–192, 199, 226, 235
SchemaIntegrator, 259
scope, *see* perspective
SecondString, 83
SEKT, 229
  mapping language, 229–231, 234
selector (in Rondo), *236*
semantic
  bridge (in MAFRA), 221
  method, 272
  peer-to-peer system, 16
  precision and recall, 212
  technique, 62, 64, *64*, 66, 70, 110–115, 163, 164, 248
  web

browsing, 22–24
  service, 19–20, 229
Semantic Bridge Ontology, 221–222, 234
Semantic Web Rule Language, *see* SWRL
semantics
  alignment -, 51–56
  emergent -, 18–19
  ontology -, 39–40
SEMINT, 176–177, 189, 191
sequential composition, 156, 160, 167, 173, 179
service
  (in MAFRA), 221
  web -, 9, 273
    composition, 43
    semantic -, 19, 26
set, 97
sharing (alignment -), 264
sigmoïd, *146*, 179
Signal ontology, 202
silence, 207
similarity, *73*, 148, 211
  aggregation, 121–125, 240
  compound -, 121
  corpus-based -, 90
  cosine -, *81*
  cosynonymy -, *89*
  filter, *145*
  global computation, 126–133
  gloss-overlap -, *90*
  information theoretic -, 90, 92
  Jaccard -, 90, 97, *106*, 167
  match-based -, *109*
  multiplicity, 97
  n-gram -, *78*
  non symmetric -, *74*
  pre-, *74*, 78
  Resnik -, *90*, 92
  strength-based -, *208*
  structural topological -, *89*, 92, *100*
  substring -, *77*
  synonymy -, *88*

upward cotopic -, *102*
vector-based -, *81*, 175
Wu–Palmer -, 89, 92, *101*
Similarity flooding, 127–129, 132, 161, 188, 191, 236
SimPack, 83, 92
simplicity, 233
SIMS, 153
single linkage, *109*, 167
singular value decomposition, 81
SKAT, 156–157, 188, 190, 260
SKOS, 203, 223, 231–234
SMART, *see* Prompt
Smith–Waterman measure, 79, 84
Smoa measure, *80*, 84
SomeWhere, 17
soundex, 84
SPARQL, 264
SPEC, 197
specialisation, *38*
speed, 212, 216
sPLMap, 176, 189, 191, 262
SQL, 33, 36, 177, 262, 263, 265
  DLL, 199
stable marriage, 147–149
stacked generalisation, 139–141
statistical technique, 66, 70
stemmer, 91
stemming, *see* lemmatisation
stopword elimination, *85*
strength-based similarity, *208*
strengthening, 146–147
string, *74*
  -based technique, 66, 67, 76–83, 107, 154, 156, 159–161, 165, 167, 174, 176, 177, 179–181, 184, 187–189
  concatenation, *74*
  equality, *77*, 82

normalisation, 76–77
occurence
  of a string, *74*
  of a substring, *75*
  substring, *75*
stringmetrics, 83
structural
  technique, 66
  topological dissimilarity,
    89, 92, *100*
structure
  -based technique, 63, 64,
    64, 66, 92–105, 178
  internal - technique, *64*,
    66, 107
  relational - technique, 66,
    92, 98–105, 156–158,
    161, 167, 180–182,
    188
Stylus, 259
substring, *75*
  similarity, *77*
  test, *77*
subsumption, 114
  test, 114
sum (weighted -), *124*,
    122–125, 161, 162,
    167, 171, 214
SUMO, 66, 68, 111
support vector machines,
    185
surjective alignment, *49*
SWRL, 223, 225–226, 228,
    234, 240, 260, 264,
    265
symmetric precision and
    recall, 211
symmetry, *73*
  evaluation measure -, 210
  property -, 92, 98
synonym, 67, 75, 86, *86*, 88,
    156, 162, 169
synonymy similarity, 88, 89
synset, 86, *87*, 88–90
syntactic technique, 62, 64,
    64
syntax (ontology -), *39*
systematicity (evaluation -),
    194

T-tree, 169–170, 189, 191
taxonomy, *31*, 31–32,
    99–103, 171
  -based technique, 66, 69
term, *83*, 86
  extraction, *85*
  frequency, *82*
  -inverse document
    frequency, *82*, 84, 135,
    167, 170, 174, 182
terminological
  technique, *see* name-
    based technique
terminology, *86*
Tess, 158–159, 188, 190
TFIDF, *see* term frequency-
    inverse document
    frequency
thesaurus, *86*, 91, 231, 232,
    239
threshold, 144–146, 148,
    156, 157, 159–161,
    167–171, 173, 179,
    183–185, 207, 240
  delta -, *146*
  hard -, *145*, 158
  percentage -, *146*
  proportional -, *146*
time
  consumption, 272
  processing -, 200
tokenisation, 84, 166, 238
ToMAS, 162–163, 188, 191,
    262
total
  alignment, *49*, 145, 148,
    200, 260, 262
  autonomy, 16, 17
trace (process -), 247
TranScm, 187, 190
transformation
  compatibility, 263
  data -, 262
  ontology -, 25, *43*, 226,
    234, *261*, 261, 263,
    265
  query -, 25, 262
transitivity (property -), 92
translation

data -, 2, 25, 26, 43,
    43, 222, 235, *261*,
    261–265
  message -, 26
  ontology -, *43*, 163, 242
TransScm, 155, 262
TREC, 195–196, 208
tree distance, 101
triangular
  inequality, *74*
  norm, *121*, 121–122, 125,
    128
true
  negative, *205*
  positive, 86, 205, *205*,
    206
Tsimmis, 153
tuning (parameter -), 200,
    272

ultrametric, *74*
  inequality, *74*
UML, 29, 35, 96, 199, 221,
    234
  models, 235
update (alignment -), 199
upper level ontology, 68,
    261
  technique based on -, 66
upward cotopic similarity,
    102
URI, 227, 229, 233
usability, 213
user
  -driven composition, *143*
  input, 110, 199, 200, 210,
    245
  interaction, 142–144, 161,
    166, 169, 185, 237
  interface, 274

validity (alignement -), *53*
value, *37*
variable, 225
vector-based similarity, *81*,
    175
version, 242
  ontology -, 10–11, *43*

W3C, 222

weakening, 146–147
web
  compatibility, 233
  deep -, 24
  service, 9, 19–20, 273
    composition, 43, 235
    semantic -, 26, 229
weighted
  average, *125*, 124–125, 140, 179, 214
  ordered, *126*, 126
  harmonic mean, *215*
  product, *121*, 214
  sum, *124*, 122–125, 161, 162, 167, 171, 214
WHIRL, 139

learner, 135, 171
Wise-Integrator, 180–181, 264
Wise-integrator, 189, 192
word, *83*
  sense disambiguation, 87, 92, 167
WordNet, 66, 68, *87*, 101, 102, 111–113, 122, 138, 156, 157, 162, 164–167, 180, 187–189, 199, 231, 239, 246, 251, 252
WSML, 234
WSMO, 229
WSMX, 243

Wu–Palmer similarity, 89, 92, *101*

XClust, 162, 188, 191
XML, 33, 62, 164, 223, 233, 259
  /RDF, 233
  schema, 17, 29, 33, 94, 162–164, 171, 177, 179, 190–192, 199, 226, 235
  matching, 2
Xpath, 221
XQuery, 177
XSLT, 177, 240, 262, 265

Printed in the United States
209868BV00005B/6/P